Steven Roman

Fundamentals of Group Theory

An Advanced Approach

Steven Roman
Irvine, CA
USA

ISBN 978-0-8176-8300-9 e-ISBN 978-0-8176-8301-6
DOI 10.1007/978-0-8176-8301-6
Springer New York Dordrecht Heidelberg London

Library of Congress Control Number: 2011941115

Mathematics Subject Classification (2010): 20-01

Printed on acid-free paper

Birkhäuser Boston is part of Springer Science+Business Media (www.birkhauser.com)

To Donna

Preface

This book is intended to be an advanced look at the basic theory of groups, suitable for a graduate class in group theory, part of a graduate class in abstract algebra or for independent study. It can also be read by advanced undergraduates. Indeed, I assume no specific background in group theory, but do assume some level of mathematical sophistication on the part of the reader.

A look at the table of contents will reveal that the overall topic selection is more or less standard for a book on this subject. Let me at least mention a few of the perhaps less standard topics covered in the book:

1) An historical look at how Galois viewed groups.
2) The problem of whether the commutator subgroup of a group is the same as the *set* of commutators of the group, including an example of when this is not the case.
3) A discussion of xY-groups, in particular,
   a) groups in which all subgroups have a complement
   b) groups in which all normal subgroups have a complement
   c) groups in which all subgroups are direct summands
   d) groups in which all normal subgroups are direct summands.
4) The subnormal join property, that is, the property that the join of two subnormal subgroups is subnormal.
5) Cancellation in direct sums: A group G is **cancellable in direct sums** if

   $$A \boxplus G \approx B \boxplus H, \quad G \approx H \quad \Rightarrow \quad A \approx B$$

   (The symbol $\boxplus$ represents the external direct sum.) We include a proof that any finite group is cancellable in direct sums.
6) A complete proof of the theorem of Baer that a nonabelian group G has the property that all of its subgroups are normal if and only if

   $$G = Q \bowtie A \bowtie B$$

   where Q is a quaternion group, A is an elementary abelian group of exponent 2 and B is an abelian group all of whose elements have odd order.

7) A somewhat more in-depth discussion of the structure of p-groups, including the nature of conjugates in a p-group, a proof that a p-group with a unique subgroup of any order must be either cyclic (for $p > 2$) or else cyclic or generalized quaternion (for $p = 2$) and the nature of groups of order p^n that have elements of order p^{n-1}.
8) A discussion of the Sylow subgroups of the symmetric group (in terms of wreath products).
9) An introduction to the techniques used to characterize finite simple groups.
10) Birkhoff's theorem on equational classes and relative freeness.

Here are a few other remarks concerning the nature of this book.

1) I have tried to emphasize universality when discussing the isomorphism theorems, quotient groups and free groups.
2) I have introduced certain concepts, such as subnormality and chain conditions perhaps a bit earlier than in some other texts at this level, in the hopes that the reader would acclimate to these concepts earlier.
3) I have also introduced group actions early in the text (Chapter 4), before giving a more thorough discussion in Chapter 7.
4) I have emphasized the role of applying certain operations, namely intersection, lifting, quotient and unquotient to a "group extension" $H \leq G$.

A couple of random notes: Unless otherwise indicated, any theorem not proved in the text is an invitation to the reader to supply a proof. Also, sections marked with an asterisk are optional, meaning that they can be skipped without missing information that will be required later.

Let me conclude by thanking my graduate students of the past five years, who not only put up with this material in manuscript form but also put up with the many last-minute changes that I made to the manuscript during those years. In any case, if the reader should find any errors, I would appreciate a heads-up. I can be contacted through my web site www.romanpress.com.

Steven Roman

Contents

Chapter 1
Preliminaries

In this chapter, we gather together some basic facts that will be useful in the text. Much of this material may already be familiar to the reader, so a light skim to set the notation may be all that is required. The chapter then can be used as a reference.

Multisets

The following simple concept is much more useful than its infrequent appearance would indicate.

Definition *Let S be a nonempty set. A* **multiset** *M with* **underlying set** *S is a set of ordered pairs*

$$M = \{(s_i, n_i) \mid s_i \in S, n_i \in \mathbb{Z}^+, s_i \neq s_j \textit{ for } i \neq j\}$$

where $\mathbb{Z}^+ = \{1, 2, \ldots\}$. The positive integer n_i is referred to as the **multiplicity** *of the element s_i in M. A multiset is* **finite** *if the underlying set is finite. The* **size** *of a finite multiset M is the sum of the multiplicities of its elements.* □

For example, $M = \{(a, 2), (b, 3), (c, 1)\}$ is a multiset with underlying set $S = \{a, b, c\}$. The element a has multiplicity 2. One often writes out the elements of a multiset according to their multiplicities, as in

$$M = \{a, a, b, b, b, c\}$$

Two multisets are equal if their underlying sets are equal and if the multiplicities of each element in the multisets are equal.

Words

We will have considerable use for the following concept.

Definition *Let X be a nonempty set. A finite sequence $\omega = (x_1, \ldots, x_n)$ of elements of X is called a* **word** *or* **string** *over X and is usually written in the form*

$$\omega = x_1 \cdots x_n$$

The number of elements in w *is the* **length** *of* w, *denoted by* $\operatorname{len}(w)$. *There is a unique word of length* 0, *called the* **empty word** *and denoted by* ϵ. *The set of all words over* X *is denoted by* X^* *and* X *is called the* **alphabet** *for* X^*. *A* **subword** *or* **substring** *of a word* ω *is a subsequence of* ω *consisting of consecutive elements of* ω. *The empty word is considered a subword of all words.* □

The set X^* of words over X has an algebraic structure. In particular, the operation of juxtaposition (also called concatenation) is associative and has identity ϵ. Any nonempty set with an associative operation that has an identity is called a **monoid**. Thus, X^* is a monoid under juxtaposition.

It is customary to allow the use of exponents other than 1 when writing words, where

$$x^n = \underbrace{x \cdots x}_{n \text{ factors}}$$

for $n > 0$. Note, however, that this is merely a shorthand notation. Also, it does not affect the length of a word; for example, the length of x^2y^3z is 6.

Partially Ordered Sets

We will need some basic facts about partially ordered sets.

Definition *A* **partially ordered set** *is a pair* $(P, \leq)$ *where* P *is a nonempty set and* $\leq$ *is a binary relation called a* **partial order**, *read "less than or equal to," with the following properties:*

1) **(Reflexivity)** *For all* $a \in P$,

$$a \leq a$$

2) **(Antisymmetry)** *For all* $a, b \in P$,

$$a \leq b, \quad b \leq a \quad \Rightarrow \quad a = b$$

3) **(Transitivity)** *For all* $a, b, c \in P$,

$$a \leq b, \quad b \leq c \quad \Rightarrow \quad a \leq c$$

Partially ordered sets are also called **posets**. □

Sometimes partially ordered sets are more easily defined using strict order relations.

Definition *A* **strict order** $<$ *on a nonempty set* P *is a binary relation that satisfies the following properties:*

1) **(Asymmetry)** *For all* $a, b \in P$,

$$a < b \quad \Rightarrow \quad b \nless a$$

2) **(Transitivity)** *For all* $a, b, c \in P$,

$$a < b, \quad b < c \quad \Rightarrow \quad a < c$$

□

Theorem 1.1 *If* $(P, \le)$ *is a partially ordered set, then the relation*

$$a < b \quad \textit{if} \quad a \le b, a \ne b$$

is a strict order on P. *Conversely, if* $<$ *is a strict order on* P, *then the relation*

$$a \le b \quad \textit{if} \quad a < b \text{ or } a = b$$

is a partial order on P.□

It is customary to use a phrase such as "Let P be a partially ordered set" when the partial order is understood. Also, it is very convenient to extend the notation a bit and define $S \le a$ for any subset S of P to mean that $s \le a$ for all $s \in S$. Similarly, $a \le S$ means that $a \le s$ for all $s \in S$ and $S \le T$ means that $s \le t$ for all $s \in S$ and $t \in T$.

Note that in a partially ordered set, it is possible that not all elements are comparable. In other words, it is possible to have $x, y \in P$ with the property that $x \nleq y$ and $y \nleq x$.

Here are some special kinds of partially ordered sets.

Definition *Let* $(P, \le)$ *be a partially ordered set.*

1) *The order* $\le$ *is called a* **total order** *or* **linear order** *if every two elements of* P *are comparable. In this case,* $(P, \le)$ *is called a* **totally ordered set** *or* **linearly ordered set**.
2) *A nonempty subset of* P *that is totally ordered is called a* **chain** *in* P. *The family of chains of* P *is ordered by set inclusion.*
3) *A nonempty subset of* P *for which no two elements are comparable is called an* **antichain** *in* P.
4) *A nonempty subset* D *of a partially ordered set* P *is* **directed** *if every two elements of* D *have an upper bound in* D.□

Definition *Let* $(P, \le)$ *be a poset and let* $a, b \in P$.

1) *The* **closed interval** $[a, b]$ *is defined by*

$$[a, b] = \{p \in P \mid a \le p \le b\}$$

2) *The* **open interval** (a, b) *is defined by*

$$(a, b) = \{p \in P \mid a < p < b\}$$

3) *The* **half open intervals** *are defined by*

$$(a,b] = \{p \in P \mid a < p \leq b\} \quad \textit{and} \quad [a,b) = \{p \in P \mid a \leq p < b\} \quad \square$$

Here are some key terms related to partially ordered sets.

Definition (Covering) *Let* $(P, \leq)$ *be a partially ordered set. If* $a, b \in P$*, then* b **covers** a*, written* $a \prec b$*, if* $a \leq b$ *and if there are no elements of* P *between* a *and* b*, that is, if*

$$a \leq x \leq b \quad \Rightarrow \quad x = a \text{ or } x = b \qquad \square$$

Definition (Maximum and minimum elements) *Let* $(P, \leq)$ *be a partially ordered set.*

1) *A* **maximal element** *is an element* $m \in P$ *with the property that there is no larger element in* P*, that is*

$$p \in P, m \leq p \quad \Rightarrow \quad m = p$$

A **maximum (largest** *or* **top) element** $m \in P$ *is an element for which*

$$P \leq m$$

2) *A* **minimal element** *is an element* $n \in P$ *with the property that there is no smaller element in* P*, that is*

$$p \in P, p \leq n \quad \Rightarrow \quad p = n$$

A **minimum (smallest** *or* **bottom) element** n *in* P *is an element for which*

$$n \leq P \qquad \square$$

Definition (Upper and lower bounds) *Let* $(P, \leq)$ *be a partially ordered set. Let* S *be a subset of* P*.*

1) *An element* $u \in P$ *is an* **upper bound** *for* S *if*

$$S \leq u$$

The smallest upper bound u *for* S*, if it exists, is called the* **least upper bound** *or* **join** *of* S *and is denoted by* $\operatorname{lub}(S)$ *or* $\bigvee S$*. Thus,* u *has the property that* $S \leq u$ *and if* $S \leq x$ *then* $u \leq x$*. The join of a finite set* $S = \{a_1, \ldots, s_n\}$ *is also denoted by* $\operatorname{lub}\{a_1, \ldots, a_n\}$ *or* $a_1 \vee \cdots \vee a_n$*.*

2) *An element* $\ell \in P$ *is a* **lower bound** *for* S *if*

$$\ell \leq S$$

The largest lower bound ℓ *for* S*, if it exists, is called the* **greatest lower bound** *or* **meet** *of* S *and is denoted by* $\operatorname{glb}(S)$ *or* $\bigwedge S$*. Thus,* ℓ *has the property that* $\ell \leq S$ *and if* $x \leq S$ *then* $x \leq \ell$*. The meet of a finite set* $S = \{a_1, \ldots, a_n\}$ *is also denoted by* $\operatorname{glb}\{a_1, \ldots, a_n\}$ *or* $a_1 \wedge \cdots \wedge a_n$*.* $\square$

Note that the join of the empty set $\emptyset$ is, by definition, the least upper bound of the elements of $\emptyset$. But every element of P is an upper bound for the elements of $\emptyset$ and so the least upper bound is the minimum element of P, if it exists. Otherwise $\emptyset$ has no join. Similarly, the meet of the empty set is the greatest lower bound of $\emptyset$ and since all elements of P are lower bounds for $\emptyset$, the meet of $\emptyset$ is the maximum element of P, if it exists.

Now we can state Zorn's lemma, which gives a condition under which a partially ordered set has a maximal element.

Theorem 1.2 (Zorn's lemma) *If P is a partially ordered set in which every chain has an upper bound, then P has a maximal element.* □

Zorn's lemma is equivalent to the axiom of choice. As such, it is not subject to proof from the axioms of ZF set theory. Also, Zorn's lemma is equivalent to the *well-ordering principle.* A **well ordering** on a nonempty set X is a total order on X with the property that every nonempty subset of X has a least element.

Theorem 1.3 (Well-ordering principle) *Every nonempty set has a well ordering.* □

Order-Preserving and Order-Reversing Maps

A function $f\colon P \to Q$ between partially ordered sets is **order preserving** (also called **monotone** or **isotone**) if

$$x \le y \quad \Rightarrow \quad fx \le fy$$

and an **order embedding** if

$$x \le y \quad \Leftrightarrow \quad fx \le fy$$

Note that an order embedding is injective, since $fx = fy$ implies both $fx \le fy$ and $fy \le fx$, which implies that $x \le y$ and $y \le x$, that is, $x = y$. A surjective order-embedding is called an **order isomorphism**.

Similarly, a function $f\colon P \to Q$ is **order reversing** (also called **antitone**) if

$$x \le y \quad \Rightarrow \quad fx \ge fy$$

and an **order anti-embedding** if

$$x \le y \quad \Leftrightarrow \quad fx \ge fy$$

An order anti-embedding is injective and if it is surjective, then it is called an **order anti-isomorphism**.

Chain Conditions and Finiteness

The chain conditions are a form of finiteness condition on a poset.

Definition *Let P be a poset.*

1) *P has the* **ascending chain condition** (**ACC**) *if it has no infinite strictly ascending sequences, that is, for any ascending sequence*

$$p_1 \le p_2 \le p_3 \le \cdots$$

there is an index n such that $p_{n+k} = p_n$ for all $k \ge 0$.

2) *P has the* **descending chain condition** (**DCC**) *if it has no infinite strictly descending sequences, that is, for any descending sequence*

$$p_1 \ge p_2 \ge p_3 \ge \cdots$$

there is an index n such that $p_{n+k} = p_n$ for all $k \ge 0$.

3) *P has* **both chain conditions** (**BCC**) *if P has the ACC and the DCC.*□

The following characterizations of ACC and DCC are very useful.

Definition *Let P be a poset.*

1) *P has the* **maximal condition** *if every nonempty subset of P has a maximal element.*
2) *P has the* **minimal condition** *if every nonempty subset of P has a minimal element.*□

Theorem 1.4 *Let P be a poset.*

1) *P has the ACC if and only if it has the maximal condition.*
2) *P has the DCC if and only if it has the minimal condition.*

Proof. Suppose P has the ACC and let $S \subseteq P$ be nonempty. Let $s_1 \in S$. If s_1 is maximal we are done. If not, then we can pick $s_2 \in S$ such that $s_2 > s_1$. Continuing in this way, we either arrive at a maximal element in S or we get a strictly increasing ascending chain that does not become constant, which contradicts the ACC. Hence, P has the maximal condition. Conversely, if P has the maximal condition then any ascending sequence in P has a maximal element, at which point the sequence becomes constant. The proof of part 2) is similar.□

A poset can express "infinitness" by spreading vertically, via an infinite chain or by spreading horizontally, via an infinite antichain. The next theorem shows that these are the only two ways that a poset can express infiniteness. It also says that if a poset has an infinite chain, then it has either an infinite ascending chain or an infinite descending chain. This theorem will prove very useful to us as we explore chain conditions on subgroups of a group.

Theorem 1.5 *Let P be a poset.*

1) *The following are equivalent:*
   a) *P has no infinite chains.*
   b) *P has both chain conditions.*

If these conditions hold, then for any $a < b$ in P, there is a maximal finite chain from a to b.

2) *The following are equivalent:*
   a) *P has no infinite chains and no infinite antichains.*
   b) *P is finite.*

Proof. It is clear that 1a) implies 1b). For the converse, suppose that P has BCC and let $\mathcal{C}$ be an infinite chain. The ACC implies that $\mathcal{C}$ has a maximal element x_1, which must be maximum in $\mathcal{C}$ since $\mathcal{C}$ is totally ordered. Then $\mathcal{C} \setminus \{x_1\}$ is an infinite chain and we may select its maximum element $x_2 < x_1$. Continuing in this way gives an infinite strictly descending chain, a contradiction to the DCC. Hence, 1a) and 1b) are equivalent.

If 1a) and 1b) hold, then since $(a, b]$ is nonempty, it has a minimal member a_1, whence $a \prec a_1$ is a maximal chain from a to a_1. If $a_1 < b$, then $(a_1, b]$ has a minimal member a_2 and so $a \prec a_1 \prec a_2$ is a maximal chain from a to a_2. This cannot continue forever and so must produce a maximal finite chain from a to b.

For part 2), assume that P has no infinite chains or infinite antichains but that P is infinite. Using the ACC, we will create an infinite descending chain, in contradiction to the DCC. Since P has the maximal condition, it has a maximal element. Let

$$\mathcal{M} = \{m_i \mid i \in I\}$$

be the set of all maximal elements of P. Denote by $\downarrow x$ the set of all elements of P that are less than or equal to x. (This is read: *down* x.) Since $\mathcal{M}$ is a nonempty antichain, it must be finite. Moreover, the ACC implies that

$$P = \bigcup (\downarrow m_i)$$

and so one of the sets, say $\downarrow m_{i_1}$, must be infinite. The infinite poset

$$P_1 = (\downarrow m_{i_1}) \setminus \{m_{i_1}\}$$

also has no infinite antichains and no infinite chains. Thus, we may repeat the above process and select an element $m_{i_2} \in P_1$ such that

$$P_2 = (\downarrow m_{i_2}) \setminus \{m_{i_2}\}$$

is infinite. Note that $m_{i_1} > m_{i_2}$. Continuing in this way, we get an infinite strictly descending chain.□

The presence of a chain condition on a poset P has consequences for meets and joins.

Theorem 1.6

1) *Let P be a poset in which every nonempty finite subset has a meet. If P has the DCC, then every nonempty subset of P has a meet.*

2) *Let P be a poset in which every nonempty finite subset has a join. If P has the ACC, then every nonempty subset of P has a join.*

Proof. For part 1), let $S \subseteq P$ be nonempty. The family of all meets of finite subsets of S has a minimal member m in P and the minimality of m implies that $m \wedge a = m$ for all $a \in P$, that is, $m \leq a$ for all $a \in P$. Hence,

$$\bigwedge\nolimits_{a \in S} a = m \in P$$

We leave proof of part 2) to the reader.□

Lattices

Many of the partially ordered sets that we will encounter have a bit more structure.

Definition

1) *A partially ordered set $(P, \leq)$ is a* **lattice** *if every two elements of P have a meet and a join.*
2) *A partially ordered set $(P, \leq)$ is a* **complete lattice** *if every subset of P has a meet and a join.*□

Thus, a complete lattice has a maximum element (the join of P) and a minimum element (the meet of P).

Note that if $f\colon P \to Q$ is an order isomorphism of the lattices P and Q, then f preserves meets and joins, that is,

$$f\left(\bigwedge p_i\right) = \bigwedge f p_i \quad \text{and} \quad f\left(\bigvee p_i\right) = \bigvee f p_i$$

However, an order *embedding* need not preserve these operations.

We will often encounter partially ordered sets P for which every subset of elements has a meet. In this case, joins also exist and P is a complete lattice.

Theorem 1.7 *Suppose that $(P, \leq)$ is a partially ordered set for which every subset of P has a meet. Then $(P, \leq)$ is a complete lattice, where the join of a subset S of P is the meet of all upper bounds for S.*

Proof. First, note that the meet of the empty set is the maximum element of P and the meet of P is the minimum element of P and so P is bounded. In particular, the join of $\emptyset$ exists.

Let S be a nonempty subset of P. The family U of upper bounds for S is nonempty, since it contains the maximum element. We need only show that the meet $m = \bigwedge U$ is the join of S. Since $s \leq U$ for any $s \in S$, that is, any $s \in S$ is a lower bound for U, it follows that $s \leq m$, that is, $m \geq S$. Moreover, if $n \geq S$ then $n \in U$ and so $m \leq n$. Hence, m is the least upper bound of S.□

The previous theorem is very useful in many algebraic contexts. In particular, suppose that X is a nonempty set and that $\mathcal{F}$ is a family of subsets of X that contains both $\emptyset$ and X and is closed under intersection. (Examples are the subspaces of a vector space, the subgroups of a group, the ideals in a ring, the subfields of a field, the sublattices of a latttice and so on.) Then $\mathcal{F}$ is a complete lattice where the join of any subfamily $\mathcal{G}$ of $\mathcal{F}$ is the intersection of all members of $\mathcal{F}$ containing the members of $\mathcal{G}$. Note that this join need not be the union of $\mathcal{G}$, since the union may not be a member of $\mathcal{F}$. However, if the union of $\mathcal{G}$ is a member of $\mathcal{F}$, then it will be the join of $\mathcal{G}$.

Example 1.8

1) The set $\mathbb{R}$ of real numbers, with the usual binary relation $\leq$, is a partially ordered set. It is also a totally ordered set. It has no maximal elements.
2) The set $\mathbb{N} = \{0, 1, \dots\}$ of natural numbers, together with the binary relation of divides, is a partially ordered set. It is customary to write $n \mid m$ to indicate that n divides m. The subset S of $\mathbb{N}$ consisting of all powers of 2 is a totally ordered subset of $\mathbb{N}$, that is, it is a chain in $\mathbb{N}$. The set $P = \{2, 4, 8, 3, 9, 27\}$ is a partially ordered set under $\mid$. It has two maximal elements, namely 8 and 27. The subset $Q = \{2, 3, 5, 7, 11\}$ is a partially ordered set in which every element is both maximal and minimal. The partially ordered set $\mathbb{N}$ is a complete lattice but the set of all positive integers under division is a lattice that is not complete.
3) Let S be any set and let $\mathcal{P}(S)$ be the power set of S, that is, the set of all subsets of S. Then $\mathcal{P}(S)$, together with the subset relation $\subseteq$, is a complete lattice.□

Sublattices

The subject of sublattices requires a bit of care, since a nonempty subset S of a lattice L inherits the order of L but not necessarily the meets and joins of L. That is, the meet of a subset T of S may be different when T is viewed as a subset of S than when T is viewed as a subset of L.

For example, let $L = \{1, 2, 3, 6, 12\}$ under division and let $S = L \setminus \{6\}$. Then L and S are both lattices under the same partial order. However, in L we have $2 \vee 3 = 6$ and in S we have $2 \vee 3 = 12$. Let us use the term L-meet to refer to the meet in L, and similarly for join.

Definition *Let L be a lattice and let $M \subseteq L$ be a nonempty subset of L.*

1) *M is a* **sublattice** *of L if the M-meet of any finite nonempty subset $S \subseteq M$ exists and is the same as the L-meet of S, and similarly for join, that is, if*
$$\bigwedge\nolimits_M S = \bigwedge\nolimits_L S \quad \text{and} \quad \bigvee\nolimits_M S = \bigvee\nolimits_L S$$
2) *If L is a complete lattice, then M is a* **complete sublattice** *of L if the M-meet of any subset $S \subseteq M$ exists and is the same as the L-meet of S, and*

similarly for join, that is, if

$$\bigwedge_M S = \bigwedge_L S \quad \textit{and} \quad \bigvee_M S = \bigvee_L S \qquad \square$$

Theorem 1.9

1) *A nonempty subset P of a lattice L is a sublattice of L if and only if the L-meet and the L-join of any finite nonempty subset $A \subseteq P$ are in P.*
2) *A nonempty subset P of a complete lattice L is a complete sublattice of L if and only if the L-meet and the L-join of any subset $A \subseteq P$ are in P.* $\square$

Equivalence Relations

The concept of an equivalence relation plays a major role in mathematics.

Definition *Let S be a nonempty set. A binary relation $\sim$ on S is called an* **equivalence relation** *on S if it satisfies the following conditions:*

1) **(Reflexivity)** *For all $a \in S$,*
$$a \sim a$$
2) **(Symmetry)** *For all $a, b \in S$,*
$$a \sim b \quad \Rightarrow \quad b \sim a$$
3) **(Transitivity)** *For all $a, b, c \in S$,*
$$a \sim b, b \sim c \quad \Rightarrow \quad a \sim c \qquad \square$$

Definition *Let $\sim$ be an equivalence relation on S. For $a \in S$, the set of all elements equivalent to a is denoted by*

$$[a] = \{b \in S \mid b \sim a\}$$

and is called the **equivalence class** *of a.* $\square$

Theorem 1.10 *Let $\sim$ be an equivalence relation on S. Then*

1) $b \in [a] \Leftrightarrow a \in [b] \Leftrightarrow [a] = [b]$
2) *For any $a, b \in S$, we have either $[a] = [b]$ or $[a] \cap [b] = \emptyset$.* $\square$

Definition *A* **partition** *of a nonempty set S is a collection $\mathcal{P} = \{A_i \mid i \in I\}$ of* nonempty *subsets of S, called the* **blocks** *of the partition, for which*

1) $A_i \cap A_j = \emptyset$ *for all $i \neq j$*
2) $S = \bigcup_{i \in I} A_i$

A **system of distinct representatives***, abbreviated* **SDR***, for a partition $\mathcal{P}$ is a set consisting of exactly one element from each block of $\mathcal{P}$. In various contexts, a system of distinct representatives is also called a* **transversal** *for $\mathcal{P}$ or a set of* **canonical forms** *for $\mathcal{P}$.* $\square$

The following theorem sheds considerable light on the concept of an equivalence relation.

Theorem 1.11

1) *Let* $\sim$ *be an equivalence relation on a nonempty set* S*. Then the set of distinct equivalence classes with respect to* $\sim$ *are the blocks of a partition of* S*.*
2) *Conversely, if* $\mathcal{P}$ *is a partition of* S*, the binary relation* $\sim$ *defined by*

$$a \sim b \text{ if } a \text{ and } b \text{ lie in the same block of } \mathcal{P}$$

is an equivalence relation on S*, whose equivalence classes are the blocks of* $\mathcal{P}$*.*

This establishes a one-to-one correspondence between equivalence relations on S *and partitions of* S*.* □

The most important problem related to equivalence relations is that of finding an efficient way to determine when two elements are equivalent. Unfortunately, in most cases, the definition does not provide an efficient test for equivalence and so we are led to the following concepts.

Definition *Let* $\sim$ *be an equivalence relation on a nonempty set* S*. A function* $f: S \to T$*, where* T *is any set, is called an* **invariant** *of the equivalence relation if it is constant on the equivalence classes, that is, if*

$$a \sim b \quad \Rightarrow \quad f(a) = f(b)$$

A function $f: S \to T$ *is called a* **complete invariant** *if it is constant and distinct on the equivalence classes, that is, if*

$$a \sim b \quad \Leftrightarrow \quad f(a) = f(b)$$

A collection $\{f_1, \ldots, f_n\}$ *of invariants is called a* **complete system of invariants** *if*

$$a \sim b \quad \Leftrightarrow \quad f_i(a) = f_i(b) \text{ for all } i = 1, \ldots, n$$ □

Definition *Let* $\sim$ *be an equivalence relation on a nonempty set* S*. A subset* $C \subseteq S$ *is said to be a set of* **canonical forms** *for the equivalence relation if* C *is a system of distinct representatives for the partition consisting of the equivalence classes, that is, if for every* $s \in S$*, there is exactly one* $c \in C$ *such that* $c \sim s$*.* □

A set of canonical forms determines equivalence since $a, b \in S$ are equivalent if and only if their corresponding canonical forms are equal. Of course, this will be a *practical* solution to the problem of equivalence only if there is a practical way to identify the canonical form associated with each element of S. Often, canonical forms provide more of a theoretical tool than a practical one.

Cardinality

Two sets S and T have the same **cardinality**, written

$$|S| = |T|$$

if there is a bijective function (a one-to-one correspondence) between the sets. If S is in one-to-one correspondence with a *subset* of T, we write $|S| \leq |T|$. If S is in one-to-one correspondence with a *proper* subset of T but not with T itself, then we write $|S| < |T|$. The second condition is necessary, since, for instance, $\mathbb{N}$ is in one-to-one correspondence with a proper subset of $\mathbb{Z}$ and yet $\mathbb{N}$ is also in one-to-one correspondence with $\mathbb{Z}$ itself. Hence, $|\mathbb{N}| = |\mathbb{Z}|$.

This is not the place to enter into a detailed discussion of cardinal numbers. The intention here is that the cardinality of a set, whatever that is, represents the "size" of the set. It is actually easier to talk about two sets having the same, or different, cardinality than it is to define explicitly the cardinality of a given set (a cardinal number is a special kind of ordinal number).

For us, it is sufficient simply to associate with each set S a special kind of set known as a **cardinal number**, denoted by $|S|$ or $\text{card}(S)$, that is intended to measure the size of the set. In the case of finite sets, the cardinality is the integer that equals the number of elements in the set.

Definition

1) *A set is* **finite** *if it can be put in one-to-one correspondence with a set of the form $\mathbb{Z}_n = \{0, 1, \ldots, n-1\}$, for some nonnegative integer n. A set that is not finite is* **infinite**.
2) *The cardinal number of the set $\mathbb{N}$ of natural numbers is $\aleph_0$ (read "aleph nought"), where $\aleph$ is the first letter of the Hebrew alphabet. Hence,*

$$|\mathbb{N}| = |\mathbb{Z}| = |\mathbb{Q}| = \aleph_0$$

3) *Any set with cardinality $\aleph_0$ is called a* **countably infinite** *set and any finite or countably infinite set is called a* **countable** *set. An infinite set that is not countable is said to be* **uncountable**. □

Theorem 1.12

1) **(Schröder–Bernstein Theorem)** *For any sets S and T,*

$$|S| \leq |T| \text{ and } |T| \leq |S| \Rightarrow |S| = |T|$$

2) **(Cantor's theorem)** *If $\mathcal{P}(S)$ denotes the power set of S then*

$$|S| < |\mathcal{P}(S)|$$

3) *If* $\mathcal{P}_0(S)$ *denotes the set of all finite subsets of* S *and if* S *is an infinite set, then*

$$|S| = |\mathcal{P}_0(S)|$$

□

Cardinal Arithmetic

If S and T are sets, the **cartesian product** $S \times T$ is the set of all ordered pairs

$$S \times T = \{(s,t) \mid s \in S, t \in T\}$$

If two sets X and Y are disjoint, their union is called a disjoint union and is denoted by

$$X \sqcup Y$$

More generally, the **disjoint union** of two arbitrary sets S and T is the set

$$S \sqcup T = \{(s,0) \mid s \in S\} \cup \{(t,1) \mid t \in T\}$$

This is just a scheme for taking the union of S and T while at the same time assuring that there is no "collapse" due to the fact that the intersection of S and T may not be empty.

Definition *Let* κ *and* λ *denote cardinal numbers. Let* S *and* T *be sets for which* $|S| = \kappa$ *and* $|T| = \lambda$.

1) *The* **sum** $\kappa + \lambda$ *is the cardinal number of the disjoint union* $S \sqcup T$.
2) *The* **product** $\kappa\lambda$ *is the cardinal number of* $S \times T$.
3) *The* **power** κ^λ *is the cardinal number of the set of all functions from* T *to* S.□

We will not go into the details of why these definitions make sense. (For instance, they seem to depend on the sets S and T, but in fact they do not.) It can be shown, using these definitions, that cardinal addition and multiplication are associative and commutative and that multiplication distributes over addition.

Theorem 1.13 *Let* κ, λ *and* μ *be cardinal numbers. Then the following properties hold:*

1) **(Associativity)**

$$\kappa + (\lambda + \mu) = (\kappa + \lambda) + \mu \text{ and } \kappa(\lambda\mu) = (\kappa\lambda)\mu$$

2) **(Commutativity)**

$$\kappa + \lambda = \lambda + \kappa \text{ and } \kappa\lambda = \lambda\kappa$$

3) **(Distributivity)**

$$\kappa(\lambda + \mu) = \kappa\lambda + \kappa\mu$$

4) **(Properties of Exponents)**
   a) $\kappa^{\lambda+\mu} = \kappa^\lambda \kappa^\mu$
   b) $(\kappa^\lambda)^\mu = \kappa^{\lambda\mu}$
   c) $(\kappa\lambda)^\mu = \kappa^\mu \lambda^\mu$ □

On the other hand, the arithmetic of cardinal numbers can seem a bit strange, as the next theorem shows.

Theorem 1.14 *Let κ and λ be cardinal numbers, at least one of which is infinite. Then*

$$\kappa + \lambda = \kappa\lambda = \max\{\kappa, \lambda\}$$ □

It is not hard to see that there is a one-to-one correspondence between the power set $\mathcal{P}(S)$ of a set S and the set of all functions from S to $\{0, 1\}$. This leads to the following theorem.

Theorem 1.15 *For any cardinal κ*
1) *If $|S| = \kappa$ then $|\mathcal{P}(S)| = 2^\kappa$*
2) $\kappa < 2^\kappa$ □

We have already observed that $|\mathbb{N}| = \aleph_0$. It can be shown that $\aleph_0$ is the smallest infinite cardinal, that is,

$$\kappa < \aleph_0 \quad \Rightarrow \quad \kappa \text{ is a natural number}$$

It can also be shown that the set $\mathbb{R}$ of real numbers is in one-to-one correspondence with the power set $\mathcal{P}(\mathbb{N})$ of the natural numbers. Therefore,

$$|\mathbb{R}| = 2^{\aleph_0}$$

The set of all points on the real line is sometimes called the **continuum** and so $2^{\aleph_0}$ is sometimes called the **power of the continuum** and denoted by c.

The previous theorem shows that cardinal addition and multiplication have a kind of "absorption" quality, which makes it hard to produce larger cardinals from smaller ones. The next theorem demonstrates this more dramatically.

Theorem 1.16
1) *Addition applied a positive countable number of times or multiplication applied a finite number of times to the cardinal number $\aleph_0$ yields $\aleph_0$. Specifically, for any nonzero $n \in \mathbb{N}$, we have*

$$\aleph_0 \cdot \aleph_0 = \aleph_0 \quad \textit{and} \quad \aleph_0^n = \aleph_0$$

2) *Addition and multiplication applied a positive countable number of times to the cardinal number* $2^{\aleph_0}$ *yields* $2^{\aleph_0}$. *Specifically,*

$$\aleph_0 \cdot 2^{\aleph_0} = 2^{\aleph_0} \quad \textit{and} \quad (2^{\aleph_0})^{\aleph_0} = 2^{\aleph_0} \qquad \square$$

Using this theorem, we can establish other relationships, such as

$$2^{\aleph_0} \leq (\aleph_0)^{\aleph_0} \leq (2^{\aleph_0})^{\aleph_0} = 2^{\aleph_0}$$

which, by the Schröder–Bernstein theorem, implies that

$$(\aleph_0)^{\aleph_0} = 2^{\aleph_0}$$

We mention that the problem of evaluating κ^λ in general is a very difficult one and would take us far beyond the scope of this book.

We conclude with the following reasonable-sounding result, whose proof is omitted.

Theorem 1.17 *Let* $\{A_k \mid k \in K\}$ *be a collection of sets with an index set of cardinality* $|K| = \kappa$. *If* $|A_k| \leq \lambda$ *for all* $k \in K$, *then*

$$\left| \bigcup_{k \in K} A_k \right| \leq \lambda\kappa \qquad \square$$

Miscellanea

The following section need not be read until it is referenced much later in the book. If p is prime, then we will have occasion to write an integer α satisfying

$$\alpha \equiv 1 \bmod p$$

in the form $\alpha = 1 + bp^t$ where $p \nmid b$. However, the case where $p = 2$ and $\alpha = 1 + 2d$ with d odd is exceptional. In this case, we will need to write $\alpha = -1 + b2^t$, where b is odd and $t \geq 2$. This will ensure that $t \geq 2$ when $p = 2$. Accordingly, it will be useful to introduce the following terminology.

Definition *Let* $\alpha \equiv 1 \bmod p$.

1) *If* $p = 2$ *and* $\alpha = 1 + 2d$ *where* d *is odd, that is, if* $\alpha \equiv 3 \bmod 4$, *then the* ***p*-standard form** *of* α *is*

$$\alpha = -1 + bp^t, \quad p \nmid b \textit{ and } t \geq 2$$

2) *In all other cases, the* ***p*-standard form** *of* α *is*

$$\alpha = 1 + bp^t, \quad p \nmid b \qquad \square$$

Theorem 1.18 Let p be a prime and let $d \geq 1$. Let $o_p(n)$ denote the largest exponent e for which p^e divides n.

1) *For* $1 \le k \le p^d$,

$$o_p\left[\binom{p^d}{k}\right] = d - o_p(k)$$

In particular,

$$p^{d-k+1} \,\Big|\, \binom{p^d}{k}$$

and if $p > 2$ *and* $k \ge 2$ *or if* $p = 2$ *and* $k \ge 3$, *then*

$$p^{d-k+2} \,\Big|\, \binom{p^d}{k}$$

2) *If the* p*-standard form of* α *is*

$$\alpha = e + bp^t$$

then for any $d \ge 0$,

$$\alpha^{p^d} = e^{p^d} + wp^{d+t}$$

where $p \nmid w$.

Proof. For part 1), write

$$\binom{p^d}{k} = \frac{p^d}{k}\frac{(p^d-1)}{1}\cdots\frac{(p^d-u)}{u}\cdots\frac{p^d-(k-1)}{k-1}$$

where $1 \le u \le k-1$. Now, if $1 \le i \le d$, then $p^i \mid u$ if and only if $p^i \mid p^d - u$ and so

$$p^n \,\Big|\, \binom{p^d}{k} \quad \Leftrightarrow \quad n \le d - o_p(k)$$

The rest follows from the fact that $p^v \mid k$ implies $v \le k-1$ and if $p > 2$ and $k \ge 2$ or if $p = 2$ and $k \ge 3$, then $p^v \mid k$ implies $v \le k-2$.

For part 2), if the p-standard form for α is $\alpha = e + bp^t$, then

$$\alpha^{p^d} = (e+bp^t)^{p^d} = e^{p^d} + e^{p^d-1}bp^{d+t} + \sum_{k=2}^{p^d}\binom{p^d}{k}e^{p^d-k}b^k p^{tk}$$

where the terms in the final sum are 0 if $d = 0$. If $p > 2$, then part 1) implies that the kth term in the final sum is divisible by p to the power

$$d - k + 2 + tk = d + t + 1 + [1 + t(k-1) - k] \ge d + t + 1$$

If $p = 2$, then $t \ge 2$ and so the kth term in the final sum is divisible by p to the power

$$d - k + 1 + tk = d + t + 1 + [t(k-1) - k] \geq d + t + 1$$

Hence, in both cases, the final sum is divisible by p^{d+t+1} and so

$$\alpha^{p^d} = e^{p^d} + e^{p^d-1}bp^{d+t} + vp^{d+t+1} = e^{p^d} + p^{d+t}(e^{p^d-1}b + vp)$$

where $p \nmid (e^{p^d-1}b + vp)$.□

Chapter 2
Groups and Subgroups

Operations on Sets

We begin with some preliminary definitions before defining our principal object of study. For a nonemtpy set X, the n-fold cartesian product is denoted by

$$X^n = \underbrace{X \times \cdots \times X}_{n \text{ factors}}$$

Definition *Let X be a nonempty set and let n be a natural number.*
1) For $n \geq 1$, an **n-ary operation** *on X is a function*

$$f: X^n \to X$$

2) A 1-ary operation $f: X \to X$ is called a **unary operation** *on X.*
3) A 2-ary operation $f: X \times X \to X$ is called a **binary operation** *on X.*
4) A **nullary operation** *on X is an element of X.*
An n-ary operation, for any natural number n, is referred to as a **finitary operation**.□

It is often the case that the result of applying a binary operation is denoted by juxtaposition, writing ab in place of $f(a,b)$.

Definition *If $f: X^n \to X$ is an n-ary operation on X and if Y is a nonempty subset of X, then the restriction of f to Y^n is a map $f|_{Y^n}: Y^n \to X$. We say that Y is* **closed** *under the operation f if $f|_{Y^n}$ maps Y^n into Y. For a nullary operation $x \in X$, this means that $x \in Y$.*□

Groups

We are now ready to define our principal object of study.

Definition *A* **group** *is a nonempty set G, called the* **underlying set** *of the group, together with a binary operation on G, generally denoted by juxtaposition, with the following properties:*

1) **(Associativity)** *For all a, b, $c \in G$,*

$$(ab)c = a(bc)$$

2) **(Identity)** *There exists an element $1 \in G$, called the* **identity** *element of the group, for which*

$$1a = a1 = a$$

for all $a \in G$.

3) **(Inverses)** *For each $a \in G$, there is an element $a^{-1} \in G$, called the* **inverse** *of a, for which*

$$aa^{-1} = a^{-1}a = 1$$

Two elements $a, b \in G$ **commute** *if*

$$ab = ba$$

A group is **abelian**, *or* **commutative**, *if every pair of elements commute. A group is* **finite** *if the underlying set G is a finite set; otherwise, it is* **infinite**. *The* **order** *of a group is the cardinality of the underlying set G, denoted by $o(G)$ or $|G|$.*□

It is customary to use the phrase "G is a group" where G is the underlying set when the group operation under consideration is understood.

We leave it to the reader to show that the identity element in a group G is unique, as is the inverse of each element. Moreover, for $a, b \in G$,

$$(a^{-1})^{-1} = a \quad \text{and} \quad (ab)^{-1} = b^{-1}a^{-1}$$

In a group G, exponentiation is defined for integral exponents as follows:

$$a^n = \begin{cases} 1 & \text{if } n = 0 \\ \underbrace{a \cdots a}_{n \text{ factors}} & \text{if } n > 0 \\ (a^{-n})^{-1} & \text{if } n < 0 \end{cases}$$

When G is abelian, the group operation is often (but not always) denoted by $+$ and is called **addition**, the identity is denoted by 0 and called the **zero** element of the group and the inverse of an element $a \in G$ is denoted by $-a$ and is called the **negative** of a. In this case, exponents are replaced by multiples:

$$na = \begin{cases} 0 & \text{if } n = 0 \\ \underbrace{a + \cdots + a}_{n \text{ terms}} & \text{if } n > 0 \\ -(-na) & \text{if } n < 0 \end{cases}$$

The Order of an Element

Definition *Let G be a group.*

1) *If $a \in G$, then any integer n for which*
$$a^n = 1$$
is called an **exponent** *of a.*
2) *The smallest positive exponent of $a \in G$, if it exists, is called the* **order** *of a and is denoted by $o(a)$. If a has no exponents, then a is said to have* **infinite order**. *An element of finite order is said to be* **periodic** *or* **torsion**.
3) An element of order 2 is called an **involution**.□

Theorem 2.1 *Let G be a group. If $a \in G$ has finite order $o(a)$, then the exponents of a are precisely the integral multiples of $o(a)$.*
Proof. Let $n = o(a)$. Any integral multiple of n is clearly an exponent of a. Conversely, if $a^m = 1$, then $m = qn + r$ where $0 \le r < n$. Hence,
$$1 = a^m = a^{qn+r} = a^{qn}a^r = a^r$$
and so the minimality of n implies that $r = 0$, whence $m = qn$ is an integral multiple of n.□

Involutions arise often in the theory of groups. Proof of the following result is left as an exercise.

Theorem 2.2 *A group in which every nonidentity element is an involution is abelian.*□

As we will see in a moment, a group may have elements of finite order and elements of infinite order.

Definition *A group G is said to be* **periodic** *or* **torsion** *if every element of G is periodic. A group that has no periodic elements other than the identity is said to be* **aperiodic** *or* **torsion free**.□

It is not hard to see that every finite group is periodic. On the other hand, as we will see, there are infinite periodic groups, that is, an infinite group need not have any elements of infinite order.

Examples

Here are some examples of groups.

Example 2.3 The simplest group is the **trivial group** $G = \{1\}$, which contains only the identity element. All other groups are said to be **nontrivial**.□

Example 2.4 The integers $\mathbb{Z}$ form an abelian group under addition. The identity is 0. The rational numbers $\mathbb{Q}$ form an abelian group under addition and the *nonzero* rational numbers $\mathbb{Q}^*$ form an abelian group under multiplication. A similar statement holds for the real numbers $\mathbb{R}$ and the complex numbers $\mathbb{C}$ (and indeed for any field F).□

Example 2.5 (Cyclic groups) If a is a formal symbol, we can define a group G to be the set of all integral powers of a:

$$C_\infty(a) = \{a^i \mid i \in \mathbb{Z}\}$$

where the product is defined by the formal rules of exponents:

$$a^i a^j = a^{i+j}$$

This group is also denoted by C_∞ or $\langle a \rangle$ and is called the **cyclic group generated by** a. The identity of $\langle a \rangle$ is $1 = a^0$.

We can also create a finite group $C_n(a)$ of positive order n by setting

$$C_n(a) = \{1 = a^0, a, a^2, \ldots, a^{n-1}\}$$

where the product is defined by addition of exponents, followed by reduction modulo n:

$$a^i a^j = a^{(i+j) \bmod n}$$

This defines a group of order n, called a **cyclic group of order** n. The inverse of a^k is $a^{(-k) \bmod n}$. The group $C_n(a)$ is also denoted by C_n or $\langle a \rangle$. Note that for any integer k, the symbol a^k refers to the element of $C_n(a)$ obtained by multiplying together k copies of a. Hence,

$$a^k = a^{k \bmod n}$$

and so for any integers k and j,

$$a^k a^j = a^{k \bmod n} a^{j \bmod n} = a^{k \bmod n + j \bmod n} = a^{(k+j) \bmod n} = a^{k+j}$$

Thus, we can feel free to represent the elements of $C_n(a)$ using all integral powers of a and the rules of exponents will hold, although we must remember that a single element of $C_n(a)$ has many representations as powers of a. The groups $C_n(a)$ or $C_\infty(a)$ are called **cyclic groups**.□

Example 2.6 The set

$$\mathbb{Z}_n = \{0, \ldots, n-1\}$$

of integers modulo a positive integer n is a cyclic group of order n under addition modulo n, generated by the element 1, since $k \in \mathbb{Z}_n$ is the sum of k ones. The notation $\mathbb{Z}/n\mathbb{Z}$ is preferred by some mathematicians since when p is a prime, the notation $\mathbb{Z}_p$ is also used for the p-adic integers. However, we will not use it in this way.

If p is a prime, then the set

$$\mathbb{Z}_p^* = \{1, \ldots, p-1\}$$

of nonzero elements of $\mathbb{Z}_p$ is an abelian group under multiplication modulo p. Indeed, by definition, the set F^* of nonzero elements of any field F is a group under multiplication. It is possible to prove (and we leave it as an exercise) that F^* is cyclic if and only if F is a *finite* field.

More generally, the set R^* of units of a commutative ring R with identity is a multiplicative group. In the case of the ring $\mathbb{Z}_n$, this group is

$$\mathbb{Z}_n^* = \{a \in \mathbb{Z}_n \mid (a, n) = 1\}$$

To see directly that this is a group, note that for each $a \in \mathbb{Z}_n^*$, there exists integers x and y such that $xa + yn = 1$. Hence, $xa \equiv 1 \bmod n$, that is, $x \in \mathbb{Z}_n^*$ is the inverse of a in $\mathbb{Z}_n^*$. The group $\mathbb{Z}_n^*$ is abelian and it is possible to prove (although with some work; see Theorem 4.43) that $\mathbb{Z}_n^*$ is cyclic if and only if $n = 2, 4, p^e$ or $2p^e$, where p is an odd prime.□

Example 2.7 (**Matrix groups**) The set $\mathcal{M}_{m,n}(F)$ of all $n \times m$ matrices over a field F is an abelian group under addition of matrices. The set $GL(n, F)$ of all nonsingular $n \times n$ matrices over F is a nonabelian (for $n > 1$) group under multiplication, known as the **general linear group**. The set $SL(n, F)$ of all $n \times n$ matrices over F with determinant equal to 1 is a group under multiplication, called the **special linear group**.□

Example 2.8 (**Functions**) Let G be a group and let X be a nonempty set. The set G^X of all functions from X to G is a group under product of functions, defined by

$$(fg)(x) = f(x)g(x)$$

for all $x \in X$. The identity in G^X is the function that sends all elements of X to the identity element $1 \in G$. This map is often referred to as the **zero map**. (We cannot call it the identity map!) Also, the set $\mathcal{F}$ of all *bijective* functions on G is a group under composition.□

Example 2.9 The set

$$G = \{(e_1, e_2, \ldots) \mid e_i \in \mathbb{Z}_2\}$$

of all infinite binary sequences, with componentwise addition modulo 2, is an infinite abelian group that is periodic, since

$$2(e_1,e_2,\ldots) = (0,0,\ldots)$$

and so every nonidentity element has order 2.□

The External Direct Product of Groups

One important method for creating a new group from existing groups is as follows. If $G_1,\ldots,G_n$ are groups, then the cartesian product

$$P = G_1 \times \cdots \times G_n$$

is a group under **componentwise product** defined by

$$(a_1,\ldots,a_n)(b_1,\ldots,b_n) = (a_1b_1,\ldots,a_nb_n)$$

where $a_i, b_i \in G_i$. The group P is called the **external direct product** of the groups $G_1,\ldots,G_n$. Although the notation $\times$ is often used for the external direct product of groups, we will use the notation

$$G_1 \boxtimes \cdots \boxtimes G_n$$

to distinguish it from the cartesian product *as a set.*

As an example, the direct product

$$V = C_2(a) \boxtimes C_2(b) = \{(1,1),(a,1),(1,b),(a,b)\}$$

of two cyclic groups of order 2 is called the **(Klein) 4-group** (and was called the **Vierergruppe** by Felix Klein in 1884). We will generalize the direct product construction to arbitrary (finite or infinite) families of groups in a later chapter.

Symmetric Groups

Let X be a nonempty set. A bijective function from X to itself is called a **permutation** of X. The set S_X of all permutations of X is a group under composition, with order $|X|!$ when X is finite. Also, S_X is nonabelian for $|X| \geq 3$. The group S_X is called the **symmetric group** or **permutation group** on the set X. The group of permutations of the set $I_n = \{1,\ldots,n\}$ is denoted by S_n and has order $n!$.

We will study permutation groups in detail in a later chapter, but we want to make a few remarks here for use in subsequent examples. (Proofs will be given later.) If $a_1,\ldots,a_k$ are distinct elements of X, the expression

$$(a_1 \cdots a_k)$$

denotes the permutation that sends a_i to a_{i+1} for $i = 1,\ldots,k-1$ and sends the last element a_k to the first element a_1. All other elements of X are held fixed. This permutation is called a k**-cycle** in S_X. For example, in $S_{\{1,2,3,4\}}$ the permutation $(1\,3\,4)$ sends 1 to 3, 3 to 4, 4 to 1 and 2 to itself. A 2-cycle $(a\,b)$ is called a **transposition**, since it simply transposes a and b, leaving all other elements of X fixed. We can now see why a permutation group with at least three elements a, b and c is nonabelian, since for example

$$(a\,b)(a\,c) \neq (a\,c)(a\,b)$$

(Composition is generally denoted by juxtaposition as above.)

The **support** of a permutation $\sigma \in S_X$ is the set of elements of X that are moved by σ, that is,

$$\text{supp}(\sigma) = \{x \in X \mid \sigma x \neq x\}$$

Two permutations $\sigma, \tau \in S_X$ are **disjoint** if their supports are disjoint. In particular, two cycles $(a_1 \cdots a_k)$ and $(b_1 \cdots b_m)$ are disjoint if the underlying sets $\{a_1, \ldots, a_k\}$ and $\{b_1, \ldots, b_m\}$ are disjoint. It is not hard to see that disjoint permutations commute, that is, if σ and τ are disjoint, then $\sigma\tau = \tau\sigma$.

It is also not hard to see that every permutation σ is a product (composition) of pairwise disjoint cycles, the product being unique except for the order of factors and the inclusion of 1-cycles. In fact, this is a direct result of the fact that the relation

$$x \equiv y \quad \text{if} \quad \sigma^k x = y \text{ for some } k \in \mathbb{Z}$$

is an equivalence relation and thereby induces a partition on I_n. (The reader is invited to write a complete proof at this time or to refer to Theorem 6.1.) This factorization is called the **cycle decomposition** of σ. The **cycle structure** of σ is the number of cycles of each length in the cycle decomposition of σ. For example, the permutation

$$\sigma = (1\,2\,3)(4\,5\,6)(7\,8)(9)$$

has cycle structure consisting of two cycles of length 3, one cycle of length 2 and one cycle of length 1.

It is easy to see that for $|X| \geq 2$, any cycle in S_X is a product of transpositions, since

$$(a_1 \cdots a_n) = (a_1\,a_n)(a_1\,a_{n-1}) \cdots (a_1\,a_3)(a_1\,a_2)$$

and

$$(a) = (a\,b)(a\,b)$$

Hence, the cycle decomposition implies that every permutation is a product of (not necessarily disjoint) transpositions. Although such a factorization is far from unique, we will show that the *parity* of the number of transpositions in the factorization is unique. In other words, if a permutation can be written as a product of an even number of transpositions, then all factorizations into a product of transpositions have an even number of transpositions. Such a permutation is called an **even permutation**. For example, since

$$(1\,3\,4) = (1\,4)(1\,3)$$

the permutation $(1\,3\,4)$ is even. Similarly, a permutation is **odd** if it can be written as a product of an odd number of transpositions. For example, the equation above for $(a_1 \cdots a_n)$ shows that a cycle of odd length is an even permutation and a cycle of even length is an odd permutation.

One of the most remarkable facts about the permutation groups is that every group has a "copy" that sits inside (is isomorphic to a subgroup of) some permutation group. For example, the Klein 4-group sits inside S_4 as follows:

$$V = \{\iota, (1\,2)(3\,4), (1\,3)(2\,4), (1\,4)(2\,3)\}$$

This is the content of *Cayley's theorem*, which we will discuss later. Thus, if we knew "everything" about permutation groups, we would know "everything" about all groups!

The Order of a Product

One must be very careful not to jump to false conclusions about the order of the product of elements in a group. For example, consider the general linear group $GL(2, \mathbb{C})$ of all nonsingular 2×2 matrices over the complex numbers. Let

$$A = \begin{pmatrix} 0 & -1 \\ 1 & 0 \end{pmatrix} \quad \text{and} \quad B = \begin{pmatrix} 0 & 1 \\ -1 & -1 \end{pmatrix}$$

We leave it to the reader to show that A and B have finite order but that their product AB has infinite order. On the other extreme, we have $o(aa^{-1}) = 1$ regardless of the value of $o(a)$. Thus, the order of a product of two nonidentity elements can be as small as 1 or as large as infinity.

On the other hand, the following key theorem relates the order of a *power* a^k of an element $a \in G$ to the order of a. It also tells us something quite specific about the order of the product of *commuting* elements.

Theorem 2.10 *Let G be a group and $a, b \in G$.*
1) If $o(a) = n$, then for $1 \le k < n$,

$$o(a^k) = \frac{n}{\gcd(n, k)}$$

In particular,

$$\langle a \rangle = \langle a^k \rangle \quad \Leftrightarrow \quad \gcd(o(a), k) = 1$$

2) *If* $o(a) = n$ *and* $d \mid n$, *then*

$$o(a^k) = d \quad \Leftrightarrow \quad k = r\frac{n}{d}, \text{ where } \gcd(r, d) = 1$$

3) *If* a *and* b *commute, then*

$$\frac{\operatorname{lcm}(o(a), o(b))}{\gcd(o(a), o(b))} \,\Big|\, o(ab) \,\Big|\, \operatorname{lcm}(o(a), o(b))$$

In particular,

$$\gcd(o(a), o(b)) = 1 \quad \Rightarrow \quad o(ab) = o(a)o(b)$$

Proof. For part 1), Theorem 2.1 implies that $(a^k)^m = 1$ if and only if $n \mid km$. But

$$n \mid km \quad \Leftrightarrow \quad \frac{n}{\gcd(n,k)} \,\Big|\, \frac{km}{\gcd(n,k)} \quad \Leftrightarrow \quad \frac{n}{\gcd(n,k)} \,\Big|\, m$$

and so the smallest positive exponent m of a^k is $n/\gcd(n,k)$. For part 2), according to part 1), the equation $o(a^k) = d$ is equivalent to

$$\frac{n}{\gcd(n,k)} = d$$

which is equivalent to $\gcd(n,k) = n/d$, or

$$\gcd(d\frac{n}{d}, k) = \frac{n}{d}$$

But this holds if and only if $k = r(n/d)$ with $\gcd(d, r) = 1$.

For part 3), let $o(a) = \alpha$ and $o(b) = \beta$. Then $(ab)^\beta = a^\beta$ has order

$$o((ab)^\beta) = o(a^\beta) = \frac{\alpha}{\gcd(\alpha, \beta)}$$

But $o((ab)^\beta)$ divides $o(ab)$ and so

$$\frac{\alpha}{\gcd(\alpha, \beta)} \,\Big|\, o(ab)$$

A symmetric argument shows that

$$\frac{\beta}{\gcd(\alpha, \beta)} \,\Big|\, o(ab)$$

and since these divisors are relatively prime, we have

$$\frac{\alpha\beta}{\gcd(\alpha,\beta)^2}\,\Big|\, o(ab)$$

But $\alpha\beta/(\alpha,\beta)^2 = \text{lcm}(\alpha,\beta)/\gcd(\alpha,\beta)$.□

Corollary 2.11 *Let G be a group and let $a \in G$. If $o(a) = uv$, where $(u,v) = 1$, then*

$$a = a_1 a_2$$

where $a_1, a_2 \in \langle a \rangle$ and $o(a_1) = u$ and $o(a_2) = v$.
Proof. Since $(u,v) = 1$, there exist integers s and t for which $su + tv = 1$. Hence,

$$a = a^{su+tv} = a^{su} a^{tv}$$

Then $(su, v) = 1$ implies that $o(a^{su}) = v$ and similarly, $o(a^{tv}) = u$.□

Orders and Exponents

Let G be a group. We have defined an exponent of $a \in G$ to be any integer k for which $a^k = 1$. Here is the corresponding concept for subsets of a group.

Definition *If $S \subseteq G$ is nonempty, then an integer k for which $s^k = 1$ for all $s \in S$ is called an* **exponent** *of S. If S has an exponent, we say that S has* **finite exponent.**□

Note that many authors reserve the term exponent for the *smallest* such positive integer k. As with individual elements, if the subset S has an exponent, then all exponents of S are multiples of the smallest positive exponent of S.

Theorem 2.12 *Let G be a group. If a nonempty set S of G has finite exponent, then the set of all exponents of S is the set of all integer multiples of the smallest positive exponent of S.*□

For a finite group G, the smallest exponent minexp(G) is equal to the least common multiple lcmorders(G) of the orders of the elements of G. Thus, if maxorder(G) denotes the maximum order among the elements of G, then

$$\text{maxorder}(G) \mid \text{minexp}(G)$$

and there are simple examples to show that equality may or may not hold. (The reader is invited to find such examples.) However, in a finite *abelian* group, equality does hold.

Theorem 2.13
1) If G is a finite group, then

$$\operatorname{maxorder}(G) \mid \operatorname{minexp}(G) = \operatorname{lcmorders}(G)$$

2) If G is a finite abelian group, then all orders divide the maximum order and so

$$\operatorname{maxorder}(G) = \operatorname{minexp}(G) = \operatorname{lcmorders}(G)$$

and G is cyclic if and only if $\operatorname{minexp}(G) = o(G)$.

Proof. For the proof of part 2), let $a \in G$ have maximum order m. Suppose to the contrary that there is a $b \in G$ for which $o(b) \nmid m$. We will find an element of G of order greater than m, which is a contradiction. Since $o(b) \nmid m$, there is a prime p for which

$$o(a) = m = p^j v \quad \text{and} \quad o(b) = p^i u$$

where $p \nmid u$, $p \nmid v$ and $i > j$. Then

$$o(a^{p^j}) = \frac{m}{p^j} \quad \text{and} \quad o(b^u) = p^i$$

Since G is abelian and these orders are relatively prime, we have

$$o(a^{p^j} b^u) = mp^{i-j} > m$$

as promised. The last statement of the theorem follows from the fact that a finite group G is cyclic if and only if it has an element of order $o(G)$.$\square$

Conjugation

Let G be a group. If $a, b \in G$, then the element

$$b^a = aba^{-1}$$

is called the **conjugate** of b by a. The **conjugacy** relation is the binary relation on G defined by

$$a \equiv b \quad \text{if} \quad b = a^x \text{ for some } x \in G$$

and if $a \equiv b$, we say that a and b are **conjugate**. Conjugacy is an equivalence relation on G, since an element is conjugate to itself and if $a = b^x$ then $b = a^{x^{-1}}$ and finally, if $b = a^x$ and $c = b^y$, then

$$c = b^y = a^{xy}$$

Note that some authors define the conjugate of b by a as $a^{-1}ba$, so care must be taken when reading other literature.

The function $\gamma_a\colon G \to G$ defined by

$$\gamma_a x = x^a$$

is called **conjugation by** a. Conjugation is very well-behaved: It is a bijection and preserves the group operation, in the sense that

$$\gamma_a(xy) = (\gamma_a x)(\gamma_a y)$$

Thus, in the language of a later chapter, γ_a is a *group automorphism*. The maps γ_a are called **inner automorphisms** and the set of inner automorphisms is denoted by $\operatorname{Inn}(G)$. We will have more to say about $\operatorname{Inn}(G)$ in later chapters.

Theorem 2.14 *Let G be a group and let $a, b, x \in G$. Then*

1) *Conjugation is a bijection that preserves the group operation, that is,*

$$(x^a)^{-1} = (x^{-1})^a \quad \textit{and} \quad (xy)^a = x^a y^a$$

for all $x, y \in G$.

2) *The conjugation map satisfies*

$$(x^b)^a = x^{ab}$$

for all $x \in G$.

3) *Conjugacy is an equivalence relation on G. The equivalence classes under conjugacy are called* **conjugacy classes.**□

We can also apply conjugation to subsets of G. If $S \subseteq G$ and $a \in G$, we write

$$\gamma_a S = S^a = \{s^a \mid s \in S\}$$

The previous rules generalize to conjugation of sets. In fact, for any $S, T \subseteq G$ and $a, b \in G$, we have

$$(S^a)^b = S^{ba} \quad \text{and} \quad S^a = T^a \Leftrightarrow S = T$$

Conjugation in the Symmetric Group

In general, it is not always easy to tell when two elements of a group are conjugate. However, in the symmetric group, it is surprisingly easy.

Theorem 2.15 *Let S_n be the symmetric group.*

1) *Let $\sigma \in S_n$. For any k-cycle $(a_1 \cdots a_k)$, we have*

$$(a_1 \cdots a_k)^\sigma = (\sigma a_1 \cdots \sigma a_k)$$

Hence, if $\tau = c_1 \cdots c_k$ is a cycle decomposition of τ, then

$$\tau^\sigma = c_1^\sigma \cdots c_k^\sigma$$

is a cycle decomposition of τ^σ.

2) *Two permutations are conjugate if and only if they have the same cycle structure.*

Proof. For part 1), we have

$$(a_1 \cdots a_k)^\sigma (\sigma a_i) = \begin{cases} \sigma a_{i+1} & i < k \\ \sigma a_1 & i = k \end{cases}$$

Also, if $b \neq \sigma a_i$ for any i, then $\sigma^{-1} b \neq a_i$ and so

$$(a_1 \cdots a_k)^\sigma b = \sigma(a_1 \cdots a_k)(\sigma^{-1} b) = \sigma(\sigma^{-1} b) = b$$

Hence, $(a_1 \cdots a_k)^\sigma$ is the cycle $(\sigma a_1 \cdots \sigma a_k)$. For part 2), if $\tau = c_1 \cdots c_m$ is a cycle decomposition of τ, then

$$\tau^\sigma = c_1^\sigma \cdots c_m^\sigma$$

and since c_i^σ is a cycle of the same length as c_i, the cycle structure of τ^σ is the same as that of τ.

For the converse, suppose that σ and τ have the same cycle structure. If σ and τ are cycles, say

$$\sigma = (a_1 \cdots a_n) \quad \text{and} \quad \tau = (b_1 \cdots b_n)$$

then any permutation λ that sends a_i to b_i satisfies $\sigma^\lambda = \tau$. More generally, if

$$\sigma = c_1 \cdots c_m \quad \text{and} \quad \tau = d_1 \cdots d_m$$

are the cycle decompositions of σ and τ, ordered so that c_k has the same length as d_k for all k, we can define a permutation λ that sends the element in the ith position of c_k to the element in the ith position of d_k. Then $\sigma^\lambda = \tau$.□

The Set Product

It is convenient to extend the group operation on a group G from elements of G to subsets of G. In particular, if S and T are subsets of G, then the **set product** ST (also called the **complex product**, since subsets of a group are called **complexes** in some contexts) is defined by

$$ST = \{st \mid s \in S, t \in T\}$$

As a special case, we write $\{a\}S$ as aS, that is,

$$aS = \{as \mid s \in S\}$$

The set product is associative and distributes over union, but not over intersection. Specifically, for $S, T, U \subseteq G$,

$$\begin{gathered} S(TU) = (ST)U \\ S(T \cup U) = ST \cup SU \quad \text{and} \quad (T \cup U)S = TS \cup US \\ S(T \cap U) \subseteq ST \cap SU \quad \text{and} \quad (T \cap U)S \subseteq TS \cap US \end{gathered}$$

Also,

$$aS = aT \quad \Leftrightarrow \quad S = T$$

Of course, we may generalize the set product to any nonempty finite collection $S_1, \ldots, S_n$ of subsets of G by setting

$$S_1 \cdots S_n = \{s_1 \cdots s_n \mid s_i \in S_i\}$$

On the other hand, if k is a positive integer, then it is customary to let

$$S^k = \{s^k \mid s \in S\}$$

Thus, in general, S^2 is a *proper* subset of the set product SS.

Subgroups

The substructures of a group are defined as follows.

Definition *A nonempty subset H of a group G is a* **subgroup** *of G, denoted by $H \leq G$, if H is a group under the restricted product on G. If $H \leq G$ and $H \neq G$, we write $H < G$ and say that H is a* **proper subgroup** *of G. If $H_1, \ldots, H_n$ are subgroups of G, we write $H_1, \ldots, H_n \leq G$.*□

For example, $\mathbb{Z} \leq \mathbb{Q}$, since $\mathbb{Z}$ is an abelian group under addition. However, $\mathbb{Z}_p$ is not a subgroup of $\mathbb{Z}$, although it is a subset of $\mathbb{Z}$ and it is a group as well: The issue is that $\mathbb{Z}_p$ is not a group under ordinary addition of integers.

However, if H is a subgroup under the first definition above, then H satisfies the second definition. To see this, multiplying the equation $1_H 1_H = 1_H$ by the inverse of 1_H in G gives $1_H = 1_G$. Thus, for all $h \in H$, we have $hh_H^{-1} = 1 = hh_G^{-1}$ and so $h_H^{-1} = h_G^{-1}$.

There is another criterion for subgroups that involves checking only closure. Proof is left to the reader.

Theorem 2.16

1) *A nonempty subset X of a group G is a subgroup of G if and only if X is closed under the operations of taking inverses and products, that is, if and only if*

$$x \in X \quad \Rightarrow \quad x^{-1} \in X$$

and

$$x, y \in X \quad \Rightarrow \quad xy \in X$$

2) *A nonempty* finite *subset X of a group G is a subgroup of G if and only if it is closed under the taking of products.*□

Theorem 2.17 *The intersection of any nonempty family of subgroups of a group G is a subgroup of G.*$\square$

Example 2.18 The set A_n of all even permutations in S_n is a subgroup of S_n. To see that A_n is closed under the product, if σ and τ are even, then they can each be written as a product of an even number of transpositions. Hence, $\sigma\tau$ is also a product of an even number of transpositions and so is in A_n. The subgroup A_n is called the **alternating subgroup** of S_n.

For $n \geq 2$, the alternating subgroup A_n has order $n!/2$, that is, A_n is exactly half the size of S_n. To see this, note that $\sigma \in S_n$ is odd if and only if $(1\,2)\sigma$ is even and τ is even if and only if $(1\,2)\tau$ is odd. Hence, the map $\sigma \mapsto (1\,2)\sigma$ is a self-inverse bijection between A_n and the set of odd permutations.$\square$

A group has many important subgroups. One of the most important is the following.

Definition *The* **center** $Z(G)$ *of a group G is the set of all elements of G that commute with all elements of G, that is,*

$$Z(G) = \{a \in G \mid ab = ba \text{ for all } b \in G\}$$

A group G is **centerless** *if $Z(G) = \{1\}$. A subgroup H of G is* **central** *if H is contained in the center of G.*$\square$

Two subgroups of a group G can never be disjoint as sets, since each contains the identity of G. However, it will be very convenient to introduce the following terminology and notation.

Definition *Two subgroups H and K of a group G are* **essentially disjoint** *if*

$$H \cap K = \{1\}$$

We introduce the notation

$$H \bullet K$$

to denote the set product of two essentially disjoint subgroups and refer to this as the **essentially disjoint product** *of H and K.*$\square$

Note that if $o(H) = n$ and $o(K) = m$, then

$$o(H \bullet K) = nm$$

The Dedekind Law

The following formula involving the intersection and set product is very handy and we will use it often. The simple proof is left to the reader.

Theorem 2.19 *Let G be a group and let $A, B, C \leq G$ with $A \leq B$.*
1) **(Dedekind law)**

$$A(B \cap C) = B \cap AC$$

2)

$$A \cap C = B \cap C \text{ and } AC = BC \quad \Rightarrow \quad A = B$$

Proof. We leave proof of the Dedekind law to the reader. For part 2),

$$A = A(A \cap C) = A(B \cap C) = B \cap AC = B \cap BC = B$$ □

We leave it to the reader to find an example to show that the condition that $A \leq B$ is necessary in Dedekind's law, that is, $A(B \cap C)$ is not necessarily equal to $AB \cap AC$ unless $A \leq B$.

Subgroup Generated by a Subset

If G is a group and $a \in G$, then the **cyclic subgroup** generated by a is the subgroup

$$\langle a \rangle = \{a^k \mid k \in \mathbb{Z}\}$$

If $o(a) = n < \infty$, then $a^k = a^{k \bmod n}$ and so

$$\langle a \rangle = \{1, a, \ldots, a^{n-1}\}$$

Note that $\langle a \rangle$ is the smallest subgroup of G containing a, since any subgroup of G containing a must contain all powers of a.

More generally, if X is a nonempty subset of G, then the **subgroup generated by** X, denoted by $\langle X \rangle$, is defined to be the smallest subgroup of G containing X and X is called a **generating set** for $\langle X \rangle$. Such a subgroup must exist; in fact, Theorem 2.17 implies that $\langle X \rangle$ is the intersection of all subgroups of G containing X. The following theorem gives a very useful look at the elements of $\langle X \rangle$.

Theorem 2.20 *Let X be a nonempty subset of a group G and let $X^{-1} = \{x^{-1} \mid x \in X\}$. Let*

$$W = (X \cup X^{-1})^*$$

be the set of all words over the alphabet $X \cup X^{-1}$.
1) If we interpret juxtaposition in W as the group product in G and the empty word as the identity in G, then $\langle X \rangle = W$ and so

$$\langle X \rangle = \{x_1^{e_1} \cdots x_n^{e_n} \mid x_i \in X, e_i \in \mathbb{Z}, n > 0\} \cup \{1\}$$

2) *If G is abelian, then we can collect like factors and so*

$$\langle X\rangle = \{x_1^{e_1}\cdots x_n^{e_n} \mid x_i \in X, x_i \neq x_j \text{ for } i \neq j, e_i \in \mathbb{Z}, n > 0\} \cup \{1\}$$

Proof. It is clear that $W \subseteq \langle X\rangle$. It is also clear that W is closed under product. As to inverses, if $w = x_1^{e_1}\cdots x_n^{e_n} \in W$ then $w^{-1} = x_n^{-e_n}\cdots x_1^{-e_1} \in W$. Hence, $W \leq \langle X\rangle$. However, since $X \subseteq W$, it follows that $\langle X\rangle \leq W$ and so $W = \langle X\rangle$.□

Although the previous description of $\langle X\rangle$ is very useful, it does have one drawback: Distinct formal words in the set W may be the same element of the group $\langle X\rangle$, when juxtaposition in W is interpreted as the group product. For instance, the distinct words x^2x^{-1} and x are the same group element of $\langle X\rangle$. We will discuss this issue in detail when we discuss free groups later in the book: The matter need not concern us further until then.

Finitely-Generated Groups

A group G is **finitely generated** if $G = \langle X\rangle$ for some finite set X. If G has a generating set of size n, then G is said to be **n-generated** or to be an **n-generator group**.

The Burnside Problem

There is a fascinating set of problems revolving around the following question. A finite group G is obviously finitely generated and periodic. In 1902, Burnside [39] asked about the converse: Is a finitely-generated periodic group finite? This is the **general Burnside problem**.

A negative answer to the general Burnside problem took 62 years, when Golod [40] showed in 1964 that there are infinite groups that are 3-generated and whose elements each have order a power of a fixed prime p (the power depending upon the element). However, this still leaves open some refinements of the general Burnside problem. For example, the Golod groups have elements of arbitrarily large order p^n, that is, they do not have finite exponent, as do finite groups.

The **Burnside problem** is the problem of deciding, for finite integers n and m, whether every n-generated group of exponent m is finite. This problem has been the subject of a great deal of research since Burnside first formulated it in 1902. For example, it has been shown that there are infinite, finitely-generated groups of every odd exponent $m > 665$ (Adjan [38], 1979) and of every exponent of the form $2^k m$, where $k \geq 48$ (Ivanov [42], 1992).

The **restricted Burnside problem**, formulated in the 1930's, asks whether or not, for integers n and m, there are a finite number (up to isomorphism) of finite n-generated groups of exponent m. In 1994, Zelmanov answered this question in the affirmative. For more on the Burnside problem, we refer the reader to the references located at the back of the book.

Subgroups of Finitely-Generated Groups

A far simpler question related to finitely-generated groups is whether every subgroup of a finitely-generated group is finitely generated. We will show when we discuss free groups that for arbitrary groups this is false: There are finitely-generated groups with subgroups that are not finitely generated. However, in the abelian case, this cannot happen.

Theorem 2.21 *Any subgroup of an n-generated abelian group A is also n-generated. In particular, a subgroup of a cyclic group is cyclic.*
Proof. Let $H \le A$. The proof is by induction on n. If $n = 1$, then $A = \langle a \rangle$ is cyclic. Let k be the smallest positive exponent for which $a^k \in H$. Then $\langle a^k \rangle \le H$. However, if $a^m \in H$, then $m = qk + r$ where $0 \le r < k$ and so

$$a^r = a^{m-qk} = a^m (a^k)^{-q} \in H$$

which can only happen if $r = 0$, whence $a^m = (a^k)^q \in \langle a^k \rangle$. Thus, $H = \langle a^k \rangle$ is also cyclic and so the result holds for $n = 1$.

Assume the result is true for any group generated by fewer than $n \ge 2$ elements. Let $A = \langle x_1, \dots, x_n \rangle$ and let $A_{n-1} = \langle x_1, \dots, x_{n-1} \rangle$. By assumption, every subgroup of A_{n-1} is $(n-1)$-generated, in particular, there exist $h_i \in H$ for which

$$H \cap A_{n-1} = \langle h_1, \dots, h_{n-1} \rangle$$

Now, every $h \in H$ has the form $h = ax_n^e$ where $a \in A_{n-1}$ and $e \in \mathbb{Z}$. If $e = 0$ for all $h \in H$, then $H \le A_{n-1}$ and the inductive hypothesis implies that H is at most $(n-1)$-generated. So let us assume that $e \ne 0$ for some $h \in H$ and let s be the smallest positive integer for which $h_n = bx_n^s \in H$ and $b \in A_{n-1}$.

For an arbitrary $h = ax_n^e \in H$, where $a \in A_{n-1}$, write $e = qs + r$ where $0 \le r < s$. Then

$$h_n^{-q} h = (bx_n^s)^{-q}(ax_n^e) = ab^{-q}x_n^{e-qs} = ab^{-q}x_n^r$$

Since $h_n^{-q}h \in H$ and $ab^{-q} \in A_{n-1}$, the minimality of s implies that $r = 0$ and so $h_n^{-q}h \in A_{n-1}$, that is, $h \in \langle A_{n-1}, h_n \rangle$. Hence,

$$H \subseteq \langle A_{n-1}, h_n \rangle = \langle h_1, \dots, h_{n-1}, h_n \rangle \subseteq H$$

and so

$$H = \langle h_1, \dots, h_{n-1}, h_n \rangle$$

is n-generated.□

The Lattice of Subgroups of a Group

Let G be a group. The collection $\mathrm{sub}(G)$ of all subgroups of G is ordered by set inclusion. Moreover, $\mathrm{sub}(G)$ is closed under arbitrary intersection and has maximum element G. Hence, Theorem 1.7 implies that $\mathrm{sub}(G)$ is a complete lattice, where the meet of a family $\mathcal{F} = \{H_i \mid i \in I\}$ is the intersection

$$\bigwedge \mathcal{F} = \bigwedge_{i \in I} H_i = \bigcap_{i \in I} H_i$$

and the join of $\mathcal{F}$ is the smallest subgroup of G that contains all of the subgroups in $\mathcal{F}$, that is,

$$\bigvee \mathcal{F} = \bigvee_{i \in I} H_i = \bigcap \{S \leq G \mid H_i \leq S \text{ for all } i\}$$

It is also clear that the join of $\mathcal{F}$ is the subgroup generated by the union of the H_i's. Specifically, since $H_i^{-1} = H_i$, it follows that the subgroup generated by $\mathcal{F}$ is the set of all words over the union $\bigcup H_i$:

$$\bigvee_{i \in I} H_i = \left(\bigcup_{i \in I} H_i \right)^* = \{a_{i_1} \cdots a_{i_n} \mid a_{i_k} \in H_{i_k}, n \geq 0\}$$

The join of $\mathcal{F}$ is also denoted by $\langle \mathcal{F} \rangle$ and $\langle H_i \mid i \in I \rangle$.

Note that, in general, the union of subgroups is not a subgroup. (The reader may find an example in the group of integers.) However, if

$$H_1 \leq H_2 \leq \cdots$$

is an increasing sequence of subgroups of G (generally referred to as an *ascending chain* of subgroups), then it is easy to see that the union $\bigcup H_i$ is a subgroup of G. More generally, the union of any directed family of subgroups is a subgroup. (A family $\mathcal{D}$ of subgroups of G is **directed** if for every $A, B \in \mathcal{D}$, there is a $C \in \mathcal{D}$ for which $A \leq C$ and $B \leq C$.)

Theorem 2.22 *Let G be a group. Then* $\mathrm{sub}(G)$ *is a complete lattice, where meet is intersection and join is given by*

$$\bigvee_{i \in I} H_i = \left\langle \bigcup_{i \in I} H_i \right\rangle$$

Also, $\mathrm{sub}(G)$ *is closed under directed unions.* □

Hasse Diagrams

For a finite group of small order, we can sometimes describe the subgroup lattice structure using a **Hasse diagram**, which is a diagram of a partially ordered set that shows the covering relation.

Example 2.23 The Hasse diagrams for the subgroup lattices of the group $C_4 = \{1, a, a^2, a^3\}$ and the 4-group $V = \{1, a, b, ab\}$ are shown in Figure 2.1.□

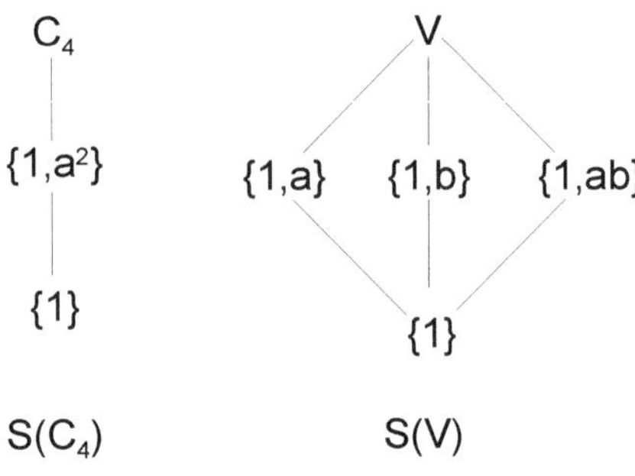

Figure 2.1

Maximal and Minimal Subgroups of a Group

The maximal and minimal subgroups of a group play an important role in the theory. At this point, we simply give the definitions.

Definition *Let G be a group and let $H \leq G$.*
1) *H is* **minimal** *if it is minimal in the partially ordered set of all nontrivial subgroups of G (under set inclusion).*
2) *H is* **maximal** *if it is maximal in the partially ordered set of all proper subgroups of G (under set inclusion).*□

We emphasize that the term maximal subgroup means maximal *proper* subgroup. Without the restriction to proper subgroups, G would be the only "maximal" subgroup of G. A similar statement can be made for minimal subgroups.

Subgroups and Conjugation

Since the conjugate of a subgroup is also a subgroup, conjugation sends $\text{sub}(G)$ to $\text{sub}(G)$. In fact, it is an order isomorphism of $\text{sub}(G)$.

Theorem 2.24 *Let G be a group and let $a \in G$. The conjugation map $\gamma_a: \text{sub}(G) \to \text{sub}(G)$ defined by $\gamma_a H = H^a$ is an order isomorphism on $\text{sub}(G)$. Hence, γ_a preserves meet and join, that is,*

$$\left[\bigcap H_i\right]^a = \bigcap H_i^a$$

and

$$\left[\bigvee H_i\right]^a = \bigvee H_i^a$$

Proof. It is clear that

$$A \leq B \quad \Leftrightarrow \quad A^a \leq B^a$$

and so γ_a is an order embedding of G into itself. But any subgroup A of G has the form $A = \gamma_a(\gamma_{a^{-1}}A)$ and so γ_a is also surjective.□

The Set Product of Subgroups

If H and K are subgroups of a group G, then the set product HK is not necessarily a subgroup of G, as the next example shows.

Example 2.25 In the symmetric group S_3, let

$$H = \{\iota, (1\,2)\} \quad \text{and} \quad K = \{\iota, (1\,3)\}$$

Then

$$HK = \{\iota, (1\,2), (1\,3), (1\,3\,2)\}$$

However, $(1\,3\,2)^2 = (1\,2\,3)$ is not in HK and so HK is not a subgroup of S_3.□

If HK is a subgroup of G, then since HK contains both H and K, we have $KH \subseteq HK$. Conversely, if $KH \subseteq HK$, then $HK \leq G$, since

$$(hk)^{-1} = k^{-1}h^{-1} \in KH \subseteq HK$$

and

$$(h_1k_1)(h_2k_2) = h_1(k_1h_2)k_2 = h_1(h_3k_3)k_2 \in HK$$

where $h_i \in H$ and $k_i \in K$. Thus,

$$HK \leq G \quad \Leftrightarrow \quad KH \subseteq HK$$

Moreover, if $KH \subseteq HK$, then equality must hold, since every $x \in HK$ has the form

$$x = (hk)^{-1} = k^{-1}h^{-1} \subseteq KH$$

for some $h \in H$ and $k \in K$. Thus $HK = KH$.

Theorem 2.26 *If $H, K \leq G$, then the following are equivalent:*
1) $HK \leq G$
2) $KH \subseteq HK$
3) $KH = HK$, *that is, H and K* **permute**.
In this case, $HK = H \vee K$ is the join of H and K in $\text{sub}(G)$.□

The Size of HK

There is a very handy formula for the size of the set product HK of two subgroups of a group G, which holds even if HK is not a subgroup of G. Consider the surjective map $f\colon H \times K \to HK$ defined by

$$f(h,k) = hk$$

The inverse map f^{-1} induces a partition $\mathcal{P}$ on $H \times K$ whose blocks are the sets $f^{-1}(x)$ for $x \in HK$. Hence, there are $|HK|$ blocks.

To detemine the size of these blocks, let $x = hk$. Then any element of $H \times K$ can be written in the form (hd, ek) for $d \in H$ and $e \in K$ and

$$f(hd, ek) = x \quad \Leftrightarrow \quad hdek = hk \quad \Leftrightarrow \quad de = 1 \quad \Leftrightarrow \quad e = d^{-1}$$

and so

$$f^{-1}(x) = \{(hd, d^{-1}k) \mid d \in H \cap K\}$$

Hence,

$$\left|f^{-1}(x)\right| = |H \cap K|$$

and so

$$|H||K| = |H \times K| = |HK||H \cap K|$$

as cardinal numbers.

Theorem 2.27 *If G is a group and $H, K \leq G$ then*

$$|H||K| = |HK||H \cap K|$$

as cardinal numbers. If H and K are finite subgroups, then

$$|HK| = \frac{|H||K|}{|H \cap K|}$$ □

The largest proper subgroups of a finite group G are the subgroups of order $o(G)/2$. The formula in Theorem 2.26 tells us something about how these large subgroups interact with other subgroups of G.

Theorem 2.28 *Let G be a finite group and let $H \leq G$ have order $o(G)/2$. Then any subgroup S of G is either a subgroup of H or else*

$$|S \cap H| = |S|/2$$

In words, S lies either completely in H or half-in and half-out of H. Also, if $a \in S \setminus H$, then

$$S = (S \cap H) \sqcup a(S \cap H)$$

Proof. If S is not contained in H, then there is an $a \in S \setminus H$ and so

$$SH \supseteq H \sqcup aH$$

But the latter has size $o(G)$ and so $SH = G$. Then

$$|H||S| = |G||H \cap S|$$

implies that $|S| = 2|H \cap S|$. The rest follows easily.□

Cosets and Lagrange's Theorem

Let $H \leq G$. For $a \in G$, the set aH is called a **left coset** of H in G. Similarly, the set Ha is called a **right coset** of H in G. The set of all left cosets of H in G is denoted by G/H and the set of all right cosets is denoted $H\backslash G$. We will refer to left cosets simply as cosets, using the adjectives "left" and "right" only to avoid ambiguity.

The map $f\colon G/H \to H\backslash G$ defined by $f(aH) = Ha^{-1}$ is easily seen to be a bijection and so

$$|G/H| = |H\backslash G|$$

Since the multiplication map $\mu_a\colon H \to aH$ defined by $\mu_a h = ah$ is a bijection, all cosets of a subgroup H have the same cardinality:

$$|aH| = |H|$$

To see that the distinct left cosets of H form a partition of G, we define an equivalence relation on G by

$$a \equiv b \quad \text{if} \quad aH = bH$$

Now, $aH = bH$ implies that $b \in aH$. Conversely, if $b \in aH$, then $b = ah$ for some $h \in H$ and so $bH = ahH = aH$. Hence,

$$a \equiv b \quad \Leftrightarrow \quad aH = bH \quad \Leftrightarrow \quad b \in aH$$

and so the equivalence class containing a is precisely the coset aH. Thus, the distinct cosets G/H form a partition of G. In particular,

$$|G| = |H| \cdot |G/H|$$

as cardinal numbers. When G is finite, this is the content of *Lagrange's theorem.*

Theorem 2.29 *Let G be a group and let $H \leq G$.*

1) *The set G/H of distinct left cosets of H in G forms a partition of G, with associated equivalence relation satisfying*

$$a \equiv b \quad \Leftrightarrow \quad aH = bH \quad \Leftrightarrow \quad b \in aH$$

This equivalence relation is called **equivalence modulo** H *and is denoted by*

$$a \equiv b \bmod H$$

or just $a \equiv b$ when the subgroup is clear. Each element $b \in aH$ is called a **coset representative** *for the coset aH, since $bH = aH$.*

2) *All cosets have the same cardinality and*

$$|G| = |H| \cdot |G/H|$$

as cardinal numbers.

3) **(Lagrange's theorem)** *If G is finite, then*

$$|G/H| = \frac{|G|}{|H|}$$

and so

$$o(H) \mid o(G)$$

In particular, the order of an element $a \in G$ divides the order of G.□

The converse of Lagrange's theorem fails: We will show later in the book that the alternating group A_4 has order 12 but has no subgroup of order 6.

Note also that

$$a \equiv b \bmod H \quad \Leftrightarrow \quad b^{-1}a \in H$$

and so, in particular,

$$a \equiv b \bmod H \quad \text{and} \quad a \equiv b \bmod K \quad \Rightarrow \quad a \equiv b \bmod (H \cap K)$$

Lagrange's Theorem and the Order of a Product

Lagrange's theorem tells us something about the order of certain products hk in a group even when h and k do not commute.

Theorem 2.30 *Let G be a group and let H and K be finite subgroups of G, with $HK \le G$.*

1) *If $h \in H$ and $k \in K$, then*

$$o(hk) \mid o(H)o(K)$$

In particular, if $\langle h \rangle \langle k \rangle \le G$ is finite, then

$$o(hk) \mid o(h)o(k)$$

2) *If $N \le G$ and $o(N)$ is relatively prime to both $o(H)$ and $o(K)$, then*

$$HN = KN \quad \Rightarrow \quad H = K$$

Proof. For part 1), Lagrange's theorem implies that

$$o(hk) \mid o(HK) \mid o(H)o(K)$$

For part 2), if $h \in H$ then $h = ka$ where $k \in K$ and $a \in N$. Hence $k^{-1}h = a \in N$ and so $o(a) \mid o(K)o(H)$ and $o(a) \mid o(N)$, whence $a = 1$, that is, $h = k \in K$. Hence, $H \leq K$ and a symmetric argument shows that $H = K$.□

Euler's Formula

The **Euler phi function** ϕ is defined, for positive integers n, by letting $\phi(n)$ be the number of positive integers less than n and relatively prime to n. The Euler phi function is important in group theory since the cyclic group C_n of order n has exactly $\phi(n)$ generators.

Theorem 2.31 (Properties of Euler's phi function)

1) The Euler phi function is **multiplicative**, *that is, if m and n are relatively prime, then*

$$\phi(mn) = \phi(m)\phi(n)$$

2) If p is a prime and $n \geq 1$, then

$$\phi(p^n) = p^{n-1}(p-1)$$

These two properties completely determine ϕ.

Proof. For part 1), consider the $m \times n$ matrix

$$A = \begin{bmatrix} 1 & m+1 & 2m+1 & \cdots & (n-1)m+1 \\ 2 & m+2 & 2m+2 & \cdots & (n-1)m+2 \\ & & & \vdots & \\ r & m+r & 2m+r & \cdots & (n-1)m+r \\ & & & \vdots & \\ m & 2m & 3m & \cdots & mn \end{bmatrix}$$

Note that the entries in the rth row of A are relatively prime to m if and only if $(r,m) = 1$. Thus, in looking for the entries that are relatively prime to both n and m, we need consider only the $\phi(m)$ rows in which $(r,m) = 1$. Note also that the difference of any two distinct entries in the same row has the form

$$(km + r) - (jm + r) = (k - j)m$$

which is not divisible by n. Hence, the entries in any row form a complete set of distinct representatives for the residue classes modulo n. Hence, modulo n, the elements of the row are $0, 1, \ldots, n-1$, of which exactly $\phi(n)$ are relatively prime to n. In other words, each of the $\phi(m)$ rows contains $\phi(n)$ elements that are relatively prime to mn.

For part 2), $\phi(p^n)$ is the number of positive integers less than or equal to p^n that are not divisible by p. The positive integers less than or equal to p^n that are divisible by p are

$$\{p \cdot 1, p \cdot 2, \ldots, p \cdot p^{n-1}\}$$

which is a set of size p^{n-1}. Hence, $\phi(p^n) = p^n - p^{n-1}$.□

Some simple group theory yields two famous old theorems from number theory.

Theorem 2.32 (Euler's theorem) *Let n be a positive integer. If a is an integer relatively prime to n, then*

$$a^{\phi(n)} \equiv 1 \bmod n$$

This formula is called **Euler's formula**.
Proof. Since $\mathbb{Z}_n^*$ is a multiplicative group of order $\phi(n)$, Lagrange's theorem implies that Euler's formula holds for $a \in \mathbb{Z}_n^*$ and since

$$(a + kn)^{\phi(n)} \equiv a^{\phi(n)} \equiv 1 \bmod n$$

Euler's formula holds for all integers that are relatively prime to n.□

Corollary 2.33 (Fermat's little theorem) *If p is a prime and $(a, p) = 1$, then*

$$a^p \equiv a \bmod p$$ □

Cyclic Groups

Let us gather some facts about cyclic groups.

Theorem 2.34 (Properties of cyclic groups)

1) **(Prime order implies cyclic)** *Every group of prime order is cyclic.*
2) **(Smallest positive exponent)** *A finite abelian group G is cyclic if and only if* $\text{minexp}(G) = o(G)$.
3) **(Subgroups)** *Every subgroup of a cyclic group is cyclic.*
   a) **(Lattice of subgroups: infinite case)** *If $\langle a \rangle$ is infinite, then each power a^k with $k \geq 0$ generates a distinct subgroup $\langle a^k \rangle$ and this accounts for all subgroups of $\langle a \rangle$.*
   b) **(Lattice of subgroups: finite case)** *If $o(a) = n$ then for each $d \mid n$, the group $\langle a \rangle$ has exactly one subgroup $S = \langle a^{n/d} \rangle$ of order d and exactly $\phi(d)$ elements of order d, all of which lie in S. This accounts for all subgroups of $\langle a \rangle$. It follows that for any positive integer n,*

$$n = \sum_{d \mid n} \phi(d)$$

4) **(Characterization by subgroups)** *If a finite group G of order n has the property that it has at most one subgroup of each order $d \mid n$, then G is cyclic (and therefore has exactly one subgroup of each order $d \mid n$).*

5) **(Direct products)** *A direct product*

$$G = G_1 \boxtimes \cdots \boxtimes G_m$$

of finite order is cyclic if and only if each G_i is cyclic and the orders of the factors G_i are pairwise relatively prime. Moreover, if $d_1, \ldots, d_n$ are pairwise relatively prime positive integers, then the following hold:

a) **(Composition)** *If $\langle a_i \rangle$ is cyclic of order d_i, then*

$$\langle a_1 \rangle \boxtimes \cdots \boxtimes \langle a_n \rangle = \langle (a_1, \ldots, a_n) \rangle$$

b) **(Decomposition)** *If $G = \langle a \rangle$ has order $d = d_1 \cdots d_n$, then*

$$\langle a \rangle = \langle a_1 \rangle \cdots \langle a_n \rangle$$

where $o(a_i) = d_i$ and

$$\langle a_i \rangle \cap \prod_{j \neq i} \langle a_j \rangle = \{1\}$$

for all i.

Proof. Part 2) follows from Theorem 2.12. For part 3b), the generators of the cyclic groups of order d are the elements of order d. However, Theorem 2.9 implies that the elements of G of order d are

$$\{a^{r(n/d)} \mid (r, d) = 1\}$$

and these all lie in the one cyclic subgroup $\langle a^{n/d} \rangle$ of order d. Hence, there can be no other cyclic subgroups of order d. For the final statement, the sum $\sum_{d|n} \phi(d)$ simply counts the elements of G by their order.

To prove 4), let D be the set of all orders of elements of G. If $a \in G$ has order $o(a) = d$, then $\langle a \rangle$ is the unique subgroup in G of order d and so all elements of order d must be in $\langle a \rangle$. It follows that there are exactly $\phi(d)$ elements in G of order $d \in D$. Hence,

$$n = \sum_{d \in D} \phi(d) \leq \sum_{d|n} \phi(d) = n$$

and so equality holds, whence D is the set of all divisors of n. In particular, $n \in D$ and so G is cyclic.

For part 5), if each $G_i = \langle a_i \rangle$ is cyclic of order d_i, where the d_i's are pairwise relatively prime, then

$$o((a_1, \ldots, a_n)) = \mathrm{lcm}(d_1, \ldots, d_n) = \prod d_i = o(G)$$

and so $G = \langle (a_1, \ldots, a_n) \rangle$ is cyclic. (This also proves part 5a).) Conversely, suppose that G is cyclic and for each $a_i \in G_i$, let

$$\widehat{a}_i = (1, \ldots, 1, a_i, 1, \ldots, 1)$$

where a_i is in the ith position. The subgroups

$$G^{(i)} = \{1\} \boxtimes \cdots \boxtimes \{1\} \boxtimes G_i \boxtimes \{1\} \boxtimes \cdots \boxtimes \{1\}$$

where G_i is in the ith position are cyclic and if $G^{(i)} = \langle \widehat{a}_i \rangle$ has order d_i, then $G_i = \langle a_i \rangle$ is also cyclic of order d_i. Moreover,

$$\prod d_i = o(G) = \text{minexp}(G) = \text{lcm}(d_1, \ldots, d_n)$$

and so the orders d_i are pairwise relatively prime.

For part 5b), since $G = \langle a \rangle$ is abelian, the product $P = \langle a_1 \rangle \cdots \langle a_n \rangle$ is a subgroup of G and since $a_1 \cdots a_n \in P$ has order

$$o(a_1 \cdots a_n) = \text{lcm}(d_1, \ldots, d_n) = d = o(\langle a \rangle)$$

it follows that $P = G$. Finally, if

$$\alpha \in \langle a_i \rangle \cap \prod_{j \neq i} \langle a_j \rangle$$

then $o(\alpha)$ divides d_i as well as the product $\prod_{j \neq i} d_j$, which are relatively prime and so $\alpha = 1$.□

Homomorphisms of Groups

We will discuss the structure-preserving maps between groups in detail in a later chapter, but we wish to introduce a few definitions here for immediate use.

Definition *Let G and H be groups. A function $\sigma: G \to H$ is called a* **group homomorphism** *(or just* **homomorphism***) if*

$$\sigma(ab) = (\sigma a)(\sigma b)$$

for all $a, b \in G$. A bijective homomorphism is an **isomorphism**. *When $\sigma: G \to H$ is an isomorphism, we write $\sigma: G \approx H$ or simply $G \approx H$ and say that G and H are* **isomorphic**.□

A property of a group is **isomorphism invariant** if whenever a group G has this property, then so do all groups isomorphic to G. For example, the properties of being finite, abelian and cyclic are isomorphism invariant.

More Groups

Let us look at a few classes of groups. As we have seen, some groups are given names, for example, the cyclic groups, the symmetric groups, the quaternion group and the dihedral groups (to be defined below). Actually, these are really *isomorphism classes* of groups. For instance, if $G \approx S_n$, then we might reasonably refer to G as a symmetric group as well. After all, it is the algebraic structure that is important and not the labeling of the elements of the underlying set.

Cyclic Groups

Theorem 2.33 describes the subgroup lattice structure of a cyclic group. Let us take a somewhat closer look at this structure.

Let $C_n(a)$ be cyclic of finite order n and let D_n be the lattice (under division) of positive integers less than or equal to n that divide n. Then the map $\sigma: D_n \to \text{sub}(C_n(a))$ defined by $\sigma d = C_d = \langle a^{n/d} \rangle$ is an order isomorphism from D_n to $\text{sub}(C_n(a))$, since σ is surjective and

$$d_1 \mid d_2 \quad \Leftrightarrow \quad (n/d_2) \mid (n/d_1) \quad \Leftrightarrow \quad C_{d_1} \le C_{d_2}$$

Thus, $\text{sub}(C_n(a))$ is order isomorphic to D_n. For example, Figure 2.2 shows the lattices D_{24} and $\text{sub}(C_{24}(a))$.

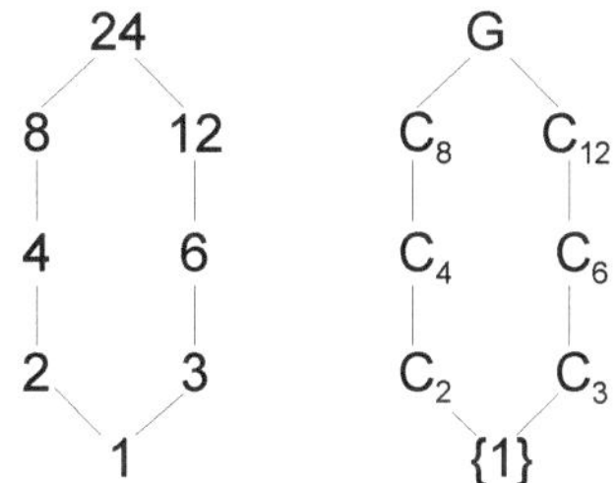

Figure 2.2

For the infinite case, we must settle for an order anti-isomorphism. Let $\mathbb{Z}^+$ be the lattice of nonnegative integers under division. If $G = \langle a \rangle$ is an infinite cyclic group, then the map $\sigma: \mathbb{Z}^+ \to \text{sub}(G)$ defined by $\sigma k = \langle a^k \rangle$ is an order anti-isomorphism, since σ is surjective and

$$k \mid n \quad \Leftrightarrow \quad \langle a^n \rangle \le \langle a^k \rangle$$

Thus, $\text{sub}(G)$ is order anti-isomorphic to $\mathbb{Z}^+$.

The Quaternion Group

Let $GL(2, \mathbb{C})$ be the general linear group of all nonsingular 2×2 matrices over the complex numbers. Let

$$I = \begin{pmatrix} 1 & 0 \\ 0 & 1 \end{pmatrix}, A = \begin{pmatrix} i & 0 \\ 0 & -i \end{pmatrix}, B = \begin{pmatrix} 0 & 1 \\ -1 & 0 \end{pmatrix}, C = \begin{pmatrix} 0 & i \\ i & 0 \end{pmatrix}$$

Then it is easy to see that

$$A^2 = B^2 = C^2 = ABC = -I$$

and

$$(-I)X = X(-I) = -X$$

for $X = A, B$ or C. These equations are sufficient to determine all products in the set

$$S = \{\pm I, \pm A, \pm B, \pm C\}$$

In particular,

$$AB = C, \quad BC = A, \quad AC = AAB = -B$$

and

$$\begin{aligned} BA &= BBC = -C \\ CB &= ABB = -A \\ CA &= ABBC = -AC = B \end{aligned}$$

Thus, S is a subgroup of $GL(2, \mathbb{C})$ of order 8. Any group isomorphic to S is called a **quaternion group**.

Note that we have defined the quaternion group in such a way that it is clearly a group, since we have defined it as a subgroup S of the *known group* $GL(2, \mathbb{C})$. However, the quaternion group is often defined without mention of matrices. In this case, it becomes necessary to verify that the definition does indeed constitute a group.

On the other hand, we can leverage our knowledge of existing groups (in particular S) by the following simple device: A bijection $f: G \to X$ from a group G to a set X can be used to transfer the group product from G to X by setting

$$f(a)f(b) = f(c) \quad \text{if} \quad ab = c$$

This makes X a group and f an isomorphism from G to X.

Now, the quaternion group is often defined as the set

$$Q = \{1, i, j, k, -1, -i, -j, -k\}$$

with multiplication defined so that 1 is the identity and

$$\begin{aligned} i^2 = j^2 = k^2 = ijk &= -1 \\ (-1)x = x(-1) &= -x \end{aligned} \tag{2.35}$$

for $x = i, j, k$. As with the set S defined above, these rules are sufficient to define all products of elements of Q. Rather than show directly that this forms a group, that is, rather than verifying directly the associative, identity and inverse properties, we can simply observe that the map $f: S \to Q$ defined by

$$\begin{aligned} f(I) &= 1, & f(-I) &= -1 \\ f(A) &= i, & f(-A) &= -i \\ f(B) &= j, & f(-B) &= -j \\ f(C) &= k, & f(-C) &= -k \end{aligned}$$

is a bijection and that

$$f(X)f(Y) = f(Z) \quad \Leftrightarrow \quad XY = Z$$

for all $X, Y, Z \in S$. Hence, the product in Q is the image of the product in S and so Q is a group because S is a group. Note that the main savings here is in the fact that we do not need to verify that the product in Q is associative. Of course, someone had to verify that matrix multiplication is associative, and we do appreciate that effort very much.

Equations (2.35) are equivalent to

$$\begin{aligned} i^2 = j^2 = k^2 &= -1 \\ ij = k, jk = i, ki &= j \\ (-1)x = x(-1) &= -x \end{aligned}$$

for $x = i, j, k$ and many authors use these equations to define the quaternion group.

In order to describe the quaternion group, it is not necessary to mention explicitly all four elements -1, i, j and k, since $-1 = i^2$ and $k = ij$. In fact, Q can also be defined as the group satisfying the following conditions:

$$Q = \langle i, j \rangle, \quad o(Q) = 8, \quad i^4 = 1, \quad i^2 = j^2, \quad ji = i^3 j$$

To see this, if

$$-1 := i^2, \quad k := ij \quad \text{and} \quad -x := (-1)x$$

for $x = 1, i, j, k$, then

$$\begin{aligned} Q &= \{1, i, i^2, i^3, j, ij, i^2 j, i^3 j\} \\ &= \{1, i, -1, -i, j, k, -j, -k\} \end{aligned}$$

Moreover, since $ji = i^3 j$, it follows that

$$k^2 = ijij = ii^3 jj = -1$$

and

$$ijk = ijij = ii^3 jj = j^2 = -1$$

Note that the conditions

$$Q = \langle i, j \rangle, \quad i \neq j, \quad o(i) = 4, \quad i^2 = j^2, \quad iji = j$$

imply that $o(G) = 8$ and so they provide perhaps the most succinct definition of the quaternion group.

Subgroups of the Quaternion Group

The quaternion group has a simple subgroup lattice. We leave it as an exercise to verify that the Hasse diagram for $\text{sub}(Q)$ is given by Figure 2.3. The subgroup $\langle -1 \rangle$ has order 2 and the other nontrivial proper subgroups have order 4.

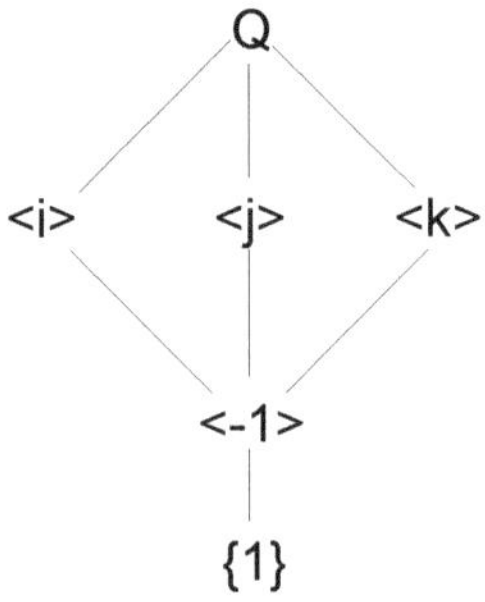

Figure 2.3

The Dihedral Groups

Many groups come from geometry. Here is one of the most famous. First, we need a bit of terminology. A **rigid motion** of the plane is a bijective distance-preserving map of the plane. If P is a nonempty subset of the plane, then a **symmetry** of P is a rigid motion of the plane that sends P onto itself.

Now, for $n \geq 2$, let P be a regular n-gon in the plane, whose center passes through the origin. Figure 2.4 shows the cases $n = 4$ and $n = 5$.

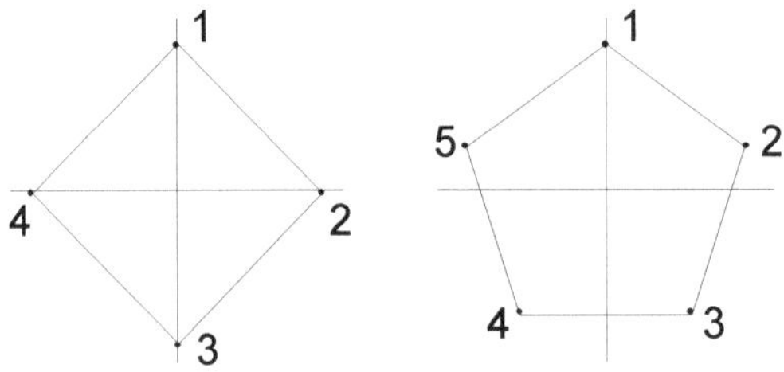

Figure 2.4

(For $n = 2$, P is a line segment.) Label the vertices $1, \ldots, n$ in clockwise order, with vertex 1 on the positive vertical axis. Let V be the set of vertices of P.

The **symmetry group** G_n of P consists of all symmetries of P. It is possible to prove that the symmetries of P are the same as the symmetries of the vertex set V and so we may regard G_n as the set of all symmetries of V.

In fact, each symmetry of V is a permutation of V and is uniquely determined by that permutation. Indeed, it is customary to think of the elements of G_n as permutations of the labeling set $I_n = \{1, \ldots, n\}$, that is, as elements of S_n. Here are a few simple facts concerning G_n, for $n \geq 2$.

1) Since any $\delta \in G_n$ preserves adjacency, if $\delta x = y$ then $\delta(x+1) = y+1$ or $\delta(x+1) = y-1$. We denote this by writing

$$\delta\colon (x, x+1) \mapsto (y, y+1)$$

or

$$\delta\colon (x, x+1) \mapsto (y, y-1)$$

In the former case, we say that δ **preserves orientation** in the pair $(x, x+1)$ and in the latter case, δ **reverses orientation**. Note that addition and subtraction are performed modulo n, but then 0 is replaced by n. Hence, for example, $1 - 1 = n$ and so 1 and n are adjacent.

2) Any $\delta \in G_n$ is uniquely determined by its value on two adjacent elements $(x, x+1)$ of I_n. To see this, we may assume that $n \geq 3$. If δ preserves orientation on $(x, x+1)$, that is, if

$$\delta\colon (x, x+1) \mapsto (y, y+1)$$

then since $\delta(x+2)$ is adjacent to $\delta(x+1) = y+1$, we have $\delta(x+2) = y$ or $\delta(x+2) = y+2$. But $\delta(x+2) \neq \delta x = y$ and so

$$\delta\colon (x+1, x+2) \mapsto (y+1, y+2)$$

Repeating this argument gives

$$\delta\colon (x+k, x+k+1) \mapsto (y+k, y+k+1)$$

for all k. A similar argument holds if δ reverses orientation, showing that

$$\delta\colon (x+k, x+k+1) \mapsto (y-k, y-k-1)$$

Note also that if δ preserves orientation for one adjacent pair, then it preserves orientation for all adjacent pairs and so we can simply say that δ either preserves orientation or reverses orientation.

The group G_n contains both the n-cycle

$$\rho = (1\,2 \cdots n)$$

which represents a clockwise *rotation* through $360/n$ degrees, and the permutation σ that represents a *reflection* across the vertical axis. If n is even,

then

$$\sigma = (2\,n)(3\,n-1)\cdots\left(\frac{n}{2}\ \frac{n}{2}+2\right)$$

and if n is odd then

$$\sigma = (2\,n)(3\,n-1)\cdots\left(\frac{n+1}{2}\ \frac{n+1}{2}+1\right)$$

To see that G_n is generated by the rotation ρ and the reflection σ, let $\delta \in G_n$. If δ preserves orientation, then

$$\delta\colon (n,1) \mapsto (k,k+1)$$

and so $\delta = \rho^k \in \langle \rho, \sigma \rangle$. On the other hand, if δ reverses orientation, then

$$\delta\colon (n,1) \mapsto (k,k-1)$$

and so the following hold:

$$\begin{aligned}\delta\sigma&\colon (2,1) \mapsto (k,k-1)\\ \delta\sigma&\colon (1,2) \mapsto (k-1,k)\\ \delta\sigma&\colon (n,1) \mapsto (k-2,k-1)\end{aligned}$$

Hence, $\delta\sigma = \rho^{k-1}$ and so $\delta = \rho^{k-1}\sigma \in \langle \rho, \sigma \rangle$. Thus $G_n = \langle \sigma, \rho \rangle$.

To examine the group structure of G_n, we have

$$o(\rho) = n \quad \text{and} \quad o(\sigma) = 2$$

Also,

$$\sigma\rho\sigma\rho\colon (1,n) \mapsto (1,n)$$

and so $\sigma\rho$ is an involution, which gives the *commutativity rule*

$$\rho\sigma = \sigma\rho^{-1} = \sigma\rho^{n-1}$$

and so every element of G_n can be written in the form $\sigma^e\rho^k$. Thus, G_n can be described succinctly by

$$G_n = \langle \sigma, \rho \rangle, \quad o(\rho) = n, \quad o(\sigma) = o(\sigma\rho) = 2$$

It is easy to see that the $2n$ elements

$$\{\iota, \rho, \ldots, \rho^{n-1}, \sigma, \sigma\rho, \ldots, \sigma\rho^{n-1}\}$$

are distinct and so $o(G_n) = 2n$.

Another description of G_n can be gleaned from the fact that $\langle \sigma, \rho \rangle = \langle \sigma, \sigma\rho \rangle$ and so G_n is generated by a pair of involutions whose product has finite order, that is, if $\pi = \sigma\rho$, then

$$G_n = \langle \sigma, \pi \rangle, \quad o(\sigma) = o(\pi) = 2, \quad o(\sigma\pi) = n$$

Without reference to geometry, a finite group G is a **dihedral group** if G is isomorphic to the symmetry group G_n, for some $n \geq 2$. Thus, a group G is a dihedral group if any of the following equivalent descriptions hold (here ρ and σ are just symbols):

1) (Common description)

$$G = \{1 = \rho^0, \rho, \ldots, \rho^{n-1}, \sigma, \sigma\rho, \ldots, \sigma\rho^{n-1}\}$$

has order $2n$ and

$$\rho^n = 1, \quad \sigma^2 = 1 \quad \text{and} \quad \rho\sigma = \sigma\rho^{n-1}$$

2) (Succinct description 1)

$$G = \langle \sigma, \rho \rangle, \quad o(\rho) = n \geq 2 \quad \text{and} \quad o(\sigma) = o(\sigma\rho) = 2$$

3) (Succinct description 2)

$$G = \langle \sigma, \pi \rangle, \quad o(\sigma) = o(\pi) = 2, \quad o(\sigma\pi) = n \geq 2$$

Note that

$$(\sigma\rho^k)(\sigma\rho^k) = \sigma\sigma\rho^{-k}\rho^k = 1$$

and so all elements of the form $\sigma\rho^k$ are involutions.

Unfortunately, the dihedral group G of order $2n$ is denoted by D_n by some authors (reflecting the fact that G consists of symmetries of n vertices) and by D_{2n} by other authors (reflecting the fact that G is a group of order $2n$). We will use the notation D_{2n}. Also, in view of the development of D_{2n}, we will often refer to ρ as a rotation and σ as a reflection, even if ρ and σ are not actually maps.

For $n = 1$, the reflection σ is the identity and the rotation ρ is the transposition $(1\,2)$ and so $D_2 = \{\iota, (1\,2)\}$ is the cyclic group C_2. For $n = 2$, the dihedral group D_4 is $C_2 \boxtimes C_2$. For $n \geq 3$, we have proved that the dihedral groups exist, namely, as certain subgroups of S_n. Note that D_{2n} is nonabelian for $n \geq 3$.

Ubiquity of the Dihedral Group

Succinct description 2 of the dihedral group shows that dihedral groups occurs quite often and reinforces the idea that a dihedral group need not consist specifically of symmetries of the plane.

Theorem 2.36 *Let G be a group. If $a, b \in G$ are distinct involutions for which $o(ab) = n < \infty$, then the subgroup $\langle a, b \rangle$ is dihedral of order $2n$.* □

Subgroups of the Dihedral Groups

Let us determine the subgroups of D_{2n}. If $S \leq D_{2n}$, then Theorem 2.28 implies that there are two possibilities. If $S \leq \langle \rho \rangle$, then $S = \langle \rho^{n/d} \rangle$ for some $d \mid n$. Otherwise, $S \cap \langle \rho \rangle = \langle \rho^{n/d} \rangle$ for some $d \mid n$ and if k is the smallest positive integer for which $\sigma\rho^k \in S$, then

$$\begin{aligned} S &= (S \cap \langle \rho \rangle) \sqcup \sigma\rho^k (S \cap \langle \rho \rangle) \\ &= \langle \rho^{n/d} \rangle \sqcup \sigma\rho^k \langle \rho^{n/d} \rangle \\ &= \langle \sigma\rho^k, \rho^{n/d} \rangle \end{aligned}$$

Note that since $\sigma\rho^k$ and $\sigma\rho^k\rho^{n/d} = \sigma\rho^{k+n/d}$ are involutions and $o(\rho^{n/d}) = d$, it follows that S is dihedral of order $2d$, generated by the "reflection" $\sigma\rho^k$ and the "rotation" $\rho^{n/d}$. Thus, the subgroups of D_{2n} fall into two categories: subgroups of $\langle \rho \rangle$, which are cyclic and the rest, which are dihedral.

Also, for distinct values of k in the range $0 \leq k < n/d$, the sets

$$S_{d,k} = \sigma\rho^k \langle \rho^{n/d} \rangle \sqcup \langle \rho^{n/d} \rangle = \langle \sigma\rho^k, \rho^{n/d} \rangle$$

are distinct subgroups of D_{2n} and this accounts for all of the dihedral subgroups of D_{2n}. We can now summarize.

Theorem 2.37 *The subgroups of the dihedral group D_{2n} are of two types. For each $d \mid n$, we have*

1) *the cyclic subgroup $\langle \rho^{n/d} \rangle$ of order d and*
2) *for each $0 \leq k < n/d$, the dihedral subgroup*

$$S_{d,k} = \langle \sigma\rho^k, \rho^{n/d} \rangle = \sigma\rho^k \langle \rho^{n/d} \rangle \sqcup \langle \rho^{n/d} \rangle$$

of order $2d$. □

The Symmetric Groups

The lattice of subgroups of the symmetric group S_n is rather complicated. Indeed, a famous theorem of Arthur Cayley [7] from 1854 (to be discussed in detail later), says that for *any* group G, the symmetric group S_G contains a subgroup that is an "exact copy" of G. Thus, a complete description of the subgroup lattice of the symmetric groups would constitute a complete description of all groups, which does not yet exist!

The Additive Rationals

Let us examine the subgroup lattice of the additive group $\mathbb{Q}$ of rational numbers. Let $\mathbb{Z}^+$ denote the set of positive integers. We say that $a/b \in \mathbb{Q}$ is **reduced** if a and b are relatively prime and $b > 0$. Let $H \leq \mathbb{Q}$ be nontrivial. Let I be the set of integers in H, let N be the set of numerators of the reduced elements of H and let D be the set of **positive** denominators of the reduced elements of H.

If $a \in N$, then $a/b \in H$ for some integer b and so $a \in I$. Thus, $I = N$. Moreover, if i is the smallest positive integer in I, then every $a \in I$ is an integral multiple of i, for if $a = qi + r$ where $0 \le r < i$, then $r = a - qi \in I$ and the minimality of i implies that $r = 0$. Thus, $N = I = \mathbb{Z}i$.

If $ai/b \in H$ is reduced, then there exist integers u and v for which $ua + vb = 1$, whence

$$\frac{i}{b} = \frac{(ua+vb)i}{b} = \frac{uai}{b} + vi \in H$$

Thus,

$$\frac{ai}{b} \in H \quad \Leftrightarrow \quad \frac{i}{b} \in H$$

for $a, b \in \mathbb{Z}$, $b > 0$. Thus,

$$H = \left\{ \frac{ai}{b} \,\middle|\, a \in \mathbb{Z}, b \in D \right\}$$

It follows that D is closed under products, since if i/b and i/c in H are reduced, then i^2/bc is also reduced and in H. Also, if $d \in D$ and $e \mid d$, then $e \in D$. Thus, D is closed under factors as well. It follows that we can describe D by describing which prime powers lie in D.

If p is prime, let D_p be the set of all powers of p that lie in D. Then D_p has one of three forms.

1) If $p \mid i$, then $D_p = \{1\}$ since if $p \in D$ then $i/p \in H$ is an integer, contradicting the fact that i is the smallest positive integer in H.
2) If $p \nmid i$ and there is a largest integer $m(p)$ for which $p^{m(p)} \in D$, then

$$D_p = \{1, \ldots, p^{m(p)}\}$$

3) If $p \nmid i$ and there is no largest integer $m(p)$ for which $p^{m(p)} \in D$, then

$$D_p = \{1, p, p^2, \ldots\}$$

In case 1) we set $m(p) = 0$ and in case 3) we set $m(p) = \infty$ and so $m(p)$ is defined for all primes. Let $p_1, p_2, \ldots$ be the sequence of all primes and let

$$m(H) = (m(p_1), m(p_2), \ldots)$$

For convenience, we say that a sequence $(a_1, a_2, \ldots)$ where a_i is a nonnegative integer or $a_i = \infty$ is **acceptable** for i if $p_k \mid i$ implies that $a_k = 0$. Thus, $m(H)$ is acceptable for i.

Let us adopt the notation $(a_1, a_2, \ldots) \leq (b_1, b_2, \ldots)$ to mean that $a_k \leq b_k$ for all k. If $b \in D$, we may write

$$b = p_1^{e_1} p_2^{e_2} \cdots$$

with the understanding that all but a finite number of exponents e_k are zero. Let

$$e(b) = (e_1, e_2, \ldots)$$

Then for any positive integer b,

$$b \in D \quad \Leftrightarrow \quad e(b) \leq m(H)$$

and so the pair $(i, m(H))$ completely determines H since

$$H = \left\{ \frac{ai}{b} \,\middle|\, a \in \mathbb{Z},\ b \in \mathbb{Z}^+, e(b) \leq m(H) \right\}$$

On the other hand, let $i \in \mathbb{Z}^+$ and let

$$s = (m_1, m_2, \ldots)$$

be acceptable for i. Then the set

$$H_{(i,s)} = \left\{ \frac{ai}{b} \,\middle|\, a \in \mathbb{Z}, b \in \mathbb{Z}^+, e(b) \leq s \right\}$$

is a nontrivial subgroup of $\mathbb{Q}$ for which $m(H) = s$. It is clear that H is closed under negatives. Also, if $ai/b \in H$ and $ci/d \in H$ are reduced, then

$$\frac{ai}{b} + \frac{ci}{d} = \frac{ui}{\operatorname{lcm}(b, d)} \in H$$

is reduced and since

$$e(\operatorname{lcm}(b, d)) \leq s$$

it follows that H is closed under addition. Thus, the subgroups $H_{(i,s)}$ of $\mathbb{Q}$ correspond bijectively to the pairs (i, s), where $i \in \mathbb{Z}^+$ and where

$$s = (m_1, m_2, \ldots)$$

is acceptable for i.

*An Historical Perspective: Galois-Style Groups

Let us conclude this chapter with a brief historical look at groups. Evariste Galois (1811–1832) was the first to develop the concept of a group, in connection with his research into the solutions of polynomial equations. However, Galois' version of a group is quite different from the modern version we see today. Here is a brief look at groups as Galois saw them (using a bit more modern terminology than Galois used).

Consider a table in which each row contains an ordered arrangement of a set X of distinct symbols (such as the roots of a polynomial), for example

$$\begin{array}{ccccc} a & b & c & d & e \\ c & a & b & d & e \\ b & c & a & d & e \\ a & b & c & e & d \\ c & a & b & e & d \\ b & c & a & e & d \end{array}$$

where $X = \{a, b, c, d, e\}$. Then each pair of rows defines a permutation of X, that is, a bijective function on X. Galois considered tables of ordered arrangements with the property that the set A_i of permutations that transform a given row r_i into the other rows (or into itself) is the same for all rows r_i, that is, $A_i = A_j$ for all i, j. Let us refer to this type of table, or list of ordered arrangements, as a **Galois-style group**.

In modern terms, it is not hard to show that a list of ordered arrangements is a Galois-style group if and only if the corresponding set $A\ (= A_i)$ of permutations is a subgroup of the symmetric group S_X. To see this, let the permutation that transforms row r_i to row r_j be $\pi_{i,j}$. Then Galois' assumption is that the sets

$$A_i = \{\pi_{i,1}, \ldots, \pi_{i,n}\}$$

are the same for all i. This implies that for each i, u and j, there is a v for which $\pi_{i,u} = \pi_{j,v}$. Hence,

$$\pi_{i,u}\pi_{i,j} = \pi_{j,v}\pi_{i,j} = \pi_{i,v} \in A_i$$

and so A_i is closed under composition and is therefore a group.

Conversely, if A_1 is a permutation group, then since

$$\pi_{i,j} = \pi_{1,j}\pi_{i,1} = \pi_{1,j}(\pi_{1,i})^{-1} \in A_1$$

it follows that $A_i = A_1$ for all i and so the ordered arrangement that corresponds to A_1 is a Galois-style group.

Galois appears not to be entirely clear about a precise meaning of the term group, but for the most part, he uses the term for what we are calling a Galois-style group. Galois also worked with subgroups and recognized the importance of what we now call normal subgroups (defined in the next chapter), although his definition is quite different from what we would see today.

When Galois' work was finally published in 1846, fourteen years after he met his untimely death in a duel at the age of 21, the theory of finite permutation groups had already been formalized by Augustin Louis Cauchy (1789–1857), who likewise required only closure under product, but who clearly recognized the importance of the other axioms by introducing notations for the identity and for inverses.

Arthur Cayley (1821–1895) was the first to consider, in 1854, the possibility of more abstract groups and the need to axiomatize associativity. He also axiomatized the identity property, but still assumed that each group was a finite set and so had no need to axiomatize inverses (only the validity of cancellation). It was not until 1883 that Walther Franz Anton von Dyck (1856–1934), in studying the relationship between groups and geometry, made explicit mention of inverses.

It is also interesting to note that Cayley's famous theorem (to be discussed in Chapter 4), to the effect that every group is isomorphic to a permutation group, completes a full circle back to Galois (at least for finite groups)!

Exercises

1. Let G be a group.
   a) Prove that G has exactly one identity.
   b) Prove that each element has exactly one inverse.
2. Prove that any group G in which every nonidentity element has order 2 is abelian.
3. Let $H \leq G$ and let $g \in G$ have order n. Show that if $g^k \in H$ for $(k,n)=1$ then $g \in H$.
4. Show that a finite subset S of a group G is a subgroup if and only if it is closed under products.
5. Show that if $a,b \in G$ then $o(ab)=o(ba)$.
6. Let G be a group with center $Z < G$. Prove that if every element of G that is not in Z has finite order, then G is periodic.
7. Show that the center of Q is equal to $\{1,-1\}$.
8. Show that in a group of even order, there is an element of order 2. (Do *not* use Cauchy's theorem, if you know it.) *Hint*: such an element is equal to its own inverse.
9. Find an example of a group G and three distinct primes p,q and r for which G has elements a and b satisfying $o(a)=p$, $o(b)=q$ and $o(ab)=r$.
10. Prove that a group with only finitely many subgroups must be finite.
11. Let G be a finite group of order m and let $(n,m)=1$. Show that every element g of G has a unique nth root h, where $h^n=g$.
12. Prove that the group $\mathbb{Q}$ of rational number has no minimal subgroups.
13. a) Find a group G and subgroups H and K for which HK is not a subgroup of G.
    b) If $G=HKL$ where H,K and L are subgroups of G, does it follow that H,K and L commute (under set product)?
14. Let F be a field and let F^* be the multiplicative group of nonzero elements of F.

a) If F is a finite field, show that F^* is cyclic. *Hint*: Use the fact that a polynomial equation of degree n has at most n distinct solutions in F.
b) Prove that if F is an infinite field, then F^* is not cyclic. *Hint*: What are the orders of the nonidentity elements?

15. Let S be a nonempty set that has an associative binary operation, denoted by juxtaposition. Show that S is a group if and only if

$$aS = S = Sa$$

for all $a \in S$.

16. Let G be a nonempty set with an associative binary operation. Assume that there is a *left* identity 1_L, that is, $1_L a = a$ for all $a \in G$ and that each element a has a *left* inverse a_L, that is, $a_L a = 1_L$. Prove that G is a group under this operation and that $1_L = 1$ and $a_L = a^{-1}$.
17. Let G be a finite abelian group of order n. Show that the product of all of the elements of G is equal to the product of all involutions in G (or 1 if G has no involutions). Apply this to the multiplicative group $\mathbb{Z}_p^*$ where p is prime to deduce that

$$(p-1)! \equiv -1 \bmod p$$

which is known as **Wilson's theorem**.

18. Draw a Hasse diagram of the subgroup lattice of
    a) the symmetric group S_3
    b) the dihedral group D_8.
19. (**Dihedral group**) Show that $\sigma\rho\sigma$ is the same as counterclockwise rotation ρ^{-1}.
20. (**Dihedral group**) Let P be a regular $2n$-gon. Show that we get the same dihedral group if we use an axis of symmetry that goes through two vertices or through the midpoint of opposite sides of P.
21. (**Dihedral group**) Find the center of the dihedral group D_{2n}.
22. Let G be a group and suppose that $a, b \in G$ satisfy $a^2 = 1$ and $ab^2a = b^3$. Prove that $b^5 = 1$.
23. Let $\mathbb{Q}$ be the additive group of rational numbers. Let $(i, (n_k))$ describe $H \le \mathbb{Q}$ and let $(j, (m_k))$ describe $K \le \mathbb{Q}$.
    a) Under what conditions is $K \le H$?
    b) Under what conditions is H cyclic?
24. Let G be a finite group and S and T be subsets of G. Prove that either $|S| + |T| \le |G|$ or $G = ST$.
25. A group G is **locally finite** if every finitely-generated subgroup is finite.
    a) Prove that a locally finite group is periodic.
    b) Prove that if G is abelian and periodic, then it is locally finite.
26. A group G is said to be **locally cyclic** if every finitely-generated subgroup of G is cyclic.
    a) Prove that G is locally cyclic if and only if every pair of elements of G generates a cyclic subgroup.
    b) Prove that a locally cyclic group is abelian.

c) Prove that any subgroup of a locally cyclic group is locally cyclic.
d) Prove that a finitely-generated locally cyclic group is cyclic.
e) Find an example of a finitely-generated group that is not locally cyclic.
f) Show that the subgroup G of the nonzero complex numbers (under multiplication) defined by

$$G = \left\{ e^{2\pi in/p^k} \mid n, k \in \mathbb{Z} \right\}$$

where p is a fixed prime is locally cyclic but not cyclic.
g) Prove that if G is locally cyclic, then all nonidentity elements have infinite order or else all elements have finite order.
h) Prove that the additive group of rational numbers is locally cyclic.
i) Show that any locally cyclic group whose nonidentity elements have infinite order is isomorphic to a subgroup of the additive group $\mathbb{Q}$.
j) The **distributive laws** are

$$A \vee (B \cap C) = (A \vee B) \cap (A \vee C)$$
$$A \cap (B \vee C) = (A \cap B) \vee (A \cap C)$$

for $A, B, C \leq G$. Prove that each distributive law implies the other. A lattice that satisfies the distributive laws is said to be a **distributive lattice**. (It is possible to prove that sub(G) is a distributive lattice if and only if G is locally cyclic. This is difficult. A complete solution can be found in Marshall Hall's book *The Theory of Groups* [16].)

Ascending Chain Condition

A group G satisfies the **ascending chain condition (ACC) on subgroups** if every ascending sequence

$$H_1 \leq H_2 \leq \cdots$$

of subgroups must eventually be constant, that is, if there is an $n > 0$ such that $H_{n+k} = H_n$ for all $k > 0$.

27. A group G satisfies the **maximal condition on subgroups** if every nonempty collection of subgroups has a maximal member. Prove that a group G satisfies the maximal condition on subgroups if and only if it satisfies the ascending chain condition on subgroups.
28. A group G satisfies the ACC on subgroups if and only if every subgroup of G is finitely generated.

Chapter 3
Cosets, Index and Normal Subgroups

We begin this chapter with a more careful look at subgroups, cosets and indices.

Cosets and Index

The number of cosets of a subgroup plays an important role in group theory.

Definition *Let G be a group. The* **index** *of $H \leq G$, denoted by $(G : H)$, is the cardinality of the set G/H of all distinct left cosets of H in G, that is,*

$$(G : H) = |G/H| \qquad \square$$

Recall that $|H\backslash G| = |G/H|$ and so the index is also the cardinality of the set of right cosets of H in G.

It is convenient to extend the quotient and index notation as follows: If $H \leq G$ and if X is any nonempty subset of G, then we write

$$X/H = \{xH \mid x \in X\}$$

and denote the cardinality of X/H by $(X : H)$. This is not entirely standard notation, but it is useful. For example, if K is a subgroup of G, but HK is not a subgroup, we may still want to consider the set $H/K = HK/K$ and the index $(HK : K)$. We will also have use for the following concept.

Definition *Let $H \leq G$.*

1) *A set consisting of exactly one element from each coset in G/H is called a* **left transversal** *for H in G (or for G/H).*
2) *A set consisting of exactly one element from each right coset in $H\backslash G$ is called a* **right transversal** *for H in G (or for $H\backslash G$).* $\square$

Now we can give some important properties of the index.

Theorem 3.1 *Let G be a group.*

1) *If $H \leq G$, then*
$$(G:H)=1 \quad \Leftrightarrow \quad G=H$$

2) *If $X \subseteq G$ is a union of cosets of $K \leq G$, then*
$$|X|=|K|\cdot(X:K)$$
Hence, if $H,K \leq G$, then
$$|HK|=|K|\cdot(HK:K)$$
In particular,
$$|G|=|K|\cdot(G:K)$$
and so if G is finite, then
$$(G:K)=\frac{|G|}{|K|}$$

3) **(Multiplicativity)** *If $H \leq K \leq G$ then*
$$(G:H)=(G:K)(K:H)$$
as cardinal numbers. Hence,
$$H \leq K \quad \text{and} \quad (G:H)=(G:K)<\infty \quad \Rightarrow \quad H=K$$

4) *Let $H,K \leq G$. If G is finite or if $HK \leq G$, then*
$$(G:K)=(HK:K)<\infty \quad \Rightarrow \quad G=HK$$

5) *If $H,K \leq G$, then*
$$(HK:K)=(H:H\cap K)$$

6) **(Poincaré's theorem)** *Let $H_1,\ldots,H_n \leq G$ and $(G:H_i)<\infty$ for all i and let $I_k=H_1\cap\cdots\cap H_k$. Then* **Poincaré's inequality** *holds:*
$$(G:H_1\cap\cdots\cap H_n)\leq(G:H_1)\cdots(G:H_n)$$
and so, in particular, $(G:H_1\cap\cdots\cap H_n)$ is also finite.
   a) *The inequality above can be replaced by division if*
   $$I_kH_{k+1}\leq G \tag{3.2}$$
   for all $k=1,\ldots,n-1$.
   b) *Equality holds in Poincaré's inequality if and only if*
   $$(I_kH_{k+1}:H_{k+1})=(G:H_{k+1})$$
   for all $k=1,\ldots,n-1$. Hence, if G is finite or if (3.2) holds for all k, then equality holds in Poincaré's inequality if and only if

$$I_k H_{k+1} = G$$

for all $k = 1, \ldots, n-1$.

7) *If a finite group* G *has subgroups* H *and* K *for which* $(G:H)$ *and* $(G:K)$ *are relatively prime, then* $G = HK$ *and equality holds in Poincaré's inequality, that is,*

$$(G : H \cap K) = (G : H)(G : K)$$

Proof. We leave proof of part 1) and part 2) to the reader. For part 3), let I be a left transversal for G/K and let J be a left transversal for K/H. If $a \in G$, then $a = ik$ for some $k \in K$ and $k = jh$ for some $j \in J$, whence

$$aH = ikH = ijhH = ijH$$

Moreover,

$$ijH = i'j'H \quad \Rightarrow \quad iK \cap i'K \neq \emptyset \quad \Rightarrow \quad i = i'$$

and so $jH = j'H$, whence $j = j'$. Thus, the set $\{ij \mid i \in I, j \in J\}$ is a left transversal for G/H. We leave proof of part 4) for the reader.

For part 5), let $I = H \cap K$ and consider the function $f : H/I \to HK/K$ defined by $f(hI) = hK$. This map is well defined and injective since if $h_1, h_2 \in H$, then $h_2^{-1}h_1 \in H$ implies that

$$h_1 I = h_2 I \quad \Leftrightarrow \quad h_2^{-1}h_1 \in I \quad \Leftrightarrow \quad h_2^{-1}h_1 \in K \quad \Leftrightarrow \quad h_1 K = h_2 K$$

Also, f is surjective since for any $h \in H$ and $k \in K$,

$$f(hI) = hK = hkK$$

Hence, $(H : H \cap K) = (HK : K)$.

For part 6), we proceed by induction on n. For $n = 2$, we have

$$\begin{aligned}(G : H_1 \cap H_2) &= (G : H_1)(H_1 : H_1 \cap H_2)\\ &= (G : H_1)(H_1H_2 : H_2)\\ &\leq (G : H_1)(G : H_2)\end{aligned}$$

and the last inequality can be replaced by division if $H_1H_2 \leq G$. Moreover, equality holds if and only if $(H_1H_2 : H_2) = (G : H_2)$. Assume the result is true for $H_1, \ldots, H_{n-1}$ and let $I_{n-1} = H_1 \cap \cdots \cap H_{n-1}$. Then

$$\begin{aligned}(G : I_{n-1} \cap H_n) &= (G : I_{n-1})(I_{n-1} : I_{n-1} \cap H_n)\\ &\leq (G : H_1)\cdots(G : H_{n-1})(I_{n-1} : I_{n-1} \cap H_n)\\ &= (G : H_1)\cdots(G : H_{n-1})(I_{n-1}H_n : H_n)\\ &\leq (G : H_1)\cdots(G : H_n)\end{aligned}$$

and both inequalities can be replaced with division signs if

$$I_k H_{k+1} \le G$$

for all $k = 1, \ldots, n-1$. Moreover, equality holds if and only if

$$(I_{k-1}H_k : H_k) = (G : H_k)$$

for all k.

For part 7), since each of $(G : H)$ and $(G : K)$ divides $(G : H \cap K)$ and since these factors are relatively prime, we have

$$(G : H)(G : K) \mid (G : H \cap K)$$

Hence, Poincaré's inequality is an equality:

$$(G : H \cap K) = (G : H)(G : K)$$

and so the finiteness of G implies that $G = HK$.□

We have seen (Theorem 2.21) that any subgroup of a finitely-generated *abelian* group is finitely generated. We can now prove that if G is a finitely-generated group, then any subgroup of finite index is also finitely generated.

Theorem 3.3 *Let G be a finitely-generated group. If H is a subgroup of G of finite index, then H is also finitely generated.*
Proof. Let $T = \{t_1, \ldots, t_m\}$ be a left transversal for G/H, with $t_1 = 1$. Let $\{x_1, \ldots, x_n\}$ be a generating set for G and let

$$W = \{x_1, \ldots, x_n\} \cup \{x_1^{-1}, \ldots, x_n^{-1}\}$$

Thus, if $a \in G$, then

$$a = w_p \cdots w_1$$

for some $w_i \in W$. But $w_1 = ts_1$ for some $t \in T$ and $s_1 \in H$ and so

$$a = w_p \cdots w_2 t s_1$$

We are now prompted to consider how the t's and w's commute. For each $t \in T$ and $w \in W$, there exist unique $t' \in T$ and $h \in H$ for which

$$wt = t'h$$

Let $S \subseteq H$ be the finite set of all such h's, as w varies over W and t varies over T. Note that $s_1 \in S$ since $w_1 t_1 = w_1 = ts_1$. By continually moving the element belonging to T forward in the product expression for a, we get

$$a = t' s_p \cdots s_1 \in t'\langle S \rangle$$

for some $t' \in T$ and $s_i \in S$. But if $a \in H$, then $a \in t'\langle S \rangle \subseteq t'H$ implies that $t' = 1$. Hence, $H \le \langle S \rangle$ and since $S \subseteq H$, we have $H = \langle S \rangle$.□

Quotient Groups and Normal Subgroups

Let G be a group and let $H \leq G$. We have seen that the equivalence relation corresponding to the partition G/H is equivalence modulo H:

$$a \equiv b \bmod H \quad \text{if} \quad aH = bH$$

Now, there seems to be a natural way to "raise" the group operation from G to G/H by defining

$$aH * bH = abH \tag{3.4}$$

Of course, for this operation to make sense, it must be well defined, that is, we must have for all $a, a_1, b, b_1 \in G$,

$$a_1^{-1}a \in H \quad \text{and} \quad b_1^{-1}b \in H \quad \Rightarrow \quad (a_1b_1)^{-1}(ab) \in H$$

Taking $a_1 = 1$ and $b_1 = b$ gives the necessary condition

$$a \in H \quad \Rightarrow \quad b^{-1}ab \in H \tag{3.5}$$

However, this condition is also sufficient, since if it holds, then

$$b_1^{-1}a_1^{-1}ab = (b_1^{-1}a_1^{-1}b_1)(b_1^{-1}ab_1)(b_1^{-1}b) \in H$$

Note that (3.5) is equivalent to each of the following conditions:

1) $aHa^{-1} \subseteq H$ for all $a \in G$
2) $aHa^{-1} = H$ for all $a \in G$
3) $aH = Ha$ for all $a \in G$.

Definition *A subgroup H of G is* **normal** *in G, written $H \trianglelefteq G$, if*

$$aH = Ha$$

for all $a \in G$. If $H \trianglelefteq G$ and $H \neq G$, we write $H \triangleleft G$. The family of all normal subgroups of a group G is denoted by $\operatorname{nor}(G)$.□

Thus, the product (3.4) is well defined if and only if $H \trianglelefteq G$. Moreover, if $H \trianglelefteq G$, then for any $a, b \in G$,

$$abH = abHH = aHbH$$

that is,

$$aH * bH = aHbH$$

In particular, the set product of two cosets of H is a coset of H. Moreover, if the set product of cosets is a coset, that is, if

$$aHbH = cH$$

for some $c \in G$, then $ab \in cH$ and so $cH = abH$, that is,

$$aHbH = abH$$

Let us refer to this as the **coset product rule**. Finally, if the coset product rule holds, then $H \trianglelefteq G$, since

$$aHa^{-1} \subseteq aHa^{-1}H = H$$

for all $a \in G$. Thus, the following are equivalent:

1) The binary operation

$$aH * bH = abH$$

is well defined on G/H.
2) $H \trianglelefteq G$.
3) The set product of cosets is a coset.
4) The coset product rule holds.

Moreover, if these conditions hold, then G/H is actually a group under the set product, for it is easy to verify that the set product is associative, G/H has identity element H and that the inverse of aH is $a^{-1}H$. Thus, we can add a fifth equivalent condition to the list above:

5) G/H is a group under set product.

Before summarizing, let us note that the following are equivalent:

$$\begin{gathered}
H \trianglelefteq G \\
aH \subseteq Ha \text{ for all } a \in G \\
b \in aH \Rightarrow b \in Ha \text{ for all } a, b \in G \\
b \in aH \Rightarrow b^{-1} \in a^{-1}H \text{ for all } a, b \in G \\
a \equiv b \bmod H \Rightarrow a^{-1} \equiv b^{-1} \bmod H \text{ for all } a, b \in G
\end{gathered}$$

Also, the following are equivalent:

$$\begin{gathered}
\text{The coset product rule holds} \\
aHbH \subseteq abH \text{ for all } a, b \in G \\
a' \in aH, b' \in bH \Rightarrow a'b' \in abH \text{ for all } a, b \in H \\
a' \equiv a \bmod H, b' \equiv b \bmod H \Rightarrow a'b' \equiv ab \bmod H
\end{gathered}$$

Now we can summarize.

Theorem 3.6 *Let $H \leq G$. The following are equivalent:*
1) The set product on G/H is a well-defined binary operation on G/H.
2) The coset product rule

$$aHbH = abH$$

holds for all $a, b \in H$.

3) *H is a normal subgroup of G.*
4) *G/H is a group under set product, called the* **quotient group** *or* **factor group** *of G by H.*
5) *The inverse preserves equivalence modulo H, that is,*

$$a \equiv b \bmod H \quad \Rightarrow \quad a^{-1} \equiv b^{-1} \bmod H$$

for all $a, b \in G$.
6) *The product preserves equivalence modulo H, that is,*

$$a' \equiv a \bmod H, \quad b' \equiv b \bmod H \quad \Rightarrow \quad a'b' \equiv ab \bmod H$$

for all $a, a', b, b' \in G$.□

When we use a phrase such as "the group G/H" it is the with the tacit understanding that H is normal in G. Note finally that statements 5) and 6) say that equivalence modulo H is a *congruence relation* on G. A **congruence relation** θ on an algebraic structure, such as a group, is an equivalence relation that preserves the (nonnullary) algebraic operations. Thus, a congruence relation θ on a group G must satisfy the conditions

$$a\theta b \quad \Rightarrow \quad a^{-1}\theta b^{-1}$$

and

$$a\theta b, \quad c\theta d \quad \Rightarrow \quad (ac)\theta(bd)$$

Theorem 3.6 shows that these two conditions are actually equivalent for groups.

More on Normal Subgroups

There are several slight variations on the definition of normality that are often useful. We leave proof of the following to the reader.

Theorem 3.7 *Let $H \leq G$. The following are equivalent:*
1) *$H \trianglelefteq G$*
2) *$H^a \subseteq H$ for all $a \in G$*
3) *$H^a \supseteq H$ for all $a \in G$*
4) *Every right coset of H is a left coset, that is, for all $a \in G$, there is a $b \in G$ such that $Ha = bH$*
5) *Every left coset is a right coset.*
6) *For all $a, b \in G$,*

$$ab \in H \quad \Rightarrow \quad ba \in H$$

7) *If $a \in G$ and $h \in H$, then $ah = h'a$ for some $h' \in H$.*□

Theorem 3.7 implies that a normal subgroup permutes with all subgroups of G. Hence, the normality of either factor guarantees that the set product HK is a subgroup.

Theorem 3.8 *Let $H, K \leq G$.*

1) *If either H or K is normal in G, then H and K permute and*
$$HK = H \vee K$$
In this case, we refer to HK as the **seminormal join** *of H and K.*
2) *If both H and K are normal in G, then HK is also normal in G and we refer to HK as the* **normal join** *of H and K.*
3) *The fact that $HK \trianglelefteq G$ does not imply that either subgroup need be normal.*

Proof. For part 3), let $G = S_4$. Let
$$H = \{\iota, (12)(34), (13)(24), (14)(23), (24), (13), (1234), (1432)\}$$
and
$$K = S_3 = \{\iota, (12), (13), (23), (123), (132)\}$$
Then $H \cap K = \{\iota, (13)\}$ has size 2 and so $|HK| = (8 \times 6)/2 = 24 = |S_4|$, which implies that $HK = S_4$. But neither subgroup is normal: H is not normal since $(14)(13)(14) = (43) \notin H$ and K is not normal since $(14)(12)(14) = (42) \notin S_3$.$\square$

Example 3.9 (The normal subgroups of D_{2n}) We have seen that for $d \mid n$, the subgroups of the dihedral group D_{2n} are

1) the cyclic subgroup $\langle \rho^{n/d} \rangle$ of order d,
2) for each $0 \leq k < n/d$, the dihedral subgroup
$$S = \sigma\rho^k \langle \rho^{n/d} \rangle \sqcup \langle \rho^{n/d} \rangle = \langle \sigma\rho^k, \rho^{n/d} \rangle$$
of order $2d$.

Subgroups of type 1) are normal, since conjugation gives
$$\sigma\rho^i (\rho^{rn/d})\rho^{-i}\sigma = \rho^{-rn/d} \in \langle \rho^{n/d} \rangle$$
Let
$$S = \langle \sigma\rho^k, \rho^{n/d} \rangle$$
be a subgroup of type 2). Then since $\langle \rho^{n/d} \rangle \trianglelefteq S$, it follows that S is normal if and only if the conjugates of the other generator $\sigma\rho^k$ are in S. Conjugation by ρ gives
$$\rho(\sigma\rho^k)\rho^{-1} = \sigma\rho^{-2+k}$$
which is in S if and only if $\rho^{-2+k} = \rho^{k+mn/d}$ for some integer m, that is, if and only if

$$-2 \equiv m\frac{n}{d} \bmod n$$

Multiplying both sides of this by d gives $-2d \equiv 0 \bmod n$, that is, $n \mid 2d$ and since $d \mid n$, we must have $n = d$ or $n = 2d$. Conversely, if $n = d$ or $n = 2d$, then the congruence holds and S is closed under conjugation by ρ.

If $n = d$, then $S = D_{2n}$. If $n = 2d$, then $o(S) = n$ and $k = 0$ or $k = 1$. If $k = 0$, then $S = \langle \sigma, \rho^2 \rangle$ and if $k = 1$, then $S = \langle \sigma\rho, \rho^2 \rangle$. Moreover, since $(D_{2n} : S) = 2$, both subgroups are normal, as we will prove in Theorem 3.17. Thus, the proper normal subgroups of D_{2n} are the subgroups of $\langle \rho \rangle$ and, for n even, the two subgroups $\langle \sigma, \rho^2 \rangle$ and $\langle \sigma\rho, \rho^2 \rangle$ of order n.□

Special Classes of Normal Subgroups

There are two very important special classes of normal subgroups. Note that $H \trianglelefteq G$ if and only if H is invariant under all inner automorphisms γ_a of G.

Definition *Let G be a group.*

1) *A subgroup H of G is* **characteristic** *in G if it is invariant under all automorphisms of G. If H is characteristic in G, we write $H \sqsubseteq G$. (This is not a standard notation, there being none.) We also write $H \sqsubset G$ if $H \sqsubseteq G$ and $H \neq G$.*
2) *A subgroup H of G is* **fully invariant** *in G if it is invariant under all endomorphisms of G.*□

Some of the most important subgroups of a group are characteristic. For example, the center $Z(G)$ of a group G is characteristic in G.

The Lattice of Normal Subgroups of a Group

If G is a group, then nor(G) is a subfamily of sub(G) and is partially ordered by set inclusion as well. Moreover, the intersection of any family $\mathcal{F}$ of normal subgroups of G is normal in G and so the meet of $\mathcal{F}$ in nor(G) is the same as the meet of $\mathcal{F}$ in sub(G).

As to join, if $\mathcal{F} = \{N_i \mid i \in I\}$ is a nonempty family of normal subgroups of G, then the join of $\mathcal{F}$ in the lattice sub(G) is the subgroup

$$\bigvee_{i \in I} N_i = \{a_{i_1} \cdots a_{i_n} \mid a_{i_k} \in N_{i_k}, n \geq 0\}$$

Actually, Theorem 3.7 implies that we can collect factors from the same subgroup N_i and so

$$\bigvee_{i \in I} N_i = \{a_{i_1} \cdots a_{i_n} \mid a_{i_k} \in N_{i_k}, i_k \neq i_j \text{ for } k \neq j, n \geq 0\}$$

In particular, if $\mathcal{F} = \{N_1, \ldots, N_m\}$ is a finite family, then the join takes the particularly simple form

$$\bigvee \mathcal{F} = \{a_1 \cdots a_m \mid a_i \in N_i\}$$

Now, if $a \in G$, then

$$\left(\bigvee_{i \in I} N_i\right)^a = \bigvee_{i \in I} N_i^a = \bigvee_{i \in I} N_i$$

and so the join of $\mathcal{F}$ in $\operatorname{sub}(G)$ is normal in G. It follows that the join of $\mathcal{F}$ in $\operatorname{sub}(G)$ is equal to the join of $\mathcal{F}$ in $\operatorname{nor}(G)$. Thus,

$$\bigcap_{\operatorname{nor}(G)} \mathcal{F} = \bigcap_{\operatorname{sub}(G)} \mathcal{F} \quad \text{and} \quad \bigvee_{\operatorname{nor}(G)} \mathcal{F} = \bigvee_{\operatorname{sub}(G)} \mathcal{F}$$

which holds also when $\mathcal{F}$ is the empty family. Hence, $\operatorname{nor}(G)$ is a complete sublattice of $\operatorname{sub}(G)$.

Theorem 3.10 *Let G be a group.*

1) *The subgroups $\{1\}$ and G are normal in G.*
2) *If $\{N_i \mid i \in I\}$ is a family of normal subgroups of G, then*

$$\bigcap_{i \in I} N_i \quad \textit{and} \quad \bigvee_{i \in I} N_i$$

are normal subgroups of G. Hence, $\operatorname{nor}(G)$ is a complete sublattice of $\operatorname{sub}(G)$. □

The maximal and minimal normal subgroups of a group play an important role in the theory. We state the definitions here for future use.

Definition *Let G be a group and let $H \trianglelefteq G$.*

1) *H is* **minimal normal** *if it is minimal in the partially ordered set of all nontrivial normal subgroups of G (under set inclusion).*
2) *H is* **maximal normal** *if it is maximal in the partially ordered set of all proper normal subgroups of G (under set inclusion).* □

The Quasicyclic Groups

For each prime p, we can now describe an infinite abelian group $\mathbb{Z}(p^\infty)$, called the ***p*-quasicyclic group** that has a very interesting subgroup lattice. Specifically, the lattice of proper subgroups of $\mathbb{Z}(p^\infty)$ consists entirely of a single ascending chain of *finite* cyclic subgroups

$$\{1\} < \langle a_1 \rangle < \langle a_2 \rangle < \cdots$$

We begin by looking at the quotient group

$$\frac{\mathbb{Q}}{\mathbb{Z}} = \left\{ \frac{m}{n}\mathbb{Z} \,\middle|\, m, n \in \mathbb{Z}, 0 \le m < n, (m, n) = 1 \right\}$$

The set

$$S = \left\{ \frac{m}{n} \,\middle|\, m,n \in \mathbb{Z}, 0 \le m < n, (m,n) = 1 \right\}$$

is a left transversal for $\mathbb{Q}/\mathbb{Z}$ and so we can simply identify $\mathbb{Q}/\mathbb{Z}$ with S, under addition modulo 1. Note that the order of $m/n \in S$ is n.

Let p be a prime and let $\mathbb{Z}(p^\infty)$ be the subgroup of S consisting of those elements of order a power of p, that is,

$$\mathbb{Z}(p^\infty) = \left\{ \frac{m}{p^k} \,\middle|\, m,k \in \mathbb{Z}, 0 \le m < p^k, p \nmid m \right\}$$

If $H < \mathbb{Z}(p^\infty)$ and $m/p^k \in H$ for $m > 0$, then there are integers a and b for which $am + bp^k = 1$ and so in $\mathbb{Z}(p^\infty)$,

$$\frac{1}{p^k} = \frac{am + bp^k}{p^k} = \frac{am}{p^k} \in H$$

Hence,

$$0 \ne m/p^k \in H \quad \Leftrightarrow \quad 1/p^k \in H \quad \Leftrightarrow \quad 1/p^j \in H \text{ for all } j \le k$$

Thus, since H is proper, there must be a largest integer n for which $1/p^n \in H$. Then

$$1/p^k \in H \quad \Leftrightarrow \quad k \le n \quad \Leftrightarrow \quad 1/p^k \in \langle 1/p^n \rangle$$

and so $H = \langle 1/p^n \rangle$ is cyclic of order p^n. Hence, the proper subgroups of $\mathbb{Z}(p^\infty)$ are the subgroups

$$\{0\} < \langle 1/p \rangle < \langle 1/p^2 \rangle < \cdots$$

In a later chapter, we will ask the reader to prove that the quasicyclic groups $\mathbb{Z}(p^\infty)$ are the only infinite groups (up to isomorphism) with the property that their proper subgroups consist entirely of a single ascending chain

$$\{0\} < S_1 < S_2 < \cdots$$

The Normal Closure of a Set

If X is a subset of a group G, then the smallest normal subgroup of G that contains X is the intersection of all normal subgroups of G that contain X. This subgroup is called the **normal closure** of X in G and we will find it useful to use the following notations for this subgroup:

$$X^G, \quad \langle X \rangle_{\text{nor}} \quad \text{and} \quad \text{nc}(X, G)$$

The normal closure has a simple characterization as follows.

Theorem 3.11 *If X is a nonempty subset of a group G, then the normal closure of X is the subgroup*

$$\mathrm{nc}(X,G)=\langle x^a \mid x\in X, a\in G\rangle$$

generated by the conjugates of X in G.$\square$

We can extend the notation and define for any $X,Y\subseteq G$,

$$X^Y=\langle x^y \mid x\in X, y\in Y\rangle$$

This allows us to describe the join of two subgroups as a set product.

Theorem 3.12 *If $H,K\leq G$, then*

$$\langle H,K\rangle = H^K K$$

Proof. Since H^K and K permute, it follows that $H^K K\leq G$. Since $H\leq H^K K$ and $K\leq H^K K$, it follows that $\langle H,K\rangle\leq H^K K$. The reverse inclusion is clear.$\square$

Internal Direct Products

Strong Disjointness of a Family of Normal Subgroups

If $\mathcal{F}=\{H_i \mid i\in I\}$ is a nonempty family of normal subgroups of a group G, we write

$$H_{(i)}:=\bigvee\{H_j \mid i\in I, j\neq i\}$$

for the join of all members of $\mathcal{F}$ *except* H_i. The members of such a family $\mathcal{F}$ can enjoy two levels of disjointness. The members of $\mathcal{F}$ can be **pairwise essentially disjoint**, that is,

$$H_i\cap H_j=\{1\}$$

for all $i\neq j$. Note that $\mathcal{F}$ is pairwise essentially disjoint if and only if $h_i h_j=1$ for $h_i\in H_i$ and $h_j\in H_j$ with $i\neq j$ imply that $h_i=h_j=1$. Also, the members of a pairwise essentially disjoint family $\mathcal{F}$ **commute elementwise**, that is, $h_i h_j=h_j h_i$ for all $h_i\in H_i$ and $h_j\in H_j$, where $i\neq j$.

A stronger level of disjointness comes when each H_i is essentially disjoint from the *join* of the other members of the family. The following useful definition is not standard in the literature.

Definition *We will say that a nonempty family $\mathcal{F}=\{H_i \mid i\in I\}$ of normal subgroups of a group G is* **strongly disjoint** *if*

$$H_i \cap H_{(i)} = \{1\}$$

for all $i \in I$.□

The property of being strongly disjoint can be characterized as follows.

Theorem 3.13 *Let* $\mathcal{F} = \{H_i \mid i \in I\}$ *be a nonempty family of normal subgroups of a group* G. *Then the following are equivalent:*
1) $\mathcal{F}$ *is strongly disjoint*
2) If

$$h_{i_1} \cdots h_{i_n} = 1$$

where $h_{i_j} \in H_{i_j}$ *and* $i_j \neq i_k$ *for* $j \neq k$, *then* $h_{i_j} = 1$ *for all* j.
3) Every nonidentity $a \in \bigvee \mathcal{F}$ *can be written, in a unique way except for the order of the factors, as a product*

$$a = h_{i_1} \cdots h_{i_n}$$

where $1 \neq h_{i_j} \in H_{i_j}$ *and* $i_j \neq i_k$ *for* $j \neq k$. *The element* h_{i_j} *is called the* i_j*th* **component** *of* a. *For* $i \in I \setminus \{i_1, \ldots, i_n\}$, *the* i*th component of* a *is* 1.

Proof. If $\mathcal{F}$ is strongly disjoint and $h_{i_1} \cdots h_{i_n} = 1$, where the factors are from different subgroups and $n \geq 2$, then

$$h_{i_1} = (h_{i_2} \cdots h_{i_n})^{-1} \in H_{i_1} \cap H_{(i_1)} = \{1\}$$

and so $h_{i_1} = 1$. Repeating this argument gives $h_{i_j} = 1$ for all j and so 1) implies 2).

If 2) holds, then the H_i's commute elementwise. To see that 3) holds, it is clear that every nonidentity $a \in \bigvee \mathcal{F}$ has such a product representation. Moreover, if a has two such product representations, then we may include additional factors equal to 1 so that

$$a = h_{i_1} \cdots h_{i_n} = k_{i_1} \cdots k_{i_n}$$

where $h_{i_j}, k_{i_j} \in H_{i_j}$ and $i_j \neq i_k$ for $j \neq k$ and at least one factor on each side is not equal to the identity. Then

$$(k_{i_1}^{-1} h_{i_1}) \cdots (k_{i_n}^{-1} h_{i_n}) = 1$$

and so 2) implies that $h_{i_j} = k_{i_j}$ for all j. Hence 2) implies 3). Finally if 3) holds and

$$a \in H_i \cap H_{(i)}$$

for some i, then the uniqueness condition implies that $a = 1$.□

The next theorem says that strong disjointness is a finitary condition and that if $\mathcal{F}$ is strongly disjoint, then it is relatively easy to check that $\mathcal{F} \sqcup \{K\}$ is also strongly disjoint.

Theorem 3.14 *Let $\mathcal{F} = \{H_i \mid i \in I\}$ be a nonempty family of normal subgroups of a group G.*
1) *$\mathcal{F}$ is strongly disjoint if and only if every nonempty finite subset of $\mathcal{F}$ is strongly disjoint.*
2) *Let $K \in \operatorname{nor}(G) \setminus \mathcal{F}$. If $\mathcal{F}$ is strongly disjoint, then $\mathcal{F} \sqcup \{K\}$ is strongly disjoint if and only if*

$$\left(\bigvee \mathcal{F}\right) \cap K = \{1\}$$ □

The following result can be quite useful.

Theorem 3.15 *Let $\mathcal{F} = \{H_i \mid i \in I\}$ be a nonempty family of normal subgroups of a group G. For any $K \trianglelefteq G$, there is a $J \subseteq I$ that is maximal with respect to the property that the family*

$$\mathcal{F}_J = \{H_j \mid j \in J\} \cup \{K\}$$

is strongly disjoint.

Proof. Write

$$\mathcal{I} = \{J \subseteq I \mid \mathcal{F}_J \text{ is strongly disjoint}\}$$

Then $\mathcal{I}$ is nonempty and the union of any chain in $\mathcal{I}$ is in $\mathcal{I}$. Hence, Zorn's lemma implies that $\mathcal{I}$ has a maximal member.□

Internal Direct Products

We have already discussed the external direct product $G \boxtimes H$ of two groups G and H. The internal direct product is defined as follows.

Definition *A group G is the* **(internal) direct product** *of two normal subgroups H and K if $G = H \bullet K$. We use the notation $G = H \bowtie K$ to denote the internal direct product.*□

The internal direct product $G = H \bowtie K$ is a decomposition of G into an essentially disjoint product of *normal* subgroups. Since the factors H and K commute elementwise, the product in G takes the form

$$(h_1 k_1)(h_2 k_2) = (h_1 h_2)(k_1 k_2)$$

where $h_i \in H$ and $k_i \in K$. Thus, the groups H and K have the same level of independence as the factors in an external direct product. Indeed, the map

$$hk \mapsto (h, k)$$

is an isomorphism from $H \bowtie K$ to $H \boxplus K$.

Definition *A nontrivial group G is said to be* **indecomposable** *if G cannot be written as an internal direct product of two proper subgroups, that is,*

$$G = H \bowtie K \quad \Rightarrow \quad H = G \text{ or } K = G \qquad \square$$

The internal direct product can easily be generalized to arbitrary nonempty families of normal subgroups.

Definition *A group G is the* **(internal) direct sum** *or* **(internal) direct product** *of a family $\mathcal{F} = \{H_i \mid i \in I\}$ of normal subgroups if $\mathcal{F}$ is strongly disjoint and $G = \bigvee \mathcal{F}$. We denote the internal direct product of $\mathcal{F}$ by*

$$\bowtie H_i \quad \text{or} \quad \bowtie \mathcal{F}$$

or when $\mathcal{F} = \{H_1, \ldots, H_n\}$ is a finite family,

$$H_1 \bowtie \cdots \bowtie H_n$$

Each factor H_i is called a **direct summand** *or* **direct factor** *of G. We denote the family of all direct summands of G by $\mathcal{DS}(G)$.* $\square$

Theorem 3.13 implies the following.

Theorem 3.16 *Let $\mathcal{F} = \{H_i \mid i \in I\}$ be a nonempty family of normal subgroups of a group G. Then the following are equivalent:*

1) $G = \bowtie \mathcal{F}$
2) *Every nonidentity $a \in G$ can be written, in a unique way except for the order of the factors, as a product*

$$a = h_{i_1} \cdots h_{i_n}$$

where $1 \neq h_{i_j} \in H_{i_j}$ and $i_j \neq i_k$ for $j \neq k$. $\square$

A note on terminology is also in order. If $\mathcal{F} = \{H_i \mid i \in I\}$ is a family of normal subgroups of a group G, to say that the join $\bigvee H_i$ **is direct** in G or to say that the direct sum $\bowtie H_i$ *exists* in G is the same as saying that $\mathcal{F}$ is strongly disjoint and that the join is the direct sum.

Projection Maps

Associated with an internal direct product

$$G = \bowtie \{H_i \mid i \in I\}$$

is a family of projection maps. Specifically, the ith **projection map** $\rho_i : G \to H_i$ is defined by setting $\rho_i(a)$ to be the ith component of a. In this case, ρ_i can be thought of as an endomorphism of G and then the following hold:

1) $(\rho_i)|_{H_i} = \iota_{H_i}$

2) $\rho_i\rho_j = 0$ if $i \neq j$, where 0 is the zero map
3) ρ_i is **idempotent**, that is, $\rho_i^2 = \rho_i$.

Note also that if $i \neq j$, then the images $H_i = \operatorname{im}(\rho_i)$ and $H_j = \operatorname{im}(\rho_j)$ commute elementwise.

We will have much more to say about the direct product in a later chapter.

Chain Conditions and Subnormality

Sequences of Subgroups and the Chain Conditions

Ordered sequences of subgroups play a key role in group theory. For infinite ascending and descending sequences of subgroups, the issue centers around the chain conditions. Generally speaking, chain conditions are considered a form of *finiteness condition* on a group and we will study the consequences of the chain conditions on various families of subgroups, such as the family of all subgroups, all normal subgroups or all subnormal subgroups throughout the book. For the record, here is the definition, which will be repeated later.

Definition *Let G be a group and let $\mathcal{S}$ be a family of subgroups of G.*

1) A group G satisfies the **ascending chain condition (ACC) on** *$\mathcal{S}$ if every ascending sequence*

$$H_1 \leq H_2 \leq \cdots$$

of subgroups in $\mathcal{S}$ must eventually be constant, that is, if there is an $n > 0$ such that $H_{n+k} = H_n$ for all $k \geq 0$. In this case, we also say that $\mathcal{S}$ has the ACC.

2) A group G satisfies the **descending chain condition (DCC) on** *$\mathcal{S}$ if every descending sequence*

$$H_1 \geq H_2 \geq \cdots$$

of subgroups in $\mathcal{S}$ must eventually be constant, that is, if there is an $n > 0$ such that $H_{n+k} = H_n$ for all $k \geq 0$. In this case, we also say that $\mathcal{S}$ has the DCC.

3) A group G satisfies **both chain conditions (BCC) on** *$\mathcal{S}$ if G has the ACC and the DCC on $\mathcal{S}$. In this case, we also say that $\mathcal{S}$ has BCC.*□

Finite Series and Subnormality

Finite ordered sequences of subgroups of a group are just as important as infinite sequences, but rather than conveying any finiteness condition about a group, they convey *structural information* about the group and are used to classify groups via this structure. For example, a group G that has a finite sequence

$$\{1\} = H_0 \trianglelefteq H_1 \trianglelefteq \cdots \trianglelefteq H_n = G$$

of subgroups for which each quotient H_{k+1}/H_k is abelian is called a **solvable** group. Solvable groups play a key role in the Galois theory of fields and we will study them in detail later in the book.

Unfortunately, the terminology surrounding finite ordered sequences of subgroups is not at all standardized. For example, consider the following types of finite sequences of subgroups of a group G:

1) An arbitrary nondecreasing sequence of subgroups of G:
$$G_0 \leq G_1 \leq \cdots \leq G_n$$
2) A sequence of subgroups of G of the form
$$G_0 \trianglelefteq G_1 \trianglelefteq \cdots \trianglelefteq G_n$$
in which each subgroup is normal in its immediate successor.
3) A sequence of subgroups of G of the form
$$G_0 \leq G_1 \leq \cdots \leq G_n$$
in which each subgroup is normal in the parent group G.

Some authors refer to 1) as a series, 2) as a subnormal series and 3) as a normal series. Some authors refer to 2) as a series and 3) as a normal series. Some authors refer to 2) as a normal series and 3) as an invariant series.

Since arbitrary finite nondecreasing sequences of subgroups are a bit too general to be really useful, it seems reasonable not to give them a special name and simply refer to them as sequences, thus reserving the term series for more useful types of sequences. Accordingly, we choose the following terminology.

Definition *Let G be a group and let $G_0, G_n \leq G$. A* **series** *in G from G_0 to G_n is a sequence of subgroups of G of the form*
$$G_0 \trianglelefteq G_1 \trianglelefteq \cdots \trianglelefteq G_n$$
where G_k is normal in G_{k+1} for each k. Each group G_k is a **term** *in the series; G_0 is the lower* **endpoint** *of the series and G_n is the upper endpoint. Each extension $G_k \trianglelefteq G_{k+1}$ is a* **step** *in the series. A series is* **proper** *if each inclusion is proper. The* **length** *of a series is the number of proper inclusions.*

1) A **normal series** *in G is a series in G in which each term is normal in the parent group G.*
2) A **characteristic series** *in G is a series in G in which each term is characteristic in the parent group G.*
3) A **fully-invariant series** *in G is a series in G in which each term is fully invariant in the parent group G.*□

The following generalization of normality is extremely important and we shall have much to say about it in this book.

Definition *A subgroup $H \leq G$ is* **subnormal** *in G, written $H \trianglelefteq \trianglelefteq G$, if there is a series*

$$H \trianglelefteq H_1 \trianglelefteq \cdots \trianglelefteq H_n = G$$

from H to G. If $H \trianglelefteq \trianglelefteq G$ and $H < G$, we write $H \lhd\lhd G$. The family of subnormal subgroups of G is denoted by $\operatorname{subn}(G)$.□

Thus, $H \trianglelefteq \trianglelefteq G$ if there is a sequence of "normal steps" from H to G.

Subgroups of Index 2

The largest proper subgroups of a group are the subgroups of index 2. We can now improve upon Theorem 2.28.

Theorem 3.17 Let H be a subgroup of G of index 2.

1) $H \trianglelefteq G$.
2) *If $a \in G$, then $a^2 \in H$.*
3) *If G is finite, then any subgroup S of G is either a subgroup of H or else*

$$|S \cap H| = |S|/2$$

In words, S lies completely in H or else S lies half-in and half-out of H. Also, if $a \in S \setminus H$, then

$$S = (S \cap H) \sqcup a(S \cap H)$$

where $\sqcup$ is the disjoint union.

Proof. For part 1), if $a \notin H$, then $\{H, aH\}$ and $\{H, Ha\}$ are both partitions of G and so $aH = Ha$ for all $a \in G$. Hence, $H \trianglelefteq G$. For part 2), if $a \notin H$ then $aH \neq H$ has order 2 in G/H and so $a^2H = (aH)^2 = H$, which implies that $a^2 \in H$. For part 3), if S is not contained in H, the normality of H implies that $SH = G$ and so

$$(S : H \cap S) = (HS : H) = (G : H) = 2$$ □

Example 3.18 For $n \geq 2$, the alternating group A_n has index 2 in the symmetric group S_n and so $A_n \trianglelefteq S_n$.□

We can now show that the converse of Lagrange's theorem fails.

Example 3.19 The alternating group A_4 has order $4!/2 = 12$, but has no subgroups of order 6. For if $H \leq A_4$ has order 6, then H has index 2 and so $\sigma^2 \in H$ for all $\sigma \in A_4$. But there are eight 3-cycles σ in A_4 and each one is a square:

$$\sigma = \sigma^4 = (\sigma^2)^2 \in H$$

Since these 8 elements cannot fit into a subgroup of size 6, there can be no such subgroup.□

Cauchy's Theorem

We have seen that the converse of Lagrange's theorem fails to hold. However, there are some *partial* converses to Lagrange's theorem. For example, there are certain classes of groups for which the converse of Lagrange's theorem does hold. In particular, we will see later in the book that if G is a finite abelian group or if $o(G) = p^n$ where p is prime, then the converse of Lagrange's theorem holds, that is, G has a subgroup of any order k that divides $o(G)$.

On the other hand, it is true that if a *prime* p divides $o(G)$, then G has an element of order p and hence a subgroup of order p. This key theorem is called *Cauchy's theorem.*

Cauchy's theorem has a very colorful history, beginning with its discovery by Cauchy in 1845 as the main conclusion of a 101-page paper ([6]). Here is a quotation from the abstract of an article on the history of Cauchy's theorem by M. Meo [24]:

> The initial proof by Cauchy, however, was unprecedented in its complex computations involving permutational group theory and contained an egregious error. A direct inspiration to Sylow's theorem, Cauchy's theorem was reworked by R. Dedekind, G. F. Frobenius, C. Jordan, and J. H. McKay in ever more natural, concise terms.

The proof we give below is essentially the proof of J. H. McKay [23], which first appeared in 1959 and reminds us that we should never stop looking for "better" proofs of even the most basic results.

Theorem 3.20 (Cauchy's theorem) *Let G be a finite group. If p is a prime dividing $o(G)$, then G has an element of order p.*
Proof. The key to the proof is to examine the set

$$X = \{(a_1, \dots, a_p) \mid a_i \in G, a_1 \cdots a_p = 1\}$$

which is nonempty, since $(1, \dots, 1) \in X$. In fact, the size of X is easily computed by observing that the first $p-1$ coordinates any $x \in X$ can be assigned arbitrarily and this uniquely determines the final coordinate. Hence,

$$|X| = |G|^{p-1}$$

which is divisible by p.

Now, if we can find a constant p-tuple $(a,\ldots,a)$ in X where $a\neq 1$, then $a^p=1$ and so $o(a)=p$, which proves the theorem. Note that a p-tuple $x=(a_1,\ldots,a_p)$ is constant if and only if rotation one position to the right (with wrap around) has no effect, that is, if and only if

$$(a_p,a_1,\ldots,a_{p-1})=(a_1,\ldots,a_p)$$

We can put this in group-theoretic language as follows. Let each $\sigma\in S_p$ act on X by permuting the coordinates. Thus, if σ is the p-cycle $(1\,2\cdots p)$, then

$$\sigma(a_1,\ldots,a_p)=(a_p,a_1,\ldots,a_{p-1})$$

and so x is constant if and only if $\sigma x=x$, that is, if and only if σ *fixes* x.

Let us consider how the powers of σ act on an element $x\in X$. The p-tuple $\sigma^k x$ comes from x by a k-fold rotation. The **orbit** of $x\in X$ is the collection

$$\mathcal{O}(x)=\{x,\sigma x,\ldots,\sigma^{p-1}x\}$$

of the various rotated versions of x. Now, the primeness of p implies that the elements of $\mathcal{O}(x)$ are either all distinct or all the same. For if $\sigma^i x=\sigma^j x$ for $i>j$, then $\sigma^{i-j}x=x$ where $0<i-j<p$. If $k=i-j$, then $(k,p)=1$ implies that $uk+vp=1$ for some $u,v\in\mathbb{Z}$ and so

$$\sigma x=\sigma^{uk+vp}x=\sigma^{uk}\sigma^{up}x=(\sigma^k)^u x=x$$

whence $\sigma^i x=x$ for all i. Thus, $\mathcal{O}(x)$ has size 1 or p for all $x\in X$.

Now, the orbits $\mathcal{O}(x)$ are the equivalence classes of the equivalence relation on X defined by

$$x\equiv y \quad\text{if}\quad y=\sigma^k x \text{ for some } k\in\mathbb{Z}$$

and so the distinct orbits form a *partition* of X. If b is the number of blocks of size 1 and c is the number of blocks of size p, then

$$|X|=b+cp$$

where $b\geq 1$, since $\{(1,\ldots,1)\}$ is a block of size 1. Hence, p divides $|X|$ implies that $p\mid b$ and so, in particular, $b\geq p$, which implies that there are at least $p-1$ constant p-tuples $(a,\ldots,a)$ with $a\neq 1$.□

Application to p-Groups

The following types of groups play a key role in the study of the structure of finite groups.

Definition *Let G be a nontrivial group and let p be a prime.*
1) *An element $a\in G$ is called a* **p-element** *if $o(a)=p^k$ for some $k\geq 0$.*

2) *G is a* **p-group** *if every element of G is a p-element.*
3) *A nontrivial subgroup S of G is called a* **p-subgroup** *of G if S is a p-group.*□

Lagrange's theorem and Cauchy's theorem combine to describe finite p-groups quite succinctly.

Theorem 3.21 *A finite group G is a p-group if and only if the order of G is a power of p.*
Proof. If $o(G) = p^k$, then Lagrange's theorem implies that every element of G is a p-element and so G is a p-group. Conversely, if G is a finite p-group but $q \mid o(G)$ where $q \neq p$ is prime, then Cauchy's theorem implies that G has an element of order q, which is false. Hence, $o(G) = p^k$ for some $k \geq 1$.□

The Center of a Group; Centralizers

We briefly mentioned the following concept earlier.

Definition *The* **center** *$Z(G)$ of a group G is the set of all elements of G that commute with all elements of G, that is,*

$$Z(G) = \{a \in G \mid ab = ba \text{ for all } b \in G\}$$

A group G is **centerless** *if $Z(G) = \{1\}$. A subgroup H of G is* **central** *if H is contained in the center of G.*□

It is easy to see that the center of G is a normal subgroup of G.

Example 3.22
a) The center of the quaternion group Q is $Z(Q) = \{1, -1\}$.
b) For $n \geq 3$ odd, the dihedral group D_{2n} is centerless and for $n \geq 3$ even,

$$Z(D_{2n}) = \{\iota, \rho^{n/2}\}$$

c) For $n \geq 3$, the symmetric group S_n is rather large and it should come as no surprise that S_n is centerless. To see this, suppose that $\sigma \in S_n$ is not the identity. If the cycle decomposition of σ contains a k-cycle with $k \geq 3$, that is, if

$$\sigma = \cdots(a\,b\,c\cdots)\cdots$$

then $\sigma^{(a\,b)} \neq \sigma$ and so $(a\,b)\sigma \neq \sigma(a\,b)$. On the other hand, if the cycle decomposition of σ is a product of disjoint transpositions:

$$\sigma = (a\,b)\cdots$$

then $\sigma^{(b\,c)} \neq \sigma$ and so $(b\,c)\sigma \neq \sigma(b\,c)$ for $c \notin \{a, b\}$. In either case, $\sigma \notin Z(S_n)$ and so S_n is centerless.

d) We will prove in a later chapter that for $n \neq 4$, the alternating group A_n is *simple*, that is, A_n has no nontrivial proper normal subgroups. Hence, A_n is centerless for $n \geq 4$.□

The Number of Conjugates of an Element

Definition *Let G be a group. The* **centralizer** *of an element $b \in G$ is the set of all elements of G that commute with b:*

$$C_G(b) = \{a \in G \mid ab = ba\}$$

The **centralizer** *of a subgroup $H \leq G$ is the set*

$$C_G(H) = \bigcap_{a \in H} C_G(a)$$

of all elements of G that commute with every element of H.□

It is easy to see that centralizers are subgroups of the parent group.

Let G be a group and let $a \in G$. Then

$$a^x = a^y \;\Leftrightarrow\; a^{y^{-1}x} = a \;\Leftrightarrow\; y^{-1}x \in C_G(a) \;\Leftrightarrow\; xC_G(a) = yC_G(a)$$

and so the element a has precisely $(G : C_G(a))$ distinct conjugates. This formula is of considerable importance in the study of finite groups.

Theorem 3.23 *Let G be a group and let $a \in G$. Let $\mathrm{conj}_G(a)$ denote the set of conjugates of a in G. Then*

$$|\mathrm{conj}_G(a)| = (G : C_G(a))$$

which divides $o(G)$ when G is finite. The set $\mathrm{conj}_G(a)$ is called a **conjugacy class** *in G.*□

The Normalizer of a Subgroup

Suppose that $H \leq G$. Then of course H is normal in itself. But it may also be normal in a larger subgroup of G.

Definition *Let G be a group and let $H \leq G$. The largest subgroup $N_G(H)$ of G for which $H \trianglelefteq N_G(H)$ is called the* **normalizer** *of H in G. A subset $X \subseteq G$ is said to* **normalize** *H if $X \subseteq N_G(H)$.*□

Theorem 3.24 *Let G be a group. The normalizer of $H \leq G$ is*

$$N_G(H) = \{a \in G \mid H^a = H\}$$

□

Will see in the chapter on free groups (Theorem 12.21) that it is possible to have $H^a \subset H$, in which case $a \notin N_G(H)$. However, since

$$H^a = H \quad \Leftrightarrow \quad H^a \subseteq H \quad \text{and} \quad H^{a^{-1}} \subseteq H$$

if a subset $S \subseteq G$ is closed under inverses and has the property that $H^s \subseteq H$ for all $s \in S$, then $S \subseteq N_G(H)$. In particular, since

$$S = \{a \in G \mid H^a \subseteq H\}$$

is closed under products, if S is finite then S is a subgroup of G and so S is closed under inverses, whence $S = N_G(H)$. In particular, if G is finite, then $S = N_G(H)$.

The normalizer of H should not be confused with the *normal closure* of H, which is the smallest normal subgroup of G that contains H. Note that if $K \leq G$ normalizes $H \leq G$, then HK is also a subgroup of G.

Theorem 3.25 *If G is a group and $H \leq G$, then*

$$C_G(H) \trianglelefteq N_G(H) \qquad \square$$

Elementwise Commutativity

It is clear that H and K commute elementwise if and only if

$$H \leq C_G(K) \quad \text{and} \quad K \leq C_G(H)$$

For essentially disjoint subgroups, we need only check that each subgroup is contained in the *normalizer* of the other. Proof of the following is left to the reader.

Theorem 3.26 *Let H and K be essentially disjoint subgroups of a group G.*

1) *Then H and K commute elementwise if and only if each subgroup is contained in the normalizer of the other, that is,*

$$H \leq N_G(K) \text{ and } K \leq N_G(H)$$

In particular, if $H, K \trianglelefteq G$, then H and K commute elementwise.

2) *If $G = H \bullet K$, then H and K commute elementwise if and only if $H, K \trianglelefteq G$.* $\square$

The Number of Conjugates of a Subgroup

Let G be a group and let $H \leq G$. Then

$$H^x = H^y \Leftrightarrow H^{y^{-1}x} = H \Leftrightarrow y^{-1}x \in N_G(H) \Leftrightarrow xN_G(H) = yN_G(H)$$

Thus, we obtain a count of the number of conjugates of a subgroup (the analog of Theorem 3.23).

Theorem 3.27 *Let G be a group and let $H \leq G$. Let* $\text{conj}_G(H)$ *denote the set of conjugates of H in G. Then*

$$|\text{conj}_G(H)| = (G : N_G(H))$$

which divides $o(G)$ *when* G *is finite. The set* $\text{conj}_G(H)$ *is called a* **conjugacy class** *in* $\text{sub}(G)$.□

Simple Groups

Nontrivial groups with no nontrivial proper normal subgroups are, in some sense, simple.

Definition *A nontrivial group* G *is* **simple** *if it has no normal subgroups other than* $\{1\}$ *and* G.□

Of course, if G is an abelian group, then all of its subgroups are normal and so an abelian group is simple if and only if its only subgroups are $\{1\}$ and G. This is not easy for a group.

Theorem 3.28 *An abelian group* G *is simple if and only if it is a cyclic group of prime order.*

Proof. If G is cyclic of prime order, then Lagrange's theorem implies that G has no subgroups of order different from 1 and $o(G)$ and so G is simple. Conversely, if G is simple, then every nonidentity element g of G must generate G, that is, $G = \langle g \rangle$ is cyclic. However, if G is infinite, then $\langle g^2 \rangle$ is a nontrivial proper subgroup of G, a contradiction. Hence, G is finite and cyclic. But if $p \mid o(G)$ where p is prime, then Cauchy's theorem implies that G has a subgroup S of order p and so $G = S$ has prime order.□

We will show later in the book that the alternating group A_n is simple for all $n \neq 4$. Also, a famous result of **Feit–Thompson** (1963, [11]), whose proof runs 255 pages, says that every *nonabelian* finite simple group has even order. Thus:

1) The *abelian* simple groups are the cyclic groups of prime order.
2) All *nonabelian* finite simple groups have even order.

Commutators of Elements

The elements of a group of the form $aba^{-1}b^{-1}$ play a special role.

Definition *Let* G *be a group. The* **commutator** *of* $a, b \in G$ *is the element*

$$[a, b] = aba^{-1}b^{-1}$$

We denote the set *of commutators of* G *by* $\mathcal{C}(G)$. *The* subgroup generated *by all commutators of* G *is usually denoted (unfortunately) by* G' *and is called the* **commutator subgroup**, *or* **derived subgroup** *of* G. *A group* G *is* **perfect** *if* $G = G'$.□

We should note that authors who define the conjugate of two elements by $a^b = b^{-1}ab$ also define the commutator by $[a,b] = a^{-1}b^{-1}ab$.

It is easy to see that G' is fully invariant in G; in particular, $G' \sqsubseteq G$. Also, we leave it as an exercise to prove that if $N \trianglelefteq G$, then

$$\left(\frac{G}{N}\right)' = \frac{G'N}{N}$$

Commutators have some very nice algebraic properties. Here are the most basic of these properties.

Theorem 3.29 *Let G be a group and let $a,b,c \in G$.*
1) $[a,b] = 1 \Leftrightarrow ab = ba$
2) $[a,b]^{-1} = [b,a]$
3) $[a,b]^c = [a^c, b^c]$
4) *If $H \trianglelefteq G$, then in G/H,*

$$[aH, bH] = [a,b]H$$

5) *If $a,b \in G$ commute with $c,d \in G$ and vice versa, then*

$$[a,b][c,d] = [ca, db]$$ □

Note that since $[a,b]^{-1} = [b,a]$, every element of G' is a product of commutators. The following characterization of commutator subgroups explains why these subgroups are so important.

Theorem 3.30 (R. Dedekind, c. 1880) *Let G be a group. Then for any subgroup $H \le G$,*

$$H \trianglelefteq G \text{ and } G/H \text{ is abelian} \quad \Leftrightarrow \quad G' \le H$$

In particular, G' is the smallest normal subgroup of G whose quotient is abelian.

Proof. If $H \trianglelefteq G$ and G/H is abelian, then in G/H, we have

$$[a,b]H = [aH, bH] = H$$

and so $[a,b] \in H$, whence $G' \le H$. Conversely, if $G' \le H$, then H is normal in G, since for any $h \in H$ and $a \in G$,

$$h^a = [a,h]h \in H$$

Also, G/H is abelian since

$$[aH, bH] = [a,b]H = H$$ □

Example 3.31 We leave it to the reader to show that the commutator subgroup of the quaternion group Q is $Q' = \{1, -1\}$, which is also the *set* of commutators, that is,

$$Q' = \mathcal{C}(Q)$$

Similarly, the commutator subgroup of the dihedral group D_{2n} is $D'_{2n} = \langle \rho^2 \rangle$ and so again the commutator subgroup of D_{2n} is also the *set* of commutators:

$$D'_{2n} = \mathcal{C}(D_{2n})$$ □

Example 3.32 For the symmetric group S_n on $n \geq 3$ symbols, we have

$$S'_n = A_n = \mathcal{C}(S_n)$$

Since all commutators are even permutations, we have $S'_n \leq A_n$.

Now we make a few observations. First the product of any finite number of disjoint commutators is a commutator, since if $\alpha_i\beta_i$ and $\alpha_j\beta_j$ are disjoint, then

$$[\alpha_i, \beta_i][\alpha_j, \beta_j] = [\alpha_j\alpha_i, \beta_j\beta_i]$$

Second, if $\sigma, \mu \in S_n$, then

$$\sigma^\mu \sigma^{-1} = [\mu, \sigma]$$

and so if σ and τ have the same cycle structure, then the product $\tau\sigma$ is a commutator.

Hence, we need only show that any $\sigma \in A_n$ can be written as a product of disjoint permutations, each of which is a product $\alpha_i\beta_i$ where α_i and β_i have the same cycle structure.

But since a cycle of odd length is even and a cycle of even length is odd, the cycle decomposition of σ must have an even number of cycles of even length. In other words, the cycle decomposition of σ is a product of odd cycles and *pairs* of even cycles.

Now, every odd cycle $\sigma = (a_1 \cdots a_{2m+1})$ can be written as a product of two equal-length cycles as follows:

$$\sigma = (a_1 \cdots a_{2m+1}) = (a_1 \cdots a_{m+1})(a_{m+1} \cdots a_{2m+1})$$

Similarly, every pair of disjoint even cycles can be written as a product of two equal-length cycles as follows (where $k \geq m$):

$$\sigma = (a_1 \cdots a_{2m})(b_1 \cdots b_{2k}) = (a_1 \cdots a_{2m} b_1 \cdots b_{k-m+1})(a_{2m}\, b_{k-m+1} \cdots b_{2k})$$

Hence, σ can be written in the form

$$\sigma = (\alpha_1\beta_1) \cdots (\alpha_m\beta_m)$$

where α_i and β_i are equal-length cycles for each i and $\alpha_i\beta_i$ and $\alpha_j\beta_j$ are disjoint for $i \neq j$. Hence, $\sigma \in \mathcal{C}(S_n)$.□

When is $\mathcal{C}(G) = G'$?

We have seen that for the quaternion, dihedral and symmetric groups,

$$G' = \mathcal{C}(G)$$

Despite these examples, however, this is not always true. In fact, the question of precisely when $G' = \mathcal{C}(G)$ is an active area of current research. We give an example of a group for which $G' \neq \mathcal{C}(G)$ and then discuss a few results in this area. Readers interested in more details may wish to consult the survey article of Kappe and Morse [19].

Example 3.33 (Cassidy [5]) Let G be the set of all matrices of the form

$$m(f,g,h) := \begin{bmatrix} 1 & f(x) & h(x,y) \\ 0 & 1 & g(y) \\ 0 & 0 & 1 \end{bmatrix}$$

where $f(x)$, $g(x)$ and $h(x,y)$ are polynomials with rational coefficients. A straightforward calculation shows that

$$m(f_1,g_1,h_1)m(f_2,g_2,h_2) = m(f_1+f_2, g_1+g_2, h_1+h_2+f_1g_2)$$

and

$$m(f,g,h)^{-1} = m(-f,-g,-h+fg)$$

from which it follows that G is a group. Another computation shows that the commutators are given by

$$[m(f_1,g_1,h_1), m(f_2,g_2,h_2)] = m(0,0,f_1g_2 - f_2g_1)$$

Thus, the commutator subgroup G' is contained in the subgroup of all matrices in G of the form $m(0,0,h)$ where $h \in \mathbb{Q}[x,y]$. Moreover, any such matrix is a product of commutators, since

$$m\left(0,0,\sum_{i,j} a_{i,j}x^iy^j\right) = \prod_{i,j}\left[m(a_{i,j}x^i,0,0), m(0,y^j,0)\right]$$

which shows that

$$G' = \{m(0,0,h) \mid h \in \mathbb{Q}[x,y]\}$$

Thus, the matrix

$$A = m(0,0,1+xy+x^2y^2)$$

is in G'. However, it is not a commutator, for if

$$1+xy+x^2y^2=f_1(x)g_2(y)-f_2(x)g_1(y)$$

for some $f_1, f_2 \in \mathbb{Q}[x]$ and $g_1, g_2 \in \mathbb{Q}[y]$, then equating coefficients of x^i shows that

$$y^i=f_{1,i}g_2(y)-f_{2,i}g_1(y)$$

for $i=0,1$ and 2, where $f_{u,i}$ is the coefficient of x^i in $f_u(x)$. But this implies that the two-dimensional vector subspace of $\mathbb{Q}[y]$ spanned by $g_1(y)$ and $g_2(y)$ contains the three independent vectors 1, y and y^2, which is not possible. Hence, not all members of G' are commutators.□

Here is a small sampling of known results concerning the issue of when $G'=\mathcal{C}(G)$. Many more results are contained in Kappe and Morse [19].

Theorem 3.34 (Speigel [31], 1976) *If a group G contains a normal abelian subgroup A whose quotient G/A is cyclic, then $G'=\mathcal{C}(G)$.*
Proof. If we can find a normal subgroup $B \trianglelefteq G$ for which $B \subseteq \mathcal{C}(G)$ and G/B is abelian, then

$$G' \le B \subseteq \mathcal{C}(G) \subseteq G'$$

whence $G'=\mathcal{C}(G)$. To this end, let $G/A=\langle xA\rangle$ and let

$$B=\{[x,a] \mid a \in A\}=\{a^xa^{-1} \mid a \in A\} \subseteq \mathcal{C}(G)$$

To see that $B \trianglelefteq G$, we have

$$(a^xa^{-1})(b^xb^{-1})=a^xb^xa^{-1}b^{-1}=(ab)^x(ab)^{-1} \in B$$

and

$$(a^xa^{-1})^{-1}=(a^{-1})^xa \in B$$

and for normality,

$$(a^xa^{-1})^{x^ia}=(a^xa^{-1})^{x^i}=(a^{x^i})^x(a^{x^i})^{-1} \in B$$

To see that G/B is abelian, we have

$$[xB,aB]=[x,a]B=B$$

and so xB and aB commute for all $a \in A$, which implies that x^iaB and x^jbB commute for all $a,b \in A$.□

For small groups, we also have $G'=\mathcal{C}(G)$.

Theorem 3.35 *Let G be a group. The following conditions imply that $G'=\mathcal{C}(G)$.*
1) **(Guralnick [15], 1980)**

a) G' *is abelian and either* $o(G) < 128$ *or* $o(G') < 16$.
b) G' *is nonabelian and either* $o(G) < 96$ *or* $o(G') < 24$.

2) **(Kappe and Morse [20], 2005)**
a) $o(G) = p^n$, *where* p *is an odd prime and* $n \le 5$.
b) $o(G) = 2^n$, *where* $n \le 6$.□

On the other hand, if the center of G is large compared to the size of G', then there will be elements of G' that are not commutators.

Theorem 3.36 (MacDonald [22], 1986) *Let* G *be a group with* $Z = Z(G)$. *If*

$$(G{:}\,Z)^2 < o(G')$$

then $G' \neq C(G)$.
Proof. Since $z, w \in Z$ implies that

$$[az, bw] = [a, b]$$

it follows that the number of distinct commutators is at most the number of commutators of G/Z and that is certainly at most $(G{:}\,Z)^2$.□

Additional Properties of Commutators

Here are some additional properties of commutators. The reader may wish simply to read the statement of the theorem and move on, referring to the theorem as needed at later times. (The proof is not particularly enlightening.)

Theorem 3.37 *Let* G *be a group and let* $a, b, c \in G$.
1)

$$[b, a] = [a^{-1}, b]^a = [a, b^{-1}]^b$$
$$[a, b] = [a^{-1}, b^{-1}]^{ba}$$

2)

$$[a, bc] = [a, b][a, c]^b$$
$$[ab, c] = [b, c]^a[a, c]$$

3) *Let* $a_1, \ldots, a_n, b_1, \ldots, b_m \in G$. *Then*

$$[a_1 \cdots a_n, b_1 \cdots b_m] = \prod_{i=1}^{n}\prod_{j=1}^{m}[a_i, b_j]^{\alpha_{i,j}}$$

where $\alpha_{i,j} \in \langle a_1, \ldots, a_n, b_1, \ldots, b_m \rangle$. *Also, the order of the factors on the right can be chosen arbitrarily, although the exponents depend on that order. In particular, if* m *and* n *are positive integers, then*

$$[a^n, b^m] = \prod_{i=1}^{n}\prod_{j=1}^{m} [a,b]^{\alpha_{i,j}}$$

where $\alpha_{i,j} \in \langle a, b\rangle$.

4) If $[a,b]$ commutes with both a and b, then for any integers m and n,
   a) $[a^m, b^n] = [a,b]^{mn}$
   b) $(ab)^m = a^m b^m [b,a]^{\binom{m}{2}}$
   Hence, if $o([a,b])$ *divides* $\binom{m}{2}$, *then*

$$(ab)^m = a^m b^m$$

Proof. The proofs of part 1) and part 2) are straightforward calculations. Part 3) is proved by a double induction. For $m = 1$, we induct on n. If $n = 1$, then the result is clear. If the result holds for $n - 1$, then part 2) implies that

$$[a_1 \cdots a_n, b_1] = [a_n, b_1]^{a_1 \cdots a_{n-1}} [a_1 \cdots a_{n-1}, b_1]$$

and the inductive hypothesis completes the proof for $m = 1$. Assume the result is true for $m - 1$ and let $a = a_1 \cdots a_n$. Then

$$[a, b_1 \cdots b_m] = [a, b_1 \cdots b_{m-1}][a, b_n]^{b_1 \cdots b_{m-1}}$$

and the inductive hypothesis completes the proof. As to the statement about the order, we can rearrange the factors using the identity $xy = y^x x$.

Part 4a) follows from part 3) when $m, n \geq 0$. The result then follows for negative exponents using part 1). For part 4b), note first that $c = [b,a]$ also commutes with a and b and that

$$ba = cab = abc$$

Now,

$$(ab)^m = \underbrace{(ab) \cdots (ab)(ab)}_{m \text{ factors}}$$

and if we move the rightmost a all the way to the left, the result is

$$(ab)^m = a\underbrace{(ab) \cdots (ab)}_{m-1 \text{ factors}} bc^{m-1}$$

Repeating this process gives

$$(ab)^m = a^2 \underbrace{(ab) \cdots (ab)}_{m-2 \text{ factors}} b^2 c^{m-2} c^{m-1}$$

Continuing until all of the as are on the left gives

$$(ab)^m = a^{m-1}(ab)b^{m-1}c^{1+2+\cdots+(m-1)} = a^m b^m c^{\binom{m}{2}}$$

which proves the result for $m \geq 0$. For $m < 0$ we have, using part 4b) for positive exponents along with part 4a) and part 1),

$$\begin{aligned}(ab)^{-m} &= (b^{-1}a^{-1})^m \\ &= b^{-m}a^{-m}[a^{-1},b^{-1}]^{\binom{m}{2}} \\ &= a^{-m}b^{-m}[b^m,a^m][a^{-1},b^{-1}]^{\binom{m}{2}} \\ &= a^{-m}b^{-m}[b,a]^{m^2}[a^{-1},b^{-1}]^{\binom{m}{2}} \\ &= a^{-m}b^{-m}[b,a]^{m^2}[a,b]^{\binom{m}{2}} \\ &= a^{-m}b^{-m}[b,a]^{m^2}[b,a]^{-\binom{m}{2}} \\ &= a^{-m}b^{-m}[b,a]^{\binom{-m}{2}}\end{aligned}$$

□

Here is one application of Theorem 3.37 that will be useful later in the book.

Theorem 3.38 *If G is a nonabelian group with the property that all cyclic subgroups are normal, then*
1) G is periodic
2) Any nonabelian subgroup of G contains a quaternion subgroup
3) If $a, b \in G$ and $o(a) + o(b) < 8$, then $ab = ba$.

Proof. To see that G is periodic, we first show that any $x \notin Z(G)$ has finite order. Let $y \in G$ satisfy

$$c := [x, y] \neq 1$$

Then the normality of $\langle x \rangle$ and $\langle y \rangle$ imply that $c \in \langle x \rangle \cap \langle y \rangle$ and so

$$c = x^n = y^m$$

for some $n, m \neq 0$. Thus, c commutes with x and y and so

$$1 = [c, c] = [x^n, y^m] = [x, y]^{nm} = c^{nm}$$

Hence, c has finite order and therefore so does x. Now let $x \in Z(G)$. If $y \notin Z(G)$, then $xy \notin Z(G)$ and so xy has finite order and therefore so does x. Hence, G is periodic.

Now suppose that D is a nonabelian subgroup of G. Let $x, y \in D$ be chosen so that $o(x) + o(y)$ is minimal among all pairs of *noncommuting* elements in D. We will show that $Q = \langle x, y \rangle$ is a quaternion subgroup of D. To see that x and y have order 4, let p be a prime dividing $o(x)$ and let

$$c = [x, y] = x^n = y^m$$

as shown above. Since $o(x^p) < o(x)$, it follows that x^p and y commute and so Theorem 3.37 implies that

$$1 = [x^p, y] = [x, y]^p$$

whence $o(c) = p$. Hence, p is the only prime dividing $o(x)$ and similarly $o(y)$

and so

$$o(x) = p^{i+1} \quad \text{and} \quad o(y) = p^{j+1}$$

for some $i, j \geq 0$. Moreover,

$$p = o(c) = o(x^n) = \frac{p^{i+1}}{(p^{i+1}, n)}$$

which shows that $n = p^i u$ where $p \nmid u$. A similar argument for y gives

$$x^{p^i u} = c = y^{p^j v}$$

where $p \nmid u$ and $p \nmid v$.

To eliminate u and v from the equation above, if $\alpha = u^{-1} \bmod p$ and $\beta = v^{-1} \bmod p$, then taking the $\alpha\beta$ power gives

$$x^{p^i \beta} = c^{\alpha\beta} = y^{p^j \alpha}$$

and so

$$(x^\beta)^{p^i} = [x^\beta, y^\alpha] = (y^\alpha)^{p^j}$$

where $o(x^\beta) = p^{i+1}$ and $o(y^\alpha) = p^{j+1}$. Thus, replacing x^β by x and y^α by y gives

$$x^{p^i} = c = y^{p^j} \tag{3.39}$$

where $1 \neq c = [x, y]$ and

$$o(x) = p^{i+1}, \quad o(y) = p^{j+1} \quad \text{and} \quad o(c) = p$$

We may also assume that $i \geq j \geq 1$, for if $j = 0$, then $y = c = [x, y]$, which is false.

The next step is to show that $p = 2$ and that $i = j = 1$. The key observation here is that for any integer k, the elements $x^k y$ and x do not commute and so $o(x^k y) \geq o(y) = p^{j+1}$. But, if p is odd or if $p = 2$ and $j > 1$, then

$$o(c) = p \,\Big|\, \binom{p^j}{2}$$

and so Theorem 3.37 implies that

$$1 \neq (x^k y)^{p^j} = x^{kp^j} y^{p^j} = x^{kp^j} c$$

and taking $k = -p^{i-j}$ gives $1 \neq 1$. Hence, $p = 2$ and $i = j = 1$. Thus,

$$o(x) = o(y) = 4, \quad o(c) = 2 \quad \text{and} \quad x^2 = c = y^2$$

where c commutes with x and y and since

$$xy(xy) = cyx(xy) = c$$

it follows that $Q = \langle x, y \rangle$ is quaternion.□

Commutators of Subgroups

Commutators can be defined for subsets as well as for elements.

Definition *The* **commutator** $[X, Y]$ *of two subsets* $X, Y \subseteq G$ *is the subgroup generated by the commutators:*

$$[X, Y] = \langle [x, y] \mid x \in X, y \in Y \rangle$$ □

Note that, by definition,

$$G' = [G, G]$$

for any group G. Here are some of the basic properties of these commutators.

Theorem 3.40 *Let* $H, K \leq G$.

1)

$$[H, K] = [K, H]$$

2) H *and* K *commute elementwise if and only if* $[H, K] = \{1\}$, *in particular,*

$$H \leq Z(G) \quad \Leftrightarrow \quad [H, G] = \{1\}$$

3)

$$H \trianglelefteq G \quad \Leftrightarrow \quad [H, G] \leq H$$

4) If $H, K \trianglelefteq G$, *then*

$$[H, K] \leq H \cap K \quad \textit{and} \quad [H, K] \trianglelefteq G$$

Also,

$$H, K \sqsubseteq G \quad \Rightarrow \quad [H, K] \sqsubseteq G$$

5) H *normalizes* $[H, K]$ *and so*

$$[H, K] \trianglelefteq \langle H, K \rangle$$

6)

$$H[H, K] = H^K \trianglelefteq H^K K = \langle H, K \rangle$$

In particular,

$$\operatorname{nc}(H, G) = H[H, G]$$

Proof. We prove only part 5). Theorem 3.37 implies that if $h_1 \in H$, then

$$[h,k]^{h_1} = [h_1 h,k][h_1,k]^{-1} \in [H,K]$$

Hence, H normalizes $[H,K]$. Similarly, K normalizes $[H,K]$ and so $\langle H,K\rangle$ also normalizes $[H,K]$.□

Here are some additional properties of commutators of subgroups.

Theorem 3.41 *Let* $A,H,K \le G$.

1)

$$[A,HK] = [A,H][A,K]^H$$

where

$$[A,K]^H = \langle [a,k]^h \mid a \in A, k \in K, h \in H\rangle$$

2) If $A \trianglelefteq G$ *and if* $\mathcal{F} = \{H_i \mid i \in I\}$ *is a family of normal subgroups of* G, *then*

$$[A, \bigvee H_i] = \bigvee [A,H_i]$$

This equation still holds even if one member of $\mathcal{F}$ *is not normal. In particular, if* $A,H \trianglelefteq G$ *and* $K \le G$, *then*

$$[A,HK] = [A,H][A,K]$$

3) If $N \trianglelefteq G$, *then*

$$[HN,KN] \le [H,K]N$$

and so

$$[HN,KN]N = [H,K]N$$

4) If $N \trianglelefteq G$, *then*

$$\left[\frac{HN}{N},\frac{KN}{N}\right] = \frac{[H,K]N}{N}$$

Proof. For part 1), if $a \in A$, $h \in H$ and $k \in K$, then

$$[a,hk] = [a,h][a,k]^h \in [A,H][A,K]^H$$

and so $[A,HK] \le [A,H][A,K]^H$. For the reverse inclusion, we have $[A,H] \le [A,HK]$ and since

$$[a,k]^h = [a,h]^{-1}[a,hk] \in [A,HK]$$

we also have $[A,K]^H \le [A,HK]$.

For part 2), note that part 1) and the fact that $[A,H] \trianglelefteq G$ imply that

$$[A,HK] \le [A,H][A,K]$$

and the reverse inclusion is evident. For the general case, since $[A,H_k] \le [A,\bigvee H_i]$ for all k, it follows that

$$\bigvee [A,H_k] \le [A,\bigvee H_i]$$

For the reverse inclusion, each generator of $[A,\bigvee H_i]$ belongs to a commutator subgroup of the form $[A,H_{i_1}\cdots H_{i_n}]$, which is equal to $\bigvee [A,H_{i_j}]$, which in turn is contained in $\bigvee [A,H_i]$ and so

$$[A,\bigvee H_i] \le \bigvee [A,H_i]$$

Part 3) follows from the fact that

$$\begin{aligned}[HN,KN] &= [HN,K][HN,N]^K \\ &= [K,H][K,N]^H[HN,N]^K \\ &\le [K,H]N\end{aligned}$$

Part 4) follows from part 3).□

*Multivariable Commutators

We can extend the definition of commutators as follows. If $a,b,c \in G$, then

$$[a,b,c] = [a,[b,c]]$$

and in general, if $a_1,\dots,a_n \in G$, then

$$[a_1,\dots,a_n] = [a_1,[a_2,\dots,a_n]]$$

with $[a] := a$. Note that some authors define $[a,b,c]$ to be $[[a,b],c]$ and, in general, these are not the same.

Theorem 3.42 *Let G be a group. The following are equivalent:*

1) **(Associativity)** *For all $a,b,c \in G$,*

$$[[a,b],c] = [a,[b,c]]$$

2) **(Distributivity)** *For all $a,b,c \in G$,*

$$[a,bc] = [a,b][a,c]$$

3) **(Commutator subgroup is central)**

$$G' \le Z(G)$$

Proof. Since

$$[a, bc] = [a, b][a, c]^b$$

it follows that 2) holds if and only if $[a, c]^b = [a, c]$ for all $a, b, c \in G$, which holds if and only if $[a, c] \in Z(G)$ for all $a, c \in G$, that is, if and only if $G' \leq Z(G)$. Hence, 2) and 3) are equivalent.

It is clear that 3) implies 1). Conversely, 1) is

$$[a, b]c[a, b]^{-1}c^{-1} = a[b, c]a^{-1}[b, c]^{-1}$$

which is equivalent to

$$[a, b]c[b, a]c^{-1} = a[b, c]a^{-1}[b, c]^{-1}$$

Now, as to the last factor $[b, c]^{-1}$, note that 1) with $b = c$ implies that

$$[[a, b], b] = 1$$

for all $a, b \in G$. Hence,

$$[b, c]^{-1} = [c, b] = b[b^{-1}, c]b^{-1} = [b^{-1}, c]$$

and so 1) is equivalent to

$$[a, b]c[b, a]c^{-1} = a[b, c]a^{-1}[b^{-1}, c]$$

Expanding the commutators gives

$$aba^{-1}b^{-1}cbab^{-1}a^{-1}c^{-1} = abcb^{-1}c^{-1}a^{-1}b^{-1}cbc^{-1}$$

and cancelling gives

$$a^{-1}b^{-1}cbab^{-1}a^{-1} = cb^{-1}c^{-1}a^{-1}b^{-1}cb$$

Moving the first factor on the right side to the left side and the last factor on the left side to the right side gives

$$c^{-1}a^{-1}b^{-1}cbab^{-1} = b^{-1}c^{-1}a^{-1}b^{-1}cba$$

which is equivalent to

$$[c^{-1}, a^{-1}b^{-1}]b^{-1} = b^{-1}[c^{-1}, a^{-1}b^{-1}]$$

which shows that $[c^{-1}, a^{-1}b^{-1}]$ commutes with b^{-1}, but $c^{-1}, a^{-1}b^{-1}$ and b^{-1} represent arbitrary elements of G and so $[x, y]$ commutes with z for all $x, y, z \in G$, that is, $G' \leq Z(G)$.□

Theorem 3.43 (Hall–Witt Identity) *Let $a, b, c \in G$. Then*

$$[a, b^{-1}, c]^b[b, c^{-1}, a]^c[c, a^{-1}, b]^a = 1$$

Proof. For the first factor we have

$$\begin{aligned}[a,b^{-1},c]^b &= b[a,[b^{-1},c]]b^{-1} \\ &= ba[b^{-1},c]a^{-1}[b^{-1},c]^{-1}b^{-1} \\ &= bab^{-1}cbc^{-1}a^{-1}cb^{-1}c^{-1}bb^{-1} \\ &= bab^{-1}cbc^{-1}a^{-1}cb^{-1}c^{-1} \\ &= a^b b^c a^{-1}(b^{-1})^c\end{aligned}$$

By cycling a, b and c, we get

$$[b,c^{-1},a]^c = b^c c^a b^{-1}(c^{-1})^a$$

and

$$[c,a^{-1},b]^a = c^a a^b c^{-1}(a^{-1})^b$$

The product is easily seen to self-destruct.□

If H, K and L are subgroups of G, then we define

$$[H,K,L] = [H,[K,L]]$$

and more generally, if $H_1,\ldots,H_n \le G$, then

$$[H_1,\ldots,H_n] = [H_1,[H_2,\ldots,H_n]]$$

with $[H] := H$.

The subgroup $[H,[K,L]]$ is generated by elements of the form $[h,c_1\cdots c_n]$, where $h \in H$ and $c_i = [k_i,\ell_i]$ for $k_i \in K$ and $\ell_i \in L$. Hence, Theorem 3.37 implies that

$$[h,c_1\cdots c_n] = \prod [h,c_i]^{\alpha_i}$$

where $\alpha_i \in \langle h,c_1,\ldots,c_n\rangle \le \langle H,K,L\rangle$.

Corollary 3.44 (Three subgroups lemma) *Let G be a group. Let H, K and L be subgroups of G. Then if any two of the commutators $[H,K,L]$, $[L,H,K]$ or $[K,L,H]$ are contained in a normal subgroup N of G, then so is the third commutator.*

Proof. We assume that $[H,K,L] \le N$ and $[L,H,K] \le N$ and use the Hall–Witt Identity. Since $[h,k^{-1},\ell] \in [H,K,L]$ and $[\ell,h^{-1},k] \in [L,H,K]$ for $h \in H$, $k \in K$ and $\ell \in L$, it follows from the Hall-Witt identity that

$$[k,\ell^{-1},h]^\ell = ([h,k^{-1},\ell]^k[\ell,h^{-1},k]^h)^{-1} \in N$$

and since N is normal in G, we deduce that $[k,\ell^{-1},h] \in N$. Hence, $[K,L,H] \le N$.□

Exercises

1. Let G be a group and let $H \trianglelefteq G$. Prove that if H and G/H are centerless, then G is centerless.
2. Let $G = S_3 \boxtimes C_2$. Show that G has two subgroups H and K of order 6 that are centerless but that $G = HK$ is not centerless. Thus, the set product of centerless subgroups may be a group that is not centerless. What about the direct product of centerless groups?
3. If a group G has the property that all of its subgroups are normal, must G be abelian?
4. Let G be a finite group and let $N_1, \ldots, N_m$ be normal subgroups of G. Show that $|N_1 \cdots N_m|$ divides $|N_1| \cdots |N_m|$.
5. A subgroup H of a group G is **permutable** if $HK = KH$ for all subgroups K of G. Prove that a maximal subgroup that is permutable is normal.
6. Show that if a normal subgroup H of a group G contains no nonidentity commutators, then H is central in G.
7. Use the Feit–Thompson Theorem to prove that a nontrivial group G of odd order is not perfect, that is, $G' < G$.
8. Suppose that G is a finite group and that $G' < G$. Prove that G has a normal subgroup K of prime index.
9. Show that the set product of subnormal subgroups need not be a subgroup. *Hint*: Check the dihedral group D_8.
10. Let G be a finite group and let H be a subgroup for which $(G : H)$ is prime. Show that if there is at least one left coset aH of H other than H itself that is also equal to some right coset Hb, then $H \trianglelefteq G$.
11. a) Prove that the commutator subgroup of the quaternion group is $\{1, -1\}$.
    b) Prove that the commutator subgroup of the dihedral group D_{2n} is $D'_{2n} = \langle \rho^2 \rangle$.
12. For which values of n is it true that the dihedral group D_{2n} has a pair of proper normal subgroups H and K for which

$$D_{2n} = HK \quad \text{and} \quad H \cap K = \{1\}$$

13. Let G be a group and let $H \leq G$. Prove that $C_G(H) \trianglelefteq N_G(H)$.
14. Prove that $S'_n = A_n$ for $n \geq 4$ in the following manner. Use the fact that A_n is simple for all $n \neq 4$. Show that $S'_n \cap A_n \trianglelefteq A_n$. Deduce that $S'_n \cap A_n = \{\iota\}$ or $S'_n \cap A_n = A_n$. Show that $S'_n \cap A_n = \{\iota\}$ is impossible.
15. Prove that if $N \trianglelefteq G$, then

$$\left(\frac{G}{N}\right)' = \frac{G'N}{N}$$

16. a) Find an infinite group that is periodic. *Hint*: Look at the quotient groups of the additive group of rationals.
    b) Prove that if $H \trianglelefteq G$ and H and G/H are periodic, then so is G.

17. Let $N \leq Z(G)$, the center of G. Show that $N \trianglelefteq G$ and that if G/N is cyclic, then G is abelian.
18. Prove that a maximal subgroup M of a group G is normal if and only if $G' \leq M$.
19. Let $H \lhd G$ have prime index. Let $x \in H$ have the property that $C_H(x) < C_G(x)$. Prove that an element $y \in H$ is conjugate to x in G if and only if it is conjugate to x in H.
20. Let $\pi = \{p_1, \ldots, p_m\}$ be a nonempty set of primes. A **π-group** is a group whose order n has the property that all primes dividing n lie in the set π. For example, a group of order $2^3 \cdot 5 \cdot 7^2$ is a $\{2, 5, 7\}$-group. Let G be a finite group and let $H_1, \ldots, H_m$ be normal subgroups such that G/H_i is a π-group for all i. Prove that $G/\bigcap H_i$ is also a π-group.
21. A subgroup H of a group G is **abnormal** if

$$a \in \langle H, H^a \rangle$$

for all $a \in G$. Prove that H is abnormal if and only if it satisfies the following conditions:
a) If $H \leq K \leq G$, then $N_G(K) = K$.
b) H is not contained in distinct conjugate subgroups, that is, if $K \leq G$ and $K \neq K^a$, then $H \not\subseteq K \cap K^a$.
22. Let $H, K \leq G$.
a) Prove that

$$H^K = H[H, K] \trianglelefteq \langle H, K \rangle = H^K K$$

b) If $G = \langle X \rangle$ for some subset X of G, show that

$$G' = \langle [x, y] \mid x, y \in X \rangle_{\text{nor}}$$

23. Let S be a nonempty subset of a group G and let $H, K \leq G$. Show that

$$S^{HK} = (S^K)^H$$

24. Let $A, H, K \leq G$. Prove that

$$[AH, AK] = [AH, A][AH, K]$$

25. Let $H, K, L \leq G$.
a) Show that if $[H, K, L] = \{1\}$ and $[L, H, K] = \{1\}$, then $[K, L, H] = \{1\}$.
b) Show that if H, K and L are normal in G, then

$$[H, K, L] \leq [L, H, K][K, L, H]$$

26. Let G be a group and let N be a normal subgroup. Let $x, y \in G$ and suppose that $x^n y^{-m} \in N$ and $yxy^{-t+am}x^{-1-an} \in N$. Prove that $yxy^{-t}x^{-1} \in N$.
27. Let G be a group and let $H \leq G$. Prove that if H has finite index in G, then $(G : H) = (G : H^x)$ for any $x \in G$.

28. Let G be a nonabelian group. Show that $Z(G)$ is a *proper* subgroup of any centralizer $C_G(g)$.
29. Let $H \leq G$. Prove that $C_G(H) \trianglelefteq N_G(H)$.
30. Let G be a group and let $H \leq G$.
    a) Prove that $N \leq G$ is normal in G if and only if all subsets of G normalize N.
    b) Prove that if S normalizes the subgroups H and K of G, then S normalizes their join $H \vee K$, that is,
    $$N_G(H) \cap N_G(K) \leq N_G(H \vee K)$$
    c) Prove that if S and T normalize H, then the set product ST normalizes H.
    d) Prove that a subgroup $H \leq G$ normalizes all supergroups of itself.
31. Prove that if $H \leq G$, then $N_G(H^a) = N_G(H)^a$. In particular, $C_G(h)^a = C_G(h^a)$.
32. Let G be a group and let $H \leq K \leq G$ with $H \trianglelefteq G$. Prove that
    $$N_G\left(\frac{H}{K}\right) = \frac{N_G(H)}{K}$$
33. a) Show that the subgroup $H = \{\iota, (1\,2)(3\,4), (1\,3)(2\,4), (1\,4)(2\,3)\}$ of S_4 is normal. *Hint*: Show that the conjugate of any transposition is another transposition, in fact, $(a\,b)^\sigma = (\sigma a\, \sigma b)$.
    b) Show that the subgroup $K = \{\imath, (1\,2)(3\,4)\}$ is normal in H but not normal in S_4.
    c) Conclude that normality is not transitive.
34. Let G be a group.
    a) Let M be the intersection of all subgroups of G that have finite index in G. Show that M is normal in G.
    b) Let $H \leq G$ with finite index. Show that there is a normal subgroup $N \trianglelefteq G$ for which $N \leq H \leq G$ where N also has finite index in G.
35. Let G be a group generated by two involutions x and y. Show that G has a normal subgroup of index 2.
36. Let G be a finite group of odd order. Show that the product of all of the elements of G, taken in any order, is in the commutator subgroup G'.
37. (P. Hall) Let $A, B \leq G$. Show that if $[A, A, B] = \{1\}$, then $[A, B]$ is an abelian group. *Hint*: One possible proof is as follows: Show that A commutes with $[A, B]$ and with $[B, A]$ and that B commutes with A'. Then use a direct computation to show that $[a, b][x, y] = [y, x^{-1}][a, b]$ where $a, x \in A$ and $b, y \in B$. Finally, use Theorem 3.37.
38. Let $p \neq 2$ be prime and consider the group
    $$G = \mathbb{Z}(p^\infty) \boxplus \mathbb{Z}_2$$
    where $\mathbb{Z}(p^\infty)$ is the p-quasicyclic group. Show that G has a unique

maximal subgroup M, but that M is not maximum in the lattice of all proper subgroups of G.

39. A group G is said to be **metabelian** if its commutator subgroup G' is abelian. (Some authors define a group G to be *metabelian* if G' is central in G, which is stronger than our definition.)
    a) Prove that G is metabelian if and only if G has a normal abelian subgroup A for which G/A is also abelian.
    b) Prove that the dihedral group D_{2n} is metabelian.
    c) Let F be a finite field, such as $\mathbb{Z}_p$ where p is prime. Let $a, b \in F$ where $a \neq 0$. The map $\sigma_{a,b}: F \to F$ defined by

    $$\sigma_{a,b}: x \mapsto ax + b$$

    is called an **affine transformation** of F. Show that the set $\text{aff}(F)$ of all affine transformations of F is a subgroup of F^F. Show that $\text{aff}(F)$ is metabelian.
    d) Prove that if $G = AB$ where A and B are abelian, then G is metabelian. *Hint*: Show that $[a,b]^{b_1 a_1} = [a,b]^{a_1 b_1}$ for $a, a_1 \in A$ and $b, b_1 \in B$. Use the fact that $AB = BA$.
40. Let G be a group of order p^n, where p is prime. Suppose that for each $a \in G$, the centralizer $C_a = C_G(a)$ has index 1 or p.
    a) Prove that $C_a \trianglelefteq G$. *Hint*: Assume $(G : C_a) = p$. Let

    $$I = \bigcap_{g \in G} C_a^g \leq C_a$$

    Show that I is normal in G. Then show that the elements of G/I are actually permutations of G/C_a, where $gI(xC_a) = gxC_a$. Show that G/I is a subgroup of $S_{G/H}$.
    b) Prove that $G' \leq Z(G)$.
41. We have seen that the family $\text{sub}(G)$ of all subgroups of G is a complete lattice. Let $\text{nor}(G)$ be the family of normal subgroups of G.
    a) Show that the join of two normal subgroups A and B is the set product AB.
    b) Show that $\text{nor}(G)$ is a complete sublattice of $\text{sub}(G)$.
    c) Prove that the distributive laws

    $$A \vee (B \cap C) = (A \vee B) \cap (A \vee C)$$
    $$A \cap (B \vee C) = (A \cap B) \vee (A \cap C)$$

    for $A, B, C \leq G$ imply the **modular law**: For $A, B, C \leq G$ with $A \leq B$,

    $$A \vee (B \cap C) = B \cap (A \vee C)$$

    A lattice that satisfies the modular law is said to be a **modular lattice**.
    d) Prove that $\text{nor}(G)$ is a modular lattice.

e) Is $\mathrm{nor}(G)$ necessarily distributive, that is, do the distributive laws necessarily hold? *Hint*: Consider the 4-group $V = \{1, a, b, ab\}$.

f) Prove that $\mathrm{sub}(G)$ need not be modular. *Hint*: Consider the alternating group A_4 of order 12. Let $A = \langle(1\,2)(3\,4)\rangle, B = \langle(1\,2)(3\,4), (1\,3)(2\,4)\rangle$ and $C = \langle(1\,2\,3)\rangle$.

g) Prove the Dedekind law: For $A, B, C \leq G$ with $A \leq B$,

$$A(B \cap C) = B \cap (AC)$$

Find an example to show that for arbitrary subgroups A, B, C of G, $A(B \cap C)$ is not necessarily equal to $AB \cap AC$ and so the condition that $A \leq B$ is necessary.

h) Let A, B and C be subgroups of G with $A \leq B$. Prove that if

$$A \cap C = B \cap C \quad \text{and} \quad AC = BC$$

then $A = B$.

42. Let G be a group with subgroups N and H and suppose that $NH \leq G$. Prove that if $(G : H)$ and $|N|$ are finite and relatively prime, then $N \leq H$.

43. Let G be a group and $H \leq G$.

a) Let $H \subseteq X \subseteq G$. Show that the relation $x \equiv y$ if $x^{-1}y \in H$ is an equivalence relation on X. What do the equivalence classes of this relation look like? Let $(X : H)$ be the cardinality of the set of equivalence classes.

b) If $H \subseteq X \subseteq Y \subseteq G$, where $XH \subseteq X$. Show that if $(X : H) = (Y : H) < \infty$, then $X = Y$.

44. Prove that $N \trianglelefteq G$ and G/N both have the ACC on subgroups if and only if G has the ACC on subgroups.

45. Sometimes it is useful to relate the commutator subgroup $[H, K]$ of subgroups H and K to the commutator subgroup $[X, Y]$ of generating sets for H and K. Let X and Y be nonempty subsets of a group G.

a) Show that

$$[X, \langle Y\rangle] = [X, \langle Y\rangle]^{\langle Y\rangle} = [X, Y]^{\langle Y\rangle}$$

b) Show that

$$[\langle X\rangle, \langle Y\rangle] = [X, Y]^{\langle X\rangle\langle Y\rangle} = [X, Y]^{\langle Y\rangle\langle X\rangle}$$

Complex Groups

Let G be a group. Let $\mathcal{G}$ be a nonempty family of subsets of G that forms a group under set product. Such a group has been called a **complex group** based on G (see Allen [1]). Let $\bigcup\mathcal{G}$ be the union of the subsets in $\mathcal{G}$. We denote the identity of $\mathcal{G}$ by E. Also, we denote the inverse of $A \in \mathcal{G}$ by A^{-1}. In the following exercises, let $\mathcal{G}$ be an arbitrary complex group based on the group G.

46. Let $N \trianglelefteq G$. Show that G/N is a complex group in which the members form a partition of G. Thus, quotient groups are complex groups.
47. Show that if $\mathbb{Q}^+$ is the multiplicative group of all positive rational numbers and if

$$\mathcal{Q} = \{(r, \infty) \mid r \in \mathbb{R}\}$$

then $\mathcal{Q}$ is a complex group. Do the members of $\mathcal{Q}$ form a partition of $\mathbb{Q}^+$? What is the identity of $\mathcal{Q}$? Does it contain the identity $1 \in \mathbb{Q}^+$? Compare the sizes of $\mathbb{Q}^+$ and $\mathcal{Q}$. Could $\mathcal{Q}$ be a quotient group of $\mathbb{Q}^+$?
48. Let $\mathbb{Z}$ be the additive group of integers and let k be a positive integer. Show that the set

$$\mathcal{Z}_k = \{\{n\} \cup [n+k, \infty) \mid n \in \mathbb{Z}\}$$

is a complex group. What is the identity of $\mathcal{Z}_k$? What is the negative of the element $A = \{1\} \cup [1+k, \infty)$? Is this the set of negatives in $\mathbb{Z}$ of the elements of A?
49. Show that all members of $\mathcal{G}$ have the same cardinality.
50. Show that for $A, B \in \mathcal{G}$,

$$A \subseteq B \quad \Rightarrow \quad B^{-1} \subseteq A^{-1}$$

51. Show that for $A, B \in \mathcal{G}$,

$$E \leq G, AB = E \quad \Rightarrow \quad A \cup B \subseteq E \quad \text{or} \quad (A \cup B) \cap E = \emptyset$$

52. Show that for $A, B \in \mathcal{G}$,

$$e \in AB^{-1} \cap BA^{-1} \quad \Rightarrow \quad A = B$$

and

$$AB^{-1} \cup BA^{-1} \subseteq E \quad \Rightarrow \quad A = B$$

53. Suppose that the members of $\mathcal{G}$ form a partition of G. Prove that $\mathcal{G}$ is a quotient group of G, that is, $\mathcal{G}$ is the set of cosets of some normal subgroup of G. *Hint*: Show that $1 \in E$ and that $E \trianglelefteq G$.
54. Prove that $\mathcal{G}$ is a quotient group of G if and only if $E \leq G$. *Hint*: Prove that $E \leq G$ if and only if the members of $\mathcal{G}$ are pairwise disjoint and $\bigcup \mathcal{G} \leq G$.
55. Prove that

$$A \neq B \in \mathcal{G} \quad \Rightarrow \quad A \cap B = \emptyset \quad \text{or} \quad A \cap B \text{ is infinite}$$

Chapter 4
Homomorphisms, Chain Conditions and Subnormality

Homomorphisms

The structure-preserving functions between two groups are referred to as *homomorphisms*. Before giving a formal definition, let us make a few remarks about functions.

We denote the action of a function $f: S \to T$ on $s \in S$ by either fs or $f(s)$, depending on readability. If $\wp$ denotes the power set, then the **induced map** $f: \wp(S) \to \wp(T)$ is defined by

$$f(U) = \{f(u) \mid u \in U\}$$

and the **induced inverse map** $f^{-1}: \wp(T) \to \wp(S)$ is defined by

$$f^{-1}(V) = \{s \in S \mid f(s) \in V\}$$

If $f: S \to S$, then a subset $A \subseteq S$ is **invariant under** f, or $\boldsymbol{f}$**-invariant**, if $f(A) \subseteq A$. If $\mathcal{F}$ is a family of functions from S to S, then a subset $A \subseteq S$ is $\mathcal{F}$**-invariant** if A is f-invariant for all $f \in \mathcal{F}$.

Now we can define homomorphisms.

Definition *Let G and H be groups. A function $\sigma: G \to H$ is called a* **group homomorphism** (*or just* **homomorphism**) *if*

$$\sigma(ab) = (\sigma a)(\sigma b)$$

The set of all homomorphisms from G to H is denoted by $\hom(G,H)$. *The following terminology is employed:*

1) *A surjective homomorphism is an* **epimorphism**, *which we denote by* $\sigma: G \twoheadrightarrow H$.

2) *An injective homomorphism is a* **monomorphism** *or* **embedding***, which we denote by* $\sigma: G \hookrightarrow H$*. If there is an embedding from* G *to* H*, we say that* G *can be* **embedded** *in* H *and write* $G \hookrightarrow H$*.*
3) *A bijective homomorphism is an* **isomorphism***, which we denote by* $\sigma: G \approx H$*. If there is an isomorphism from* G *to* H*, we say that* G *and* H *are* **isomorphic** *and write* $G \approx H$*.*
4) *A homomorphism of* G *into itself is an* **endomorphism***. The set of all endomorphisms of* G *is denoted by* $\mathrm{End}(G)$*.*
5) *An isomorphism of* G *onto itself is an* **automorphism***. The set of all automorphisms of* G *is denoted by* $\mathrm{Aut}(G)$*.*□

If $\sigma: G \to H$ is a homomorphism, then it is easy to see that

$$\sigma 1 = 1 \quad \text{and} \quad \sigma(a^{-1}) = (\sigma a)^{-1}$$

for any $a \in G$. Also, if $\sigma: G \approx H$, then the inverse map $\sigma^{-1}: H \approx G$ is an isomorphism from H to G. The map that sends every element of G to the identity $1 \in H$ is called the **zero map**. (We cannot call it the identity map!)

In general, induced inverse maps are more well behaved than induced direct maps. Here is an example.

Theorem 4.1 *Let* $\sigma: G \to H$ *be a group homomorphism.*

1) *a)* **(Image preserves subgroups)**

$$S \leq G \quad \Rightarrow \quad \sigma S \leq H$$

b) **(Surjective image preserves normality)** *If* σ *is surjective, then*

$$S \trianglelefteq G \quad \Rightarrow \quad \sigma S \trianglelefteq H$$

2) **(Inverse image preserves subgroups and normality)**

$$T \leq H \quad \Rightarrow \quad \sigma^{-1} T \leq G$$

and

$$T \trianglelefteq H \quad \Rightarrow \quad \sigma^{-1} T \trianglelefteq G$$

□

While it is true that isomorphic groups have essentially the same group–theoretic structure, one must be careful in applying this notion to the subgroup structure of a group. For example, in the group $\mathbb{Z}$ of integers, the subgroups $2\mathbb{Z} = \{2n \mid n \in \mathbb{Z}\}$ and $3\mathbb{Z} = \{3n \mid n \in \mathbb{Z}\}$ are isomorphic but the quotients $\mathbb{Z}/2\mathbb{Z}$ and $\mathbb{Z}/3\mathbb{Z}$ have different sizes. Thus, isomorphic subgroups need not have the same index.

Example 4.2 Let $D_6 = \langle \rho, \sigma \rangle$ be the dihedral group, where $o(\rho) = 3$ and $o(\sigma) = 2$. Let S_3 be the symmetric group of order 6. The map $f: D_6 \to S_3$ defined by

$$f(\sigma^i \rho^k) = (2\,3)^i (1\,2\,3)^k$$

is an isomorphism and so $D_6 \approx S_3$. This tells us that the group of symmetries of the triangle is the group of permutations of the vertices.□

Definition *Let G be a group and let $\sigma: G \to G$ be an endomorphism of G.*

1) *σ is* **nilpotent** *if*

$$\sigma^n = 0$$

for some $n > 0$, where 0 is the zero map.

2) *σ is* **idempotent** *if*

$$\sigma^2 = \sigma$$ □

We leave it as an exercise to show that the zero map is the only endomorphism that is both nilpotent and idempotent.

Sums of Homomorphisms

If $\sigma, \tau: G \to H$ are homomorphisms, then the map $\sigma + \tau: G \to H$ defined by

$$(\sigma + \tau)(a) = (\sigma a)(\tau a)$$

is a homomorphism if and only if the images $\mathrm{im}(\sigma)$ and $\mathrm{im}(\tau)$ commute elementwise.

Definition *Let G and H be groups. The* **sum** *of a family $\sigma_1, \ldots, \sigma_n: G \to H$ of homomorphisms is the function defined by*

$$(\sigma_1 + \cdots + \sigma_n)(a) = (\sigma_1 a)\cdots(\sigma_n a)$$

for all $a \in G$.□

If the images $\mathrm{im}(\sigma_i)$ commute elementwise, then the sum $\sigma_1 + \cdots + \sigma_n$ is a homomorphism and in this case, the sum itself is commutative, that is,

$$\sigma_1 + \cdots + \sigma_n = \sigma_{\pi 1} + \cdots + \sigma_{\pi n}$$

for any $\pi \in S_n$. Moreover, composition distributes over addition,

$$\sigma(\sigma_1 + \cdots + \sigma_n) = \sigma\sigma_1 + \cdots + \sigma\sigma_n$$

and

$$(\sigma_1 + \cdots + \sigma_n)\sigma = \sigma_1\sigma + \cdots + \sigma_n\sigma$$

Hence, if the images $\mathrm{im}(\sigma_i)$ commute elementwise, then the binomial formula holds,

$$(\sigma_1 + \cdots + \sigma_n)^m = \sum_{j_1+\cdots+j_n=m} \binom{m}{j_1, \ldots, j_n} \sigma_1^{j_1} \cdots \sigma_n^{j_n}$$

for any $m > 0$.

Theorem 4.3 *Let* $\sigma_1, \ldots, \sigma_n \in \mathrm{End}(G)$ *have the property that the images* $\mathrm{im}(\sigma_i)$ *commute elementwise. If each* σ_i *is nilpotent, then so is the sum*

$$\tau = \sigma_1 + \cdots + \sigma_n$$

Proof. For any $m > 0$,

$$\tau^{nm} = \sum_{j_1+\cdots+j_n=nm} \binom{nm}{j_1, \ldots, j_n} \sigma_1^{j_1} \cdots \sigma_n^{j_n}$$

But if each σ_i is nilpotent, then there is a positive integer m for which $\sigma_i^m = 0$ for all i and so each term in the sum above is the zero map, since one of the exponents j_i is greater than or equal to m.□

Kernels and the Natural Projection

The **kernel** of a homomorphism $\sigma: G \to H$ is the subgroup

$$\ker(\sigma) = \{a \in G \mid \sigma a = 1\}$$

consisting of all elements of G that are sent to the identity of H. This is easily seen to be a *normal* subgroup of G. Conversely, any normal subgroup is a kernel.

Theorem 4.4 *Let* $N \trianglelefteq G$. *The map* $\pi_N: G \to G/N$ *defined by*

$$\pi_N(a) = aN$$

is an epimorphism with kernel N *and is called the* **natural projection** *or* **canonical projection** *modulo* N.□

It is possible to tell whether a homomorphism is injective from its kernel.

Theorem 4.5 *A group homomorphism* $\sigma: G \to H$ *is injective if and only if* $\ker(\sigma) = \{1\}$.□

Groups of Small Order

One of the most important outstanding problems of group theory is the problem of classifying various types of groups up to isomorphism. This problem is called the **classification problem** and is, in general, very difficult. The classification problem for all groups is unsolved. The classification problem for finitely-generated abelian groups is solved (see Theorem 13.4). The classification problem for finite simple groups *seems* to have been solved, and we discuss this in more detail in a later chapter.

Groups of relatively small order have been fully classified up to isomorphism. For example, one can find such a classification of groups of order 50 or less in Weinstein [36]. Of course, Lagrange's theorem gives a simple solution to the classification problem for groups of prime order, since all such groups are cyclic.

A bit later in the book, we will solve the classification problem for groups of order 15 or less. For now, we can solve the classification problem for groups of order 8 or less, which are easily analyzed by looking at the possible orders of the elements. In fact, since groups of prime order are cyclic, we need only look at groups of order 4, 6 and 8.

Groups of Order 4

Let G be a group of order 4. If G has an element of order 4, then G is cyclic and $G \approx C_4$. Otherwise, all nonidentity elements have order 2, which implies that G is abelian. In this case, if $a, b \in G$ are distinct nonidentity elements, then $G = \{1, a, b, ab\}$ is the Klein 4-group. Thus, the groups of order 4 are (up to isomorphism):

1) C_4, the cyclic group
2) $V \approx C_2 \boxtimes C_2$, the Klein 4-group.

Groups of Order 6

If $o(G) = 6$, then Cauchy's theorem implies that there exist $a, b \in G$ with $o(a) = 3$ and $o(b) = 2$. Since $\langle a \rangle$ has index 2, it is normal in G and since conjugation preserves order, we have

$$a^b = a \quad \text{or} \quad a^b = a^2$$

If $a^b = a$, then G is abelian. In this case, since $o(a)$ and $o(b)$ are relatively prime,

$$o(ab) = o(a)o(b) = 6$$

and so G is cyclic. If $a^b = a^2 = a^{-1}$, then

$$G = \langle a, b \rangle, o(a) = 3, o(b) = 2, ba = a^{-1}b$$

and so G is the dihedral group D_6. Also, since $D_6 \approx S_3$, the group G is also a symmetric group. Thus, the groups of order 6 are (up to isomorphism):

1) C_6, the cyclic group
2) $D_6 \approx S_3$, the nonabelian dihedral (and symmetric) group.

Groups of Order 8

Let $o(G)=8$. If G has an element of order 8, it is cyclic and $G\approx C_8$. If every nonidentity element of G has order 2, then G is abelian. In this case, let a,b and c be distinct elements of G with $c\neq ab$. It is easy to see that

$$G=\{1,a,b,c,ab,ac,bc,abc\}$$

and that $G\approx C_2\boxtimes C_2\boxtimes C_2$.

Now suppose that G has an element a of order 4 but no elements of order 8. Then $\langle a\rangle\trianglelefteq G$. For any $b\notin\langle a\rangle$, we have $G=\langle a,b\rangle$ and since $a^b\in\langle a\rangle$ has order 4, we must have

$$bab^{-1}=a\quad\text{or}\quad bab^{-1}=a^3$$

If $bab^{-1}=a$, then G is abelian. Moveover, G has an element c of order 2 that is not in $\langle a\rangle$. To see this, note that if $b\notin\langle a\rangle$ and $o(b)=4$, then $b^2\in\langle a\rangle$ has order 2 and so $b^2=a^2$. Hence, $(ab)^2=a^2b^2=a^4=1$ and so $c=ab$. Thus,

$$G=\langle a\rangle\bullet\langle c\rangle\approx C_4\boxtimes C_2$$

On the other hand, suppose that $bab^{-1}=a^3=a^{-1}$. If there is a involution $b\notin\langle a\rangle$, then ab is also an involution and so Theorem 2.36 implies that $\langle b,ab\rangle$ is dihedral of order $2o(a)=8$, that is, $G=\langle b,ab\rangle\approx D_8$.

Finally, if all elements $b\notin\langle a\rangle$ have order 4, then $b^2\in\langle a\rangle$ and so

$$G=\langle a,b\rangle, a\neq b, o(a)=4, b^2=a^2, bab^{-1}=a^3$$

which is the quaternion group. Thus, the groups of order 8 are (up to isomorphism):

1) C_8, the cyclic group
2) $C_4\boxtimes C_2$, abelian but not cyclic
3) $C_2\boxtimes C_2\boxtimes C_2$, abelian but not cyclic
4) D_8, the (nonabelian) dihedral group
5) Q, the (nonabelian) quaternion group.

A Universal Property and the Isomorphism Theorems

The following property is key to many other properties of group homomorphisms.

Definition *Let G be a group and let $K\trianglelefteq G$. Let $\mathcal{F}(G;K)$ be the family of all pairs $(H,\sigma\colon G\to H)$, where $\sigma\colon G\to H$ is a group homomorphism and $K\subseteq\ker(\sigma)$.*

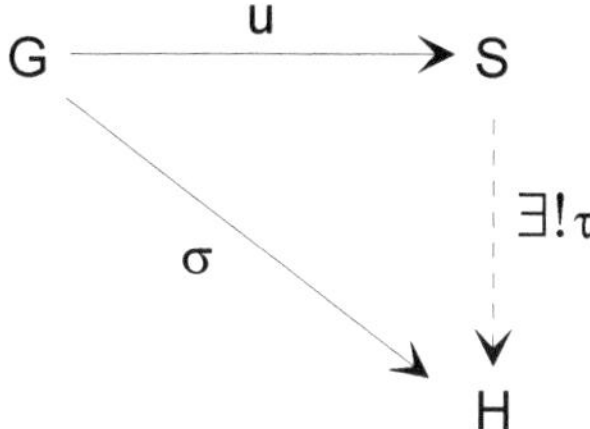

Figure 4.1

Referring to Figure 4.1, a pair $(S, u{:}\, G \to S) \in \mathcal{F}(G;K)$ *is* **universal** *in* $\mathcal{F}(G;K)$ *if for any pair* $(H, \sigma{:}\, G \to H) \in \mathcal{F}(G;K)$, *there is a unique group homomorphism* $\tau{:}\, S \to H$ *for which the diagram in Figure 4.1* **commutes**, *that is, for which*

$$\tau \circ u = \sigma$$

The map τ *is called the* **mediating morphism** *for* σ *and we say that* σ *can be* **factored uniquely** *through* u *or that* σ *can be* **lifted** *uniquely to* S.□

Existence and uniqueness (up to isomorphism) of universal pairs is given by the following theorem.

Theorem 4.6 (Universal pairs) *Let* G *be a group and let* $K \trianglelefteq G$.

1) **(Existence)** *The pair*

$$(G/K, \pi_K{:}\, G \to G/K)$$

where π_K *is the canonical projection modulo* K *is universal in* $\mathcal{F}(G;K)$. *The mediating morphism for* $\sigma{:}\, G \to H$ *is the map* $\tau{:}\, G/K \to H$ *defined by*

$$\tau(gK) = \sigma g$$

Also,

$$\operatorname{im}(\tau) = \operatorname{im}(\sigma) \quad \textit{and} \quad \ker(\tau) = \ker(\sigma)/K$$

2) **(Uniqueness)** *Referring to Figure 4.2*

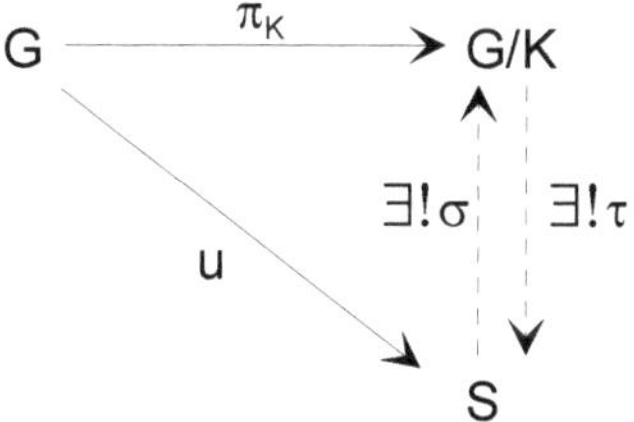

Figure 4.2

if $(S, u: G \to S)$ is also universal in $\mathcal{F}(G; K)$, then the mediating morphisms σ and τ are inverse isomorphisms, whence $S \approx G/K$.

Proof. For part 1), a mediating morphism for $\sigma: G \to H$, if it exists, must satisfy

$$\tau(gK) = \sigma g$$

for all $g \in G$ and so must be unique. But the condition $K \subseteq \ker(\sigma)$ implies that τ is a well defined map and it is easy to see that it is a homomorphism. Finally,

$$\operatorname{im}(\tau) = \tau(G/K) = \tau \circ \pi_K(G) = \sigma(G) = \operatorname{im}(\sigma)$$

and

$$\tau(aK) = 0 \quad \Leftrightarrow \quad \tau \circ \pi_K(a) = 0 \quad \Leftrightarrow \quad \sigma a = 0 \quad \Leftrightarrow \quad a \in \ker(\sigma)$$

and so $\ker(\tau) = \ker(\sigma)/K$.

For part 2), since $(G/K, \pi_K)$ and $(S, u: G \to S)$ are both universal, u can be factored uniquely through π_K, that is,

$$\tau \circ \pi_K = u$$

and π_K can be factored uniquely through u, that is,

$$\sigma \circ u = \pi_K$$

Hence,

$$(\tau \circ \sigma) \circ u = u$$

But u can be factored *uniquely* through itself and since $\iota \circ u = u$, it follows that $\tau \circ \sigma = \iota$. Similarly, $\sigma \circ \tau = \iota$ and so σ and τ are inverse isomorphisms.□

The following well-known results are direct consequences of the universal property of the pair $(G/H, \pi_H)$.

Theorem 4.7 (The isomorphism theorems) *Let G be a group.*

1) **(First isomorphism theorem)** *Every group homomorphism $\sigma: G \to H$ induces an embedding $\overline{\sigma}: G/\ker(\sigma) \hookrightarrow H$ of $G/\ker(\sigma)$ defined by*

$$\overline{\sigma}(g\ker(\sigma)) = \sigma g$$

and so

$$\frac{G}{\ker(\sigma)} \approx \operatorname{im}(\sigma)$$

2) **(Second isomorphism theorem)** *If* $H,K \le G$ *with* $K \trianglelefteq G$, *then* $H \cap K \trianglelefteq H$ *and*

$$\frac{HK}{K} \approx \frac{H}{H \cap K}$$

3) **(Third isomorphism theorem)** *If* $H \le K \le G$ *with* $H,K \trianglelefteq G$, *then* $K/H \trianglelefteq G/H$ *and*

$$\frac{G}{H} \Big/ \frac{K}{H} \approx \frac{G}{K}$$

Hence

$$(G : K) = (G/H : K/H)$$

Proof. For part 1), since $(G/K, \pi_K)$ is universal, there is a unique homomorphism $\tau: G/K \to \text{im}(\sigma)$ for which $\tau \circ \pi_K = \sigma$. The rest follows from the fact that $\text{im}(\tau) = \text{im}(\sigma)$ and $\ker(\tau) = \ker(\sigma)/K = \{1\}$.

For part 2), the map $\sigma: H \to HK/K$ defined by $\sigma(h) = hK$ is clearly an epimorphism with kernel $H \cap K$ and the first isomorphism theorem completes the proof. Proof of part 3) is left to the reader.□

Theorem 4.8 *Let* G_1 *and* G_2 *be groups and let* $H_i \trianglelefteq G_i$ *for* $i = 1, 2$. *Then*

$$\frac{G_1 \boxtimes G_2}{H_1 \boxtimes H_2} \approx \frac{G_1}{H_1} \boxtimes \frac{G_2}{H_2}$$

Proof. Let $\tau: G_1 \boxtimes G_2 \to (G_1/H_1) \boxtimes (G_2/H_2)$ be defined by

$$\tau(a_1, a_2) = (a_1H_1, a_2H_2)$$

We leave it to the reader to show that τ is an epimorphism with kernel $H_1 \boxtimes H_2$. The first isomorphism theorem then completes the proof.□

The Correspondence Theorem

The following correspondence theorem has a great many uses. We use the notation $\text{sub}(N; G)$ to denote the lattice of all subgroups of G that contain N.

Theorem 4.9 (Correspondence theorem) *Let* G *be a group, let* $N \trianglelefteq G$ *and let* $\pi: G \to G/N$ *be the natural projection. Referring to Figure 4.3,*

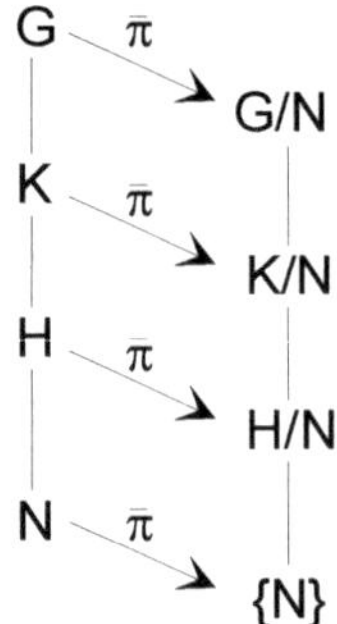

Figure 4.3

let $\overline{\pi}\colon \operatorname{sub}(N;G) \to \operatorname{sub}(G/N)$ *be the map defined by*

$$\overline{\pi}(H) = H/N$$

1) $\overline{\pi}$ *is an order isomorphism, that is,* $\overline{\pi}$ *is a bijection for which*

$$H \le K \quad \Leftrightarrow \quad H/N \le K/N$$

for all $H, K \in \operatorname{sub}(N;G)$*. In particular, every subgroup of* G/N *has the form* H/N *for a unique* $H \in \operatorname{sub}(N;G)$*.*

2) *Normality is preserved in both directions, that is,*

$$H \trianglelefteq K \quad \Leftrightarrow \quad H/N \trianglelefteq K/N$$

for all $H, K \in \operatorname{sub}(N;G)$*. Moreover, the corresponding factor groups are isomorphic, that is,*

$$\frac{K}{H} \approx \frac{K}{N} \Big/ \frac{H}{N}$$

and so $\overline{\pi}$ *also preserves index:*

$$(K : H) = (K/N : H/N)$$

It follows that subnormality is also preserved in both directions:

$$H \trianglelefteq\trianglelefteq K \quad \Leftrightarrow \quad H/N \trianglelefteq\trianglelefteq K/N$$

3) *If* $N \sqsubseteq G$*, then the property of being characteristic is preserved in only one direction, specifically,*

$$\frac{H}{N} \sqsubseteq \frac{G}{N} \quad \Rightarrow \quad H \sqsubseteq G$$

for any $H \in \operatorname{sub}(N;G)$*, but the converse fails.*

Proof. For the surjectivity of $\overline{\pi}$, any subgroup of G/N has the form

$$S = \{xN \mid x \in X\}$$

for some index set X. The set $H = \bigcup(xN)$ is a subgroup of G since $x, y \in H$ implies that $xN, yN \in S$ and so $x^{-1}N$ and xyN are in S, whence $x^{-1}, xy \in H$. Also,

$$\overline{\pi}H = H/N = \{xN \mid x \in H\} = S$$

and so $\overline{\pi}$ is surjective.

For part 3), if $\sigma \in \operatorname{Aut}(G)$, then $\pi_N \circ \sigma: G \to G/N$ is an epimorphism with kernel $\sigma^{-1}N = N$. Hence, the map $\tau: G/N \to G/N$ defined by $\tau(aN) = (\sigma a)N$ is an automorphism of G/N and so $\tau(H/N) \le H/N$. Thus, for $h \in H$,

$$(\sigma h)N = \tau(hN) \in H/N$$

and so $\sigma h \in H$, which shows that $H \sqsubseteq G$. As to the converse, let $G = K = D_8$, $H = \langle \rho \rangle$ and $N = Z(G) = \langle \rho^2 \rangle$. Then H and N are characteristic in D_8 but $H/Z(G) \approx C_2$ is not characteristic in $D_8/Z(G) \approx C_2 \boxtimes C_2$. The rest of the proof is left to the reader.□

Group Extensions

It will be convenient to make the following definition.

Definition *Let G be a group. We refer to subgroups H and K of G for which $H \le K$ as an* **extension**. *We also refer to $H \trianglelefteq G$ as a* **normal extension**. *The* **index** *of an extension $H \le K$ is $(K : H)$.*□

This use of the term *extension* is consistent with its use in field theory, where if $F \subseteq K$ are fields, then K is referred to as an extension of F. The term extension has another meaning in group theory, which we will encounter later in the book: An **extension** of a pair (N, Q) of groups is a group G that has a normal subgroup N' isomorphic to N and for which $G/N' \approx Q$. However, since $H \le G$ *is* an extension whereas (N, Q) *has* an extension, there should be no ambiguity in adopting the current definition.

Various operations can be performed on a group extension to yield another extension.

Definition *Let G be a group. Let $A \le B \le G$ and $C \le G$.*

1) The **intersection** *of $A \le B$ with C is*

$$A \cap C \le B \cap C$$

2) If $C \trianglelefteq G$, the **normal lifting** *of $A \le B$ by C is*

$$AC \le BC$$

3) *If* $N \leq A \leq B$ *with* $N \trianglelefteq B$*, then the* **quotient** *of* $A \leq B$ *by* N *is*

$$\frac{A}{N} \leq \frac{B}{N}$$

and (for want of a better term) the **unquotient** *of* $A/N \leq B/N$ *is* $A \leq B$.□

Inheritance of Group Properties

Let $\mathcal{P}$ be a property of groups, such as being cyclic, being finite or being abelian. (Technically, $\mathcal{P}$ can be thought of as a subclass of the class of all groups.) We write $G \in \mathcal{P}$ to denote the fact that the group G has property $\mathcal{P}$. A group property $\mathcal{P}$ is **isomorphism invariant** if

$$G \in \mathcal{P}, \quad H \approx G \quad \Rightarrow \quad H \in \mathcal{P}$$

We will say that a normal extension $A \trianglelefteq B$ has property $\mathcal{P}$ if the quotient B/A has property $\mathcal{P}$. For example, to say that $A \trianglelefteq B$ is cyclic is to say that B/A is cyclic. A property $\mathcal{P}$ of groups is **inherited** by subgroups if

$$G \in \mathcal{P}, \quad H \leq G \quad \Rightarrow \quad H \in \mathcal{P}$$

and $\mathcal{P}$ is inherited by quotients if

$$G \in \mathcal{P}, \quad H \trianglelefteq G \quad \Rightarrow \quad G/H \in \mathcal{P}$$

Preservation of Group Properties

The second isomorphism theorem implies that intersection preserves normality, that is,

$$A \trianglelefteq B, \quad C \leq G \quad \Rightarrow \quad A \cap C \trianglelefteq B \cap C$$

and that the quotient satisfies

$$\frac{B \cap C}{A \cap C} = \frac{B \cap C}{A \cap (B \cap C)} \approx \frac{A(B \cap C)}{A} \leq \frac{B}{A}$$

Hence, any isomorphism-invariant property of $A \trianglelefteq B$ that is inherited by subgroups, such as being finite, abelian or cyclic, is inherited by intersections. For example, if $A \trianglelefteq B$ is cyclic, then so is $A \cap C \trianglelefteq B \cap C$. Similar statements can be made about normal liftings, quotients and unquotients, as follows.

Theorem 4.10 *Let* G *be a group with* $A \leq B \leq G$*. Let* $\mathcal{P}$ *be an isomorphism-invariant property of groups.*

1) **(Intersection)** *Normality is preserved by intersection, that is, for* $C \leq G$,

$$A \trianglelefteq B \quad \Rightarrow \quad A \cap C \trianglelefteq B \cap C$$

Also, if $\mathcal{P}$ *is inherited by subgroups, then* $\mathcal{P}$ *is preserved by intersection.*

2) **(Normal lifting)** *Normality is preserved by normal lifting, that is, for* $N \trianglelefteq G$,

$$A \trianglelefteq B \quad \Rightarrow \quad AN \trianglelefteq BN$$

Also, if $\mathcal{P}$ *is inherited by quotients, then* $\mathcal{P}$ *is preserved by normal lifting.*

3) **(Quotient and unquotient)** *Normality is preserved by quotient and unquotient, that is, for* $N \trianglelefteq B$ *and* $N \le A \le B$,

$$A \trianglelefteq B \quad \Leftrightarrow \quad A/N \trianglelefteq B/N$$

Also, $\mathcal{P}$ *is preserved by quotient and unquotient.*

Proof. For part 2), to see that $AN \trianglelefteq BN$, note that B normalizes both A and N and so B normalizes AN. Also, $N \le AN$ implies that N normalizes AN and so BN normalizes AN, that is, $AN \trianglelefteq BN$. Since $BN = ABN = ANB$, it follows that

$$\frac{BN}{AN} = \frac{ANB}{AN} \approx \frac{B}{B \cap AN} = \frac{B}{A(B \cap N)} \approx \frac{B}{A} \Big/ \frac{A(B \cap N)}{A} \qquad \square$$

Centrality

Another very important property of extensions $A \le B$ in a group G that involves more than just the quotient B/A alone is *centrality*.

Definition *Let* $A \trianglelefteq G$. *The extension* $A \trianglelefteq B$ *in* G *is* **central in** G *if*

$$\frac{B}{A} \le Z\left(\frac{G}{A}\right) \qquad \square$$

Note that centrality is not a property of B/A alone, since it depends on G/A as well, which is why we use the phrase *central in* G. Theorem 3.40 implies that an extension $A \trianglelefteq B$ is central in G if and only if

$$[B, G] \le A$$

Thus, for $N \trianglelefteq G$ and $N \le A$,

$$A \trianglelefteq B \text{ central in } G \quad \Leftrightarrow \quad \frac{A}{N} \trianglelefteq \frac{B}{N} \text{ central in } \frac{G}{N}$$

and so centrality is preserved by quotient and unquotient. Also, for any $H \le G$,

$$[B \cap H, H] \le A \cap H$$

and so

$$A \trianglelefteq B \text{ central in } G \quad \Rightarrow \quad A \cap H \le B \cap H \text{ central in } H$$

Finally, for any $N \trianglelefteq G$,

$$[BN,G]=[B,G][N,G]\le AN$$

and so

$$A \trianglelefteq B \text{ central in } G \quad \Rightarrow \quad AN \le BN \text{ central in } G$$

Thus, centrality is preserved by intersection and normal lifting as well.

Theorem 4.11 *Let G be a group. The property of an extension being central in G is preserved by intersection, normal lifting, quotient and unquotient.* □

Projections and the Zassenhaus Lemma

Combining intersection with lifting allows us to *project* one normal extension in G into another. Specifically to project $A \trianglelefteq B$ into $H \trianglelefteq K$, we first intersect $A \trianglelefteq B$ with K to get

$$(A\cap K)\trianglelefteq(B\cap K)$$

and then lift by H to get

$$H(A\cap K)\trianglelefteq H(B\cap K)$$

This extension is the **projection** of $A \trianglelefteq B$ into $H \trianglelefteq K$ and is denoted by

$$(A\trianglelefteq B)\to(H\trianglelefteq K)$$

We leave it to the reader to show that the same extension is obtained by first lifting $A \trianglelefteq B$ by H and then intersecting with K.

As to the factor group of the projection, the isomorphism theorems give

$$\begin{aligned}\frac{H(B\cap K)}{H(A\cap K)}&=\frac{H(A\cap K)(B\cap K)}{H(A\cap K)}\\&\approx\frac{B\cap K}{(B\cap K)\cap H(A\cap K)}\\&=\frac{B\cap K}{(A\cap K)[(B\cap K)\cap H]}\\&=\frac{B\cap K}{(A\cap K)(B\cap H)}\end{aligned}$$

But the last quotient remains unchanged if we reverse the roles of the two extensions $A \le B$ and $H \le K$. Hence, the **reverse projection**

$$(H\trianglelefteq K)\to(A\trianglelefteq B)$$

has an isomorphic quotient, that is,

$$\frac{H(B\cap K)}{H(A\cap K)}\approx\frac{A(K\cap B)}{A(H\cap B)}$$

This result was proved by Zassenhaus in 1934 and was given the name **butterfly lemma** by Serge Lang because of the shape of a certain figure associated with an alternate proof.

Theorem 4.12 (Zassenhaus lemma [37], 1934) *Let G be a group and let*

$$A \trianglelefteq B \quad \textit{and} \quad H \trianglelefteq K$$

be normal extensions in G. Then the reverse projections

$$(A \trianglelefteq B) \to (H \trianglelefteq K) \quad \textit{and} \quad (H \trianglelefteq K) \to (A \trianglelefteq B)$$

have isomorphic factor groups, that is,

$$\frac{H(B \cap K)}{H(A \cap K)} \approx \frac{A(K \cap B)}{A(H \cap B)}$$ □

Let us make a few very simple observations about projections:

1) If $A \trianglelefteq B \trianglelefteq C$, then the projections of the contiguous extensions $A \trianglelefteq B$ and $B \trianglelefteq C$ into $H \trianglelefteq K$ are also contiguous, that is,
$$H(A \cap K) \trianglelefteq H(B \cap K) \trianglelefteq H(C \cap K)$$
2) If $A \trianglelefteq B$ is projected into $H \trianglelefteq K$ where $B \geq K$, then the projection has top group subgroup K.
3) If $A \trianglelefteq B$ is projected into $H \trianglelefteq K$ where $A \leq H$, then the projection has bottom subgroup H.

Thus, we can project a series

$$A_1 \trianglelefteq A_2 \trianglelefteq \cdots \trianglelefteq A_n$$

in G into an extension $H \trianglelefteq K$ to get a new series

$$H(A_1 \cap K) \trianglelefteq H(A_2 \cap K) \trianglelefteq \cdots \trianglelefteq H(A_n \cap K)$$

In particular, if $A_1 \leq H$ and $K \leq A_n$, then the series runs from H to K.

Inner Automorphisms

If G is a group, then the map $\gamma\colon G \to \operatorname{Aut}(G)$ defined by

$$\gamma a = \gamma_a$$

is a group homomorphism, since $\gamma_a \gamma_b = \gamma_{ab}$. The kernel of γ is $Z(G)$ and the image of γ is $\operatorname{Inn}(G)$, which is normal in $\operatorname{Aut}(G)$ since for any $\sigma \in \operatorname{Aut}(G)$ and $g \in G$,

$$(\sigma\gamma_a\sigma^{-1})g = \sigma[a(\sigma^{-1}g)a^{-1}] = (\sigma a)g(\sigma a)^{-1} = \gamma_{\sigma a}g$$

and so

$$\gamma_a^\sigma = \gamma_{\sigma a}$$

An automorphism of G that is not an inner automorphism is called an **outer automorphism** of G and the factor group $\mathrm{Aut}(G)/\mathrm{Inn}(G)$ is called the **outer automorphism group** of G, even though its elements are *not* automorphisms.

Theorem 4.13 *Let G be a group.*

1) *The map $\gamma\colon G \to \mathrm{Aut}(G)$ defined by*

$$\gamma a = \gamma_a$$

is a group homomorphism with image $\mathrm{Inn}(G) \trianglelefteq \mathrm{Aut}(G)$ and kernel $Z(G)$. Hence,

$$\mathrm{Inn}(G) \approx \frac{G}{Z(G)}$$

In particular, if G is centerless, then $G \approx \mathrm{Inn}(G)$.

2) *If $H \le G$, then*

$$a \in N_G(H) \quad \Leftrightarrow \quad \gamma_a \in \mathrm{Aut}(H)$$

Moreover, the restricted map $\gamma\colon N_G(H) \to \mathrm{Aut}(H)$ has kernel $C_G(H)$ and so $C_G(H) \trianglelefteq N_G(H)$ and

$$\frac{N_G(H)}{C_G(H)} \hookrightarrow \mathrm{Aut}(H)$$

In particular, if $H \trianglelefteq G$, then $C_G(H) \trianglelefteq G$ and

$$\frac{G}{C_G(H)} \hookrightarrow \mathrm{Aut}(H)$$ □

Characteristic Subgroups

Normality is not transitive: If $N \trianglelefteq G$, then the normal subgroups of N are not necessarily normal in G. Examples can be found in the symmetric group S_4. However, the property of being characteristic is transitive. Here are some of the basic properties associated with characteristic subgroups.

Theorem 4.14 *Let G be a group.*

1) *$H \sqsubseteq G$ if and only if $\sigma H = H$ for all $\sigma \in \mathrm{Aut}(G)$.*
2) **(Transitivity)**

$$A \sqsubseteq B \sqsubseteq C \quad \Rightarrow \quad A \sqsubseteq C$$

and

$$A \sqsubseteq B \trianglelefteq C \quad \Rightarrow \quad A \trianglelefteq C$$

3) *Any subgroup of a cyclic group is characteristic and so*

$$H \le \langle a \rangle \trianglelefteq G \quad \Rightarrow \quad H \trianglelefteq G$$

4) **(Extension property)** *For any* $N \le H \le G$*,*

$$N \sqsubseteq G, \quad \frac{H}{N} \sqsubseteq \frac{G}{N} \quad \Rightarrow \quad H \sqsubseteq G$$

Proof. For part 3), a finite cyclic group G has only one subgroup of each size and so it must be invariant under any automorphism of G. On the other hand, if $G = \langle a \rangle$ is infinite cyclic and $\sigma \in \text{Aut}(G)$, then σa must be a generator of G and so $\sigma a = a$ or $\sigma a = a^{-1}$. Hence, σ is either the identity map or the map that sends any element to its inverse. In either case, any subgroup of G is σ-invariant.□

Definition *A nontrivial group* G *is* **characteristically simple** *if it has no nontrivial proper characteristic subgroups.*□

As an example, the 4-group $V = \{1, a, b, ab\}$ is characteristically simple but not simple.

Elementary Abelian Groups

The simplest type of abelian group is a cyclic group of prime order. Perhaps the next simplest type of abelian group is an external direct product of cyclic groups of the same prime order.

Definition *An* **elementary abelian group** G *is an abelian group in which every nonidentity element has the same finite order.*□

Theorem 4.15 *Let* G *be an elementary abelian group.*
1) *Every nonidentity element of* G *has prime order* p*.*
2) *Writing the group product additively,* G *is a vector space over* $\mathbb{Z}_p$*, where if* $\alpha \in \mathbb{Z}_p$ *and* $a \in G$*, then* αa *is the sum of* α *copies of* a*. The subgroups of* G *are the same as the subspaces of* G *and the group endomorphisms of* G *are the same as linear operators on* G*.*
3) *If* G *is finite, then*

$$G \approx \mathbb{Z}_p \boxtimes \cdots \boxtimes \mathbb{Z}_p$$

(This also holds when G *is infinite, but we have not yet defined infinite direct products.)*

Proof. We use additive notation for G. For part 1), if $o(a) = mn$ where $m > 1$, then $o(na) = m$ and so $mn = m$, that is, $n = 1$. Thus, $o(a) = p$ is prime for all nonidentity $a \in G$. For part 2), let $\oplus$ and $\odot$ denote addition and multiplication in $\mathbb{Z}_p$, respectively. If $n \in \mathbb{Z}$ is nonnegative, let $n * a$ be the sum of n copies of a. Then for $\alpha, \beta \in \mathbb{Z}_p$, there exists an integer n for which

$$\alpha \oplus \beta = (\alpha + \beta) + np$$

and so

$$(\alpha \oplus \beta)a = [(\alpha + \beta) + np] * a = (\alpha + \beta) * a = \alpha a + \beta b$$

Similarily,

$$(\alpha \odot \beta)a = [\alpha\beta + np] * a = (\alpha\beta) * a = \alpha(\beta a)$$

The other requirements of scalar multiplication are also met, namely, $1a = a$ and

$$\alpha(a + b) = \alpha a + \alpha b$$

We leave proof of the remaining part of part 2) to the reader. For part 3), G is vector-space isomorphic to a direct sum of a certain number of one-dimensional subspaces of G, that is, subspaces of the form $\mathbb{Z}_p a = \langle a \rangle$ for $a \in G$. These subspaces are cyclic subgroups of G and the vector space isomorphism is also a group isomorphism.□

Theorem 4.16 *The following are equivalent for an abelian group G:*
1) *G is elementary abelian*
2) *G is characteristically simple and has at least one nonidentity torsion element.*

Proof. If G is an elementary abelian group, then the automorphisms of G are the linear automorphisms of G. It follows that if $S \leq G$ is nontrivial and proper, then for any nonzero $a \in S$ and nonzero $x \in G \setminus S$, there is an automorphism sending a to x and so S is not characteristic in G. Hence, G is characteristically simple. Of course, G has a nonidentity torsion element.

For the converse, assume that G is characteristically simple and let p be a prime dividing the order of some element $a \in A$. Then (using additive notation)

$$A_p = \{a \in A \mid pa = 0\}$$

is nontrivial and characteristic in A, whence $A = A_p$ is an elementary abelian group.□

Note that the p-quasicyclic group $\mathbb{Z}(p^\infty)$, which has proper subgroup lattice

$$\{0\} < \langle 1/p \rangle < \langle 1/p^2 \rangle < \cdots$$

is characteristically simple, since there is at most one subgroup of any given size. However, it is not an elementary abelian group since it has no nonzero torsion elements.

Multiplication as a Permutation

If G is a group, then multiplication by $a \in G$ is a permutation of G; specifically, the multiplication map $\mu_a: G \to G$ defined by

$$\mu_a x = ax$$

is bijective. In this context, multiplication is also referred to as **(left) translation**.

The map $\mu: G \to S_G$ that sends a to μ_a provides a *representation* of the elements of the group G as permutations of the set G. Moreover, the representation μ is a group homomorphism, since

$$\mu_{ab} = \mu_a \mu_b$$

This is not the only way to represent the elements of a group as permutations of some set. For example, if $H \leq G$, then the elements of G can also be represented as permutations of a quotient set G/H via the multiplication map $\sigma_a: G \to G/H$ defined by

$$\sigma_a(gH) = agH$$

It is clear that σ_a is a permutation of G/H and that the map $\sigma: G \to S_{G/H}$ sending a to σ_a is a group homomorphism.

The representation map $\mu: G \to S_G$ is described by saying that G *acts on itself by* (left) *translation* and the representation map $\sigma: G \to G/H$ is described by saying that G *acts on* G/H *by* (left) *translation*. Both of these representations fit the pattern of the following definition.

Definition *An* **action** *of a group G on a nonempty set X is a group homomorphism $\lambda: G \to S_X$, called the* **representation map** *for the action. Thus,*

$$\lambda(1) = \iota, \quad \lambda(a^{-1}) = (\lambda a)^{-1} \quad \textit{and} \quad \lambda(ab) = \lambda(a)\lambda(b)$$

for all $a, b \in G$. When λ is a group action, we say that G **acts** *on X by λ. The permutation λa is usually denoted by λ_a, or simply by a itself.*□

We should mention that translation is not the only important action of a group on a set. We will see in a later chapter that conjugation is also an important group action.

Since a representation map is assumed only to be a homomorphism, it is possible for two distinct elements of G to have the same representation in S_X. Although it may seem at first that this is not a particularly desirable quality, we will see that such representations can yield important results. An injective representation, that is, an embedding $\lambda: G \hookrightarrow S_X$ is said to be **faithful**. In this case, G is isomorphic to a subgroup of S_X. The action of G on itself by translation is faithful but the action of G on G/H by translation need not be faithful.

Here is one interesting application of left translation.

Theorem 4.17 *Let H be a proper subgroup of a group G. Then there are distinct group homomorphisms $\sigma, \tau: G \to K$ into some group K that agree on H.*

Proof. For $x \notin G/H$, we construct two distinct group actions of G on the set

$$X = G/H \cup \{x\}$$

that agree on H. Let $\sigma: G \to S_X$ be left translation on G/H that also leaves x fixed, that is, for any $a \in G$,

$$\sigma_a(bH) = abH \quad \text{and} \quad \sigma_a x = x$$

and let $\tau: G \to S_X$ be defined by

$$\tau_a = (H\,x)\sigma_a(H\,x)$$

If $h \in H$, then σ_h fixes both H and x and so σ_h and $(H\,x)$ are disjoint permutations, whence $\sigma_h = \tau_h$ for all $h \in H$. On the other hand, τ and σ are distinct since if $a \notin H$, then

$$\tau_a H = H \quad \text{but} \quad \sigma_a H \neq H$$

□

The Left Regular Representation: Cayley's Theorem

The action $\mu: G \to S_G$ of G on itself by left translation is called the **left regular representation** of G. Since the left regular representation is faithful, it follows that *every* group is isomorphic to a subgroup of some symmetric group!

Theorem 4.18 (Cayley's theorem [7], 1854) *Every group G is isomorphic to a subgroup of the symmetric group S_G, via the action of G on itself by left translation.* □

Multiplication by G on G/H; the Normal Interior

If $H < G$ has finite index, the action $\sigma: G \to S_{G/H}$ of G on G/H by left translation yields a variety of remarkable consequences, mainly due to the nature of the kernel of this action, which is the intersection of all conjugates of H:

$$\begin{aligned}\ker(\sigma) &= \{x \in G \mid \sigma_x = \iota\} \\ &= \{x \in G \mid xaH = aH \text{ for all } a \in G\} \\ &= \{x \in G \mid xa \in aH \text{ for all } a \in G\} \\ &= \{x \in G \mid x \in a^{-1}Ha \text{ for all } a \in G\} \\ &= \bigcap_{a \in G} H^a\end{aligned}$$

Thus,

$$K := \bigcap\nolimits_{a \in G} H^a$$

is both normal in G and contained in H. Moreover, if N is any normal subgroup of G that is contained in H, then

$$N = N^a \le H^a$$

for all $a \in G$ and so $N \le K$. In other words, K is the *largest* subgroup of H that is normal in G.

Theorem 4.19 *Let $H \le G$. The largest subgroup of H that is normal in G is called the* **normal interior** *or* **core** *of H, which we denote by H°. The core of H is the kernel*

$$H^\circ = \bigcap_{a \in G} H^a$$

of the action of G on G/H by left translation. □

Thus, left translation $\sigma\colon G \to S_{G/H}$ induces an embedding

$$\frac{G}{H^\circ} \hookrightarrow S_{G/H}$$

of G/H° into $S_{G/H}$ and so

$$(G : H^\circ) \mid (G : H)!$$

In particular, if G is simple, then H° is trivial and so

$$o(G) \mid (G : H)!$$

These facts have some rather interesting consequences relating to the existence of normal subgroups and subgroups of small index of a group.

Theorem 4.20 *Let G be a group and let $H \le G$ have finite index. Then*

$$G/H^\circ \hookrightarrow S_{G/H}$$

and so

$$(G : H^{\circ}) \mid (G : H)!$$

In particular, $(G : H^{\circ})$ *is also finite and*

$$(H : H^{\circ}) \mid ((G : H) - 1)!$$

1) *Any of the following imply that* $H \trianglelefteq G$*:*
   a) H *is periodic and* $(G : H) = p$ *is equal to the smallest order among the nonidentity elements of* H.
   b) G *is finite and* $o(H)$ *and* $((G : H) - 1)!$ *are relatively prime, that is, for all primes* p,

$$p \mid o(H) \quad \Rightarrow \quad p \geq (G : H)$$

   which happens, in particular, if $(G : H)$ *is equal to the smallest prime dividing* $o(G)$.
2) *If G is finitely generated, then G has at most a finite number of subgroups of any finite index* m.
3) *If G is simple, then*

$$o(G) \mid (G : H)!$$

   a) *If G is infinite, then G has no proper subgroups of finite index.*
   b) *If G is finite and* $o(G) \nmid m!$ *for some integer* m*, then G has no subgroups of index* m *or less.*

Proof. For part 1), first note that p is a prime. If q is a prime dividing $(H : H^{\circ})$, then Cauchy's theorem implies that there is an $h \in H$ for which $q = o(hH^{\circ}) \mid o(h)$, whence $q \geq p$, a contradiction to the fact that

$$q \mid (H : H^{\circ}) \mid (p - 1)!$$

Hence, $(H : H^{\circ}) = 1$ and $H = H^{\circ} \trianglelefteq G$.

For part 2), $(H : H^{\circ})$ divides both $((G : H) - 1)!$ and $o(H)$ and so must equal 1

$$(H : H^{\circ}) \mid (m - 1)!$$

But $(H : H^{\circ})$ also divides $o(H)$, which is relatively prime to $(p - 1)!$ and so $(H : H^{\circ}) = 1$.

For part 3), suppose that the result is true for all normal subgroups. Then there are only finitely many possibilities for normal interiors of subgroups of index m, since such a normal interior must have index dividing $m!$. But each normal subgroup N of finite index can be the normal interior for only finitely many subgroups, since there are only finitely many subgroups containing N.

Now, normal subgroups correspond to kernels of homomorphisms. In particular, if $(G : N) = k$, then the homomorphism

$$\sigma \circ \pi_N: G \to G/N \hookrightarrow S_k$$

where σ is the left regular representation of G/N in S_k, has kernel N. Thus, distinct normal subgroups N of index k yield distinct homomorphisms from G into S_k. However, since G is finitely generated, there are only a finite number of such homomorphisms and so there are only a finite number of normal subgroups of G of index k.□

Example 4.21 If $o(G) = p^n u$ with p prime and $p > u$, then we will prove in a later chapter that G has a subgroup H of order p^n (a Sylow p-subgroup of G). Since p^n and $(u-1)!$ are relatively prime, it follows that $H \trianglelefteq G$.□

The Frattini Subgroup of a Group

The following subgroup of a group is most interesting.

Definition *The* **Frattini subgroup** $\Phi(G)$ *of a finite group* G *is the intersection of the maximal subgroups of* G.□

Definition *Let* G *be a group. An element* $a \in G$ *is called a* **nongenerator** *of* G *if whenever the subset* $S \subseteq G$ *generates* G*, then so does the set* $S \setminus \{a\}$*. Thus, a nongenerator is an element that is not needed in any generating set.*□

Here are the basic properties of the Frattini subgroup of a finite group. But first a definition.

Definition *Let* G *be a group and let* $K \leq G$*. A subgroup* H *of* G *is called a* **supplement** *of* K *if* $G = HK$.□

Theorem 4.22 *Let* G *be a finite group.*

1) $\Phi(G)$ *is the set of all nongenerators of* G*.*
2) $\Phi(G) \sqsubseteq G$*.*
3) The following are equivalent:
   a) Every maximal subgroup of G *is normal.*
   b) $G/\Phi(G)$ *is abelian.*
4) If $K \lhd G$*, then* $K \leq \Phi(G)$ *if and only if* K *has no proper supplements.*

Proof. For part 1), if $S \subseteq G$ does not generate G, then S is contained in a maximal subgroup M, which also contains $\Phi(G)$ and so $S \cup \Phi(G) \subseteq M$ does not generate G. Hence, the elements of $\Phi(G)$ are nongenerators. Conversely, if x is a nongenerator of G and $x \notin M$ for some maximal subgroup M, then $M \cup \{x\}$ generates G and so M generates G, which is false. Thus, all nongenerators of G are in $\Phi(G)$.

For part 2), if $\sigma \in \operatorname{Aut}(G)$, then the induced map is a bijection on the family $\mathcal{M}$ of maximal subgroups of G and so

$$\sigma(\Phi(G)) = \sigma\left(\bigcap \mathcal{M}\right) = \bigcap \sigma\mathcal{M} = \bigcap \mathcal{M} = \Phi(G)$$

For part 3), assume that every maximal subgroup M of G is normal. Then G/M has no proper nontrivial subgroups and so is abelian. Hence, $G' \le M$ and so $G' \le \Phi(G)$. Conversely, if $G' \le \Phi(G)$ and M is a maximal subgroup of G, then $G' \le M$ and so $M \trianglelefteq G$.

For part 4), assume first that $K \le \Phi(G)$. If $H < G$, then $H \le M$ for some maximal subgroup M of G and so $HK \le M < G$. Conversely, if K has no proper supplements, then every maximal subgroup M satisfies $M \le MK < G$ and so $M = MK$, whence $K \le M$ and so $K \le \Phi(G)$.□

Subnormal Subgroups

We wish now to take a closer look at the concept of subnormality. As to the existence of subnormal subgroups, we have the following examples to show that all possibilities may occur, and that a subnormal subgroup need not be normal.

Example 4.23

1) A simple group has no nontrivial proper subnormal subgroups.
2) In the dihedral group $D_8 = \langle \sigma, \rho \rangle$, we have

$$\langle \sigma \rangle \triangleleft \langle \sigma, \rho^2 \rangle \triangleleft D_8$$

and so $\langle \sigma \rangle$ is subnormal in D_8 but not normal in D_8. In fact, all subgroups of D_8 are subnormal.
3) In the symmetric group S_3, the subgroup $H = \langle (1\,2) \rangle$ is maximal and so the only sequence of subgroups from H to S_3 is $H \le S_3$. Since H is not normal in S_3, it follows that H is not subnormal in S_3. Of course, $\langle (1\,2\,3) \rangle$ is subnormal in S_3, being normal in S_3.□

The following theorem outlines the simplest properties of subnormality and is a direct consequence of Theorem 4.10.

Theorem 4.24 *Let G be a group. Let $H, K \le G$.*

1) **(Transitivity)**

$$H \trianglelefteq\trianglelefteq K \quad \text{and} \quad K \trianglelefteq\trianglelefteq G \quad \Rightarrow \quad H \trianglelefteq\trianglelefteq G$$

2) **(Intersection)** *If $L \le G$, then*

$$H \trianglelefteq\trianglelefteq K \quad \Rightarrow \quad H \cap L \trianglelefteq\trianglelefteq K \cap L$$

In particular,

$$H \le K \le G, \quad H \trianglelefteq\trianglelefteq G \quad \Rightarrow \quad H \trianglelefteq\trianglelefteq K$$

and

$$H \trianglelefteq\trianglelefteq G, \quad K \trianglelefteq\trianglelefteq G \quad \Rightarrow \quad H \cap K \trianglelefteq\trianglelefteq G$$

3) **(Normal lifting)** *If* $N \trianglelefteq G$*, then*

$$H \trianglelefteq\trianglelefteq K \quad \Rightarrow \quad HN \trianglelefteq\trianglelefteq KN$$

4) **(Quotient/unquotient)** *If* $N \leq H$ *and* $N \trianglelefteq K$*, then*

$$H \trianglelefteq\trianglelefteq K \quad \Leftrightarrow \quad H/N \trianglelefteq\trianglelefteq K/N$$

□

Intersections and Subnormality

While the intersection of two, and hence any finite number, of subnormal subgroups is subnormal, this is not the case for an arbitrary family of subnormal subgroups.

Example 4.25 Let G be the **infinite dihedral group**,

$$G = \{\rho^i, \sigma\rho^i \mid i \in \mathbb{Z}\}$$

where $o(\sigma) = 2$, $o(\rho) = \infty$ and $\rho\sigma = \sigma\rho^{-1}$. If

$$H_i = \langle \sigma, \rho^{2^i} \rangle$$

then $H_{i+1} \trianglelefteq H_i$ and so $H_i \trianglelefteq\trianglelefteq G$. But, $\bigcap H_i = \langle \sigma \rangle$ is self-normalizing (equal to its own normalizer) and so is not subnormal in G.□

However, since the family $\operatorname{subn}(G)$ is closed under finite intersection, Theorem 1.6 does imply the following.

Theorem 4.26 *If a group* G *has the DCC on* $\operatorname{subn}(G)$*, then the intersection of any family of subnormal subgroups is subnormal.*□

The Sequence of Normal Closures of a Subgroup

If $H \trianglelefteq\trianglelefteq G$, then the first step down in a series for H is an extension $H_1 \trianglelefteq G$. Moreover, since $\operatorname{nc}(H, G) \leq H_1$, the largest possible first step down in a series from H to G is

$$\operatorname{nc}(H, G) \trianglelefteq G$$

By repeatedly taking normal closures of H, we get a series from H to G that descends more rapidly than any other series from H to G.

Definition *Let* G *be a group and let* $H \leq G$*. The* **sequence of normal closures** *of* H *in* G *is defined by*

$$H_{(0)} = G, H_{(1)} = \operatorname{nc}(H, H_{(0)}) \quad \textit{and} \quad H_{(i+1)} = \operatorname{nc}(H, H_{(i)})$$

for $i \geq 0$.□

To see that the sequence of normal closures descends more rapidly than any other proper series

$$H = H_r \lhd H_{r-1} \lhd \cdots \lhd H_0 = G$$

it is clear that $H_{(0)} \leq H_0$ and if $H_{(i)} \leq H_i$, then

$$H_{(i+1)} = \operatorname{nc}(H, H_{(i)}) \leq \operatorname{nc}(H, H_i) \leq H_{i+1}$$

Hence, $H_{(i)} \leq H_i$ for all i.

Theorem 4.27 *A subgroup H of G is subnormal in G if and only if the sequence of normal closures of H in G reaches H, in which case this sequence is a series of shortest length from H to G. If $H \trianglelefteq\trianglelefteq G$, then the length of the sequence of normal closures is called the* **subnormal index** *of H in G, which we denote by $s(H, G)$.*□

We gather a few facts about subnormal indices.

Theorem 4.28 *Let G be a group and let $H \trianglelefteq\trianglelefteq G$ have sequence of normal closures*

$$H = H_{(s)} \lhd H_{(s-1)} \lhd \cdots \lhd H_{(0)} = G$$

1) **(Triangle inequality)**

$$H \trianglelefteq\trianglelefteq K \trianglelefteq\trianglelefteq G \quad \Rightarrow \quad s(H, G) \leq s(H, K) + s(K, G)$$

2) If $\sigma \in \operatorname{Aut}(G)$, then

$$\sigma(H_{(i)}) = (\sigma H)_{(i)}$$

for all $i \geq 0$.

a) If $\sigma H = H$, then $\sigma H_{(i)} = H_{(i)}$ for all i. In particular,

$$N_G(H) \subseteq \bigcap_{i=0}^{s} N_G(H_{(i)})$$

and so if $K \leq G$ normalizes H, then K normalizes every $H_{(i)}$.

b) The sequence of normal closures for σH in G is

$$\sigma H = \sigma(H_{(s)}) \lhd \sigma(H_{(s-1)}) \lhd \cdots \lhd \sigma(H_{(0)}) = G$$

and so $s(H, G) = s(\sigma H, G)$.

Proof. For part 2), we have $\sigma(H_{(0)}) = (\sigma H)_{(0)}$ and if $\sigma(H_{(i)}) = (\sigma H)_{(i)}$, then

$$\sigma(H_{(i+1)}) = \text{nc}(\sigma H, \sigma(H_{(i)})) = \text{nc}(\sigma H, (\sigma H)_{(i)}) = (\sigma H)_{(i+1)} \qquad \square$$

Joins and Subnormality

The question of whether the *join* of subnormal subgroups is subnormal is much more involved than the same question for intersection. Of course, if K is subnormal in G and if H is *normal* in G, then we may lift the sequence of normal closures of K in G

$$K = K_{(t)} \lhd K_{(t-1)} \lhd \cdots \lhd K_{(0)} = G \tag{4.29}$$

by H to get a series for the join HK. Thus,

$$H \trianglelefteq G, \quad K \trianglelefteq\trianglelefteq G \quad \Rightarrow \quad HK \trianglelefteq\trianglelefteq G$$

Actually, the process of lifting by H is the same as projecting the sequence (4.29) into the extension $H \trianglelefteq G$.

If neither factor H or K is normal, we are tempted to project (4.29) into each extension $H_{(i)} \lhd H_{(i-1)}$ and concatenate the resulting series. This almost works, the problem being that it produces a series from

$$H_{(s)}(K_{(t)} \cap H_{(s-1)}) = HK \cap H_{(s-1)}$$

to G, rather than from HK to G. So a slight modification is in order. Note that the unwanted $H_{(s-1)}$ comes from the *upper endpoint* of the extension $H_{(s)} \lhd H_{(s-1)}$.

Thus, if we assume that K normalizes H, then K normalizes $H_{(i)}$ for all i and so $H_{(i)}K$ is a subgroup of G and $H_{(i)} \trianglelefteq H_{(i-1)}K$. If we now project (4.29) into the extension $H_{(i)} \trianglelefteq H_{(i-1)}K$, the result is a series $\mathcal{H}_i$ with lower endpoint

$$H_{(i)}(K_{(t)} \cap H_{(i-1)}K) = H_{(i)}K$$

and upper endpoint $H_{(i-1)}K$, that is,

$$\mathcal{H}_i\colon H_{(i)}K \trianglelefteq \cdots \trianglelefteq H_{(i-1)}K$$

Moreover, since $\mathcal{H}_{(i+1)}$ and $\mathcal{H}_{(i)}$ are contiguous, the concatenation

$$\mathcal{H}_{(s)}\mathcal{H}_{(s-1)}\cdots\mathcal{H}_{(1)}$$

is a series from HK to G and so $HK \trianglelefteq\trianglelefteq G$.

Note also that each of the series $\mathcal{H}_i$ has length at most t and so

$$s(HK, G) \leq s(H, G)s(K, G)$$

Theorem 4.30 *If $H,K \trianglelefteq\trianglelefteq G$ and if K normalizes H, then*

$$HK = \langle H,K\rangle \trianglelefteq\trianglelefteq G$$

and

$$s(\langle H,K\rangle,G) \le s(H,G)s(K,G)$$ □

***The Subnormal Join Property**

Theorem 4.30 implies the following useful characterization of the join question.

Theorem 4.31 Let G be a group and let $H,K \trianglelefteq\trianglelefteq G$. The following are equivalent:

1) $\langle H,K\rangle \trianglelefteq\trianglelefteq G$

2) $H^K \trianglelefteq\trianglelefteq G$

3) $[H,K] \trianglelefteq\trianglelefteq G$.

Proof. Recall from Theorem 3.40 that

$$[H,K] \trianglelefteq H[H,K] = H^K \trianglelefteq H^K K = \langle H,K\rangle$$

Thus, 1) $\Rightarrow$ 2) $\Rightarrow$ 3). On the other hand, if 3) holds, then since $H \trianglelefteq\trianglelefteq G$ and $[H,K] \trianglelefteq\trianglelefteq G$ and H normalizes $[H,K]$, Theorem 4.30 implies that

$$H^K = H[H,K] \trianglelefteq\trianglelefteq G$$

and so 2) holds. If 2) holds, then 1) holds since K normalizes H^K.□

Definition *If a group G has the property that the join of any two subnormal subgroups is subnormal, then G is said to have the* **subnormal join property**, *or* **SJP**.□

Theorem 4.31 gives us one simple criterion for the SJP. If the commutator subgroup G' of G has the property that all of its subgroups are subnormal, then $[H,K]$ is subnormal in G' and therefore also in G, for all $H,K \le G$ and so Theorem 4.31 implies that G has the SJP. In particular, this occurs when G' is abelian, that is, when G is **metabelian**.

Theorem 4.32 *If the commutator subgroup G' of G has the property that all of its subgroups are subnormal, in particular, if G is metabelian, then G has the subnormal join property.*□

On a different line, if G' has the ACC on subnormal subgroups (a finiteness condition), then G has the subnormal join property.

Theorem 4.33 (Robinson) *Let G be a group for which G' has the ACC on subnormal subgroups. Then G has the subnormal join property.*

Proof. Let H and K be subnormal in G. The proof is by induction on $s = s(H,G)$. If $s \le 1$, the result is clear, so assume that $s \ge 2$. If $a \in G$ then

$$H_{(1)}^a = \mathrm{nc}(H^a, G) = \mathrm{nc}(H, G) = H_{(1)}$$

and so H and H^a are subnormal subgroups of $H_{(1)}$. But $H_{(1)}$ also has the property that its commutator subgroup has the ACC on subnormal subgroups. To see this, note that $H_{(1)}' \sqsubseteq H_{(1)} \trianglelefteq G$ implies that $H_{(1)}' \trianglelefteq G$ and so $\mathrm{subn}(H_{(1)}')$ is contained in $\mathrm{subn}(G')$. Thus, since $s(H, H_{(1)}) = s - 1$, the inductive hypothesis implies that $\langle H, H^a \rangle \trianglelefteq\trianglelefteq H_{(1)}$ and so $\langle H, H^a \rangle \trianglelefteq\trianglelefteq G$.

Moreover, this can be extended to more than one conjugate of H. For example, if $b \in G$, then $\langle H, H^a \rangle^b = \langle H^b, H^{ba} \rangle$ and so the join of $\langle H, H^a \rangle$ and $\langle H, H^a \rangle^b$ is the subnormal subgroup $\langle H, H^a, H^b \rangle$. Note also that we can replace H by any conjugate of H and so

$$\langle H, H^{a_1}, \ldots, H^{a_n} \rangle \trianglelefteq\trianglelefteq G$$

for all $a_1, \ldots, a_n \in G$.

Next, we show that

$$\langle [H, a_1], \ldots, [H, a_n] \rangle \trianglelefteq\trianglelefteq G$$

for all $a_1, \ldots, a_n \in G$. First, note that

$$\langle H, [H, a] \rangle = \langle H, H^a \rangle$$

and so

$$\langle H, [H, a_1], \ldots, [H, a_r] \rangle = \langle H, H^{a_1}, \ldots, H^{a_r} \rangle \trianglelefteq\trianglelefteq G$$

But Theorem 3.37 implies that H normalizes each $[H, a_i]$ and so

$$\langle [H, a_1], \ldots, [H, a_r] \rangle \trianglelefteq \langle H, [H, a_1], \ldots, [H, a_r] \rangle \trianglelefteq\trianglelefteq G$$

Now we can complete the proof. The ACC on $\mathrm{subn}(G')$ implies that there is a finite subset $I = \{a_1, \ldots, a_n\} \subseteq K$ for which

$$[H, K] = M_I := \langle [H, a_1], \ldots, [H, a_n] \rangle$$

for if not, then there is a strictly increasing sequence

$$M_{I_1} < M_{I_2} < \cdots$$

of subnormal subgroups of G'. Hence, $[H, K] \trianglelefteq\trianglelefteq G$.□

We refer the reader to Robinson [26], page 389, for an example of a group that does not have the SJP.

***The Generalized Subnormal Join Property**

It is not necessarily the case that a group with the SJP also has the property that the join of *any* family of subnormal subgroups of G is subnormal. This is called the **generalized subnormal join property** or **GSJP**.

Theorem 4.34 (Wielandt [35], 1939) *Let G be a group.*

1) *If G has the ACC on* $\operatorname{subn}(G)$, *then G has the GSJP.*
2) *If G has BCC on* $\operatorname{subn}(G)$, *in particular, if G is finite, then* $\operatorname{subn}(G)$ *is a complete sublattice of* $\operatorname{sub}(G)$.□

Proof. For part 1), since G' also has the ACC on subnormal subgroups, G has the SJP. Hence, $\operatorname{subn}(G)$ is closed under finite join and so Theorem 1.6 implies that G has the GSJP.□

Robinson has provided the following characterization of the GSJP, whose proof uses ordinal numbers.

Theorem 4.35 (Robinson) *A group G has the generalized subnormal join property if and only if the union of every chain of subnormal subgroups is subnormal.*

Proof. If G has the GSJP, then the union of a chain of subnormal subgroups is the join of that chain and so is subnormal. For the converse, we first show that G has the SJP by induction on the smaller of the two subnormal indices. Let $H,K \trianglelefteq\trianglelefteq G$ and let

$$s = \min\{s(H,G), s(K,G)\}$$

If $s \le 1$, then $\langle H,K\rangle \trianglelefteq\trianglelefteq G$. Assume that $s \ge 2$ and that the result holds when the subnormal index is less than s.

Well order the set K so that

$$K = \{k_\alpha \mid \alpha < \delta\}$$

where δ is an ordinal number and $k_0 = 1$. Let

$$L_\beta = \langle H^{k_\alpha} \mid \alpha < \beta\rangle$$

be the join of the first β conjugates of H by elements of K. Then

$$L_\beta \le H_{(1)} \quad \text{and} \quad L_\delta = H^K = H[H,K]$$

To show that L_δ is subnormal in G, we use transfinite induction on β.

If $\beta = 0$, then $L_\beta = H \trianglelefteq\trianglelefteq G$. For successor ordinals, if $L_\beta \trianglelefteq\trianglelefteq G$, then L_β and H^{k_β} are contained in $H_{(1)}$ and $s(H^{k_\beta}, H_{(1)}) = s-1$. Also, $H_{(1)}$ has the property that the union of any chain of subnormal subgroups is subnormal. Hence, the induction hypothesis implies that

$$L_{\beta+1} = \langle L_\beta, H^{k_\beta} \rangle \trianglelefteq\trianglelefteq H_{(1)} \trianglelefteq G$$

Finally, if λ is a limit ordinal, then

$$L_\lambda = \bigcup \{L_\alpha \mid \alpha < \lambda\}$$

is the union of a chain of subnormal subgroups, which is subnormal in G by hypothesis. Hence, the transfinite induction is complete and $L_\delta \trianglelefteq\trianglelefteq G$, whence $\langle H, K \rangle \trianglelefteq\trianglelefteq G$ by Theorem 4.31. Thus, G has the SJP.

Now, let $\mathcal{S} \subseteq \operatorname{subn}(G)$. The join is not affected by including in $\mathcal{S}$ the join of every finite subset of $\mathcal{S}$, so we may assume that $\mathcal{S}$ is closed under finite joins. For convenience, we refer to a family $\mathcal{P} \subseteq \operatorname{subn}(G)$ as *good* if $\mathcal{P}$ contains $\mathcal{S}$ and is closed under finite joins and chain joins. Then $\operatorname{subn}(G)$ is good and the intersection $\mathcal{S}'$ of all good families is the smallest good family containing $\mathcal{S}$. Let

$$J = \bigvee \mathcal{S} \quad \text{and} \quad J' = \bigvee \mathcal{S}'$$

Then $J \subseteq J'$. But if $J \subset J'$, then there is an $S' \in \mathcal{S}'$ for which $S' \not\subseteq J$ and so if

$$\mathcal{T} = \{S \in \mathcal{S}' \mid S \subseteq J\}$$

it follows that $\mathcal{S} \subseteq \mathcal{T} \subset \mathcal{S}'$. But $\mathcal{T}$ is also good and so the minimality of $\mathcal{S}'$ implies that $J = J'$. Finally, since $\mathcal{S}'$ is closed under chain join, Zorn's lemma implies that $\mathcal{S}'$ has a maximal member M. If $M \subset J'$, then there is an $N \in \mathcal{S}'$ for which $N \not\subseteq M$ and so $M \vee N \in \mathcal{S}'$, which contradicts the maximality of M, whence $M = J'$ and

$$\bigvee \mathcal{S} = \bigvee \mathcal{S}' = M \in \mathcal{S}' \subseteq \operatorname{subn}(G)$$

Thus, G has the GSJP.□

When All Subgroups Are Subnormal

There are a variety of ways to characterize finite groups in which all subgroups are subnormal. (We will characterize groups in which all subgroups are *normal* in Theorem 5.20.) To explore this further, we need some additional terminology.

By definition, the normalizer $N_G(H)$ of a subgroup H of G is the largest subgroup of G in which H is normal. Thus, a subgroup that is equal to its own normalizer is as "unnormal" as possible, since it is normal *only* in itself.

Definition *Let G be a group.*

1) *A subgroup $H \le G$ is* **self-normalizing** *if $H = N_G(H)$.*
2) *A group G has the* **normalizer condition** *if G has no proper self-normalizing subgroups, that is, if*

$$H < G \quad \Rightarrow \quad H < N_G(H) \qquad \square$$

Note that the term *self-normalizing* is a bit misleading, since every subgroup normalizes itself. (We would prefer the term *unnormal*.)

A proper subnormal subgroup H cannot be self-normalizing, since the first step in a series shows that H is normal in some subgroup larger than itself. Hence, if all subgroups of G are subnormal, then G has the normalizer condition. The converse is also true if G is finite (or more generally, if G has the ACC on subgroups), since the proper series of normalizers

$$H \lhd N_G(H) \lhd N_G(N_G(H)) \lhd \cdots$$

must eventually reach G.

Theorem 4.36 *The following are equivalent for a finite group G:*
1) *Every subgroup of G is subnormal*
2) *G has the normalizer condition.*$\square$

If G has the normalizer condition, then any maximal subgroup M must be normal in G, for we have $M < N_G(M)$ and the maximality of M implies that $N_G(M) = G$. Thus, with respect to the following conditions on a finite group G:

1) Every subgroup of G is subnormal
2) G has the normalizer condition
3) Every maximal subgroup of G is normal
4) $G/\Phi(G)$ is abelian

we have proved (see Theorem 4.22) that

$$1) \Leftrightarrow 2) \Rightarrow 3) \Leftrightarrow 4)$$

For finite groups, it is our eventual goal to prove not only that these four conditions are equivalent, but that they are equivalent to several other conditions, one of which is that G satisfy a strong converse of Lagrange's theorem, namely, if $n \mid o(G)$, then G has a *normal* subgroup of order n.

Chain Conditions

Let us take a closer look at chain conditions for groups, beginning with the definition.

Definition *Let G be a group and let $\mathcal{S}$ be a family of subgroups of G.*
1) *A group G satisfies the* **ascending chain condition (ACC) on** $\mathcal{S}$ *if every ascending sequence*

$$H_1 \le H_2 \le \cdots$$

of subgroups in $\mathcal{S}$ must eventually be constant, that is, if there is an $n > 0$

such that $H_{n+k} = H_n$ *for all* $k \geq 0$. *In this case, we also say that* $\mathcal{S}$ *has the ACC.*

2) *A group* G *satisfies the* **descending chain condition (DCC) on** $\mathcal{S}$ *if every descending sequence*

$$H_1 \geq H_2 \geq \cdots$$

of subgroups in $\mathcal{S}$ *must eventually be constant, that is, if there is an* $n > 0$ *such that* $H_{n+k} = H_n$ *for all* $k \geq 0$. *In this case, we also say that* $\mathcal{S}$ *has the DCC.*

3) *A group* G *satisfies* **both chain condition (BCC) on** $\mathcal{S}$ *if* G *has the ACC and the DCC on* $\mathcal{S}$. *In this case, we also say that* $\mathcal{S}$ *has the BCC.*□

Theorem 1.5 implies that G has the BCC on $\mathcal{S}$ if and only if $\mathcal{S}$ has no infinite chains.

Our main interest will center on the case where G is a group, $N \trianglelefteq G$ and $\mathcal{S}$ is one of the following families:

$$\text{sub}(N;G), \quad \text{nor}(N;G), \quad \text{subn}(N;G)$$

of all subgroups, normal subgroups and subnormal subgroups, respectively, of G that contain N.

The ACC and DCC are, in general, independent of each other, that is, all four combinations are possible. For example, a finite group has both chain conditions on subgroups. An infinite cyclic group (such as the integers) has the ACC on subgroups but not the DCC on subgroups. The p-quasicyclic group $\mathbb{Z}(p^\infty)$ has the DCC on subgroups but not the ACC. Finally, the group of rational numbers $\mathbb{Q}$ has neither chain condition on subgroups.

The chain conditions can be characterized as follows.

Definition *Let* G *be a group and let* $\mathcal{S} \subseteq \text{sub}(G)$.

1) G *has the* **maximal condition** *on* $\mathcal{S}$ *if every nonempty subfamily of* $\mathcal{S}$ *contains a maximal element.*
2) G *has the* **minimal condition** *on* $\mathcal{S}$ *if every nonempty subfamily of* $\mathcal{S}$ *contains a minimal element.*□

Theorem 4.37 *Let* G *be a group and let* $\mathcal{S} \subseteq \text{sub}(G)$.

1) G *has the maximal condition on* $\mathcal{S}$ *if and only if* G *has the ACC on* $\mathcal{S}$.
2) G *has the minimal condition on* $\mathcal{S}$ *if and only if* G *has the DCC on* $\mathcal{S}$.

Proof. Suppose G satisfies the maximal condition on $\mathcal{S}$ and that

$$H_1 \leq H_2 \leq \cdots$$

is an ascending sequence of members of $\mathcal{S}$. Then the subgroups H_k have a maximal member H_n, which implies that $H_{n+k} \leq H_n$ for all $k \geq 0$. Conversely, suppose G satisfies the ACC on $\mathcal{S}$ and let $\mathcal{M}$ be a nonempty subfamily of $\mathcal{S}$. If $H_1 \in \mathcal{M}$ is not maximal, then there is an $H_2 \in \mathcal{M}$ for which $H_1 < H_2$. If H_2 is not maximal, then there is an H_3 for which $H_1 < H_2 < H_3$. This must stop after a finite number of steps and so $\mathcal{M}$ must have a maximal member. The proof of part 2) is analogous.□

In certain cases, the ACC can be characterized in another important way. If $\mathcal{S} \subseteq \text{sub}(G)$ is closed under arbitrary intersections, then every subset X of G is contained in a *smallest* element of $\mathcal{S}$, called the **$\mathcal{S}$-closure** of X, which is

$$\langle X \rangle_{\mathcal{S}} = \bigcap \{H \in \mathcal{S} \mid X \subseteq H\}$$

Note that the set on the right is nonempty, since the assumption that $\mathcal{S}$ is closed under arbitrary intersections implies that $\mathcal{S}$ contains the empty intersection, which is G itself. We say that $\langle X \rangle_{\mathcal{S}}$ is **$\mathcal{S}$-generated** by X and if X is a finite set, we say that $\langle X \rangle_{\mathcal{S}}$ is **finitely $\mathcal{S}$-generated** by X.

Theorem 4.38 *Let G be a group and let $\mathcal{S} \subseteq \text{sub}(G)$ be closed under arbitrary intersections and closed under unions of ascending sequences. Then G has the ACC on $\mathcal{S}$ if and only if every $H \in \mathcal{S}$ is finitely $\mathcal{S}$-generated.*
Proof. Suppose G satisfies the ACC on $\mathcal{S}$ and let $H \in \mathcal{S}$. If H is not finitely $\mathcal{S}$-generated, then for any $h_1 \in H$, we have

$$\langle h_1 \rangle_{\mathcal{S}} < H$$

Hence, there is an $h_2 \in H \setminus \langle h_1 \rangle_{\mathcal{S}}$ for which

$$\langle h_1 \rangle_{\mathcal{S}} < \langle h_1, h_2 \rangle_{\mathcal{S}} < H$$

We can continue to choose elements to produce an infinite strictly ascending sequence, in contradiction to the ACC on $\mathcal{S}$. Hence, H is finitely $\mathcal{S}$-generated.

Conversely, suppose every element of $\mathcal{S}$ is finitely $\mathcal{S}$-generated and let

$$H_1 \le H_2 \le \cdots$$

be an ascending sequence of subgroups in $\mathcal{S}$. Then $H = \bigcup H_k \in \mathcal{S}$ and so $H = \langle X \rangle_{\mathcal{S}}$ for some finite set X. Hence, there is an index m for which $X \subseteq H_m$ and so $H = H_m$. But then $H_{m+k} = H$ for all $k \ge 0$.□

Note that in the preceeding theorem, $\mathcal{S}$ can be $\text{sub}(G)$, $\text{nor}(G)$, the family of all characteristic subgroups of G or the family of all fully-invariant subgroups of G.

We next describe how the chain conditions are inherited.

Theorem 4.39 *Let G be a group and let $N \trianglelefteq G$. Let $\mathcal{F}(N;G)$ be one of the following families of subgroups of G:*

$$\text{sub}(N;G), \quad \text{nor}(N;G), \quad \text{subn}(N;G)$$

and let $\mathcal{F}(G) = \mathcal{F}(\{1\}, G)$. *Write* $\mathcal{F} \in$ ACC *to denote the fact that* $\mathcal{F}$ *has the* ACC, *and similarly for the* DCC.

1) **(Quotients)**

$$\mathcal{F}(N;G) \in \text{ACC} \quad \Leftrightarrow \quad \mathcal{F}(G/N) \in \text{ACC}$$

2) **(Extension)**

$$\mathcal{F}(N), \mathcal{F}(G/N) \in \text{ACC} \quad \Rightarrow \quad \mathcal{F}(G) \in \text{ACC}$$

3) **(Direct products)**

$$\mathcal{F}(G_1), \mathcal{F}(G_2) \in \text{ACC} \quad \Leftrightarrow \quad \mathcal{F}(G_1 \boxtimes G_2) \in \text{ACC}$$

Similar statements hold for the DCC *in place of the* ACC.

Proof. Part 1) follows from the correspondence theorem. For part 2), let

$$G_1 \le G_2 \le \cdots$$

be an ascending chain in $\mathcal{F}(G)$. The sequences

$$G_1 \cap N \le G_2 \cap N \le \cdots$$

and

$$G_1 N \le G_2 N \le \cdots$$

are ascending chains in $\mathcal{F}(N)$ and $\mathcal{F}(N;G)$, respectively. Since part 1) implies that $\mathcal{F}(N;G)$ has the ACC, each sequence is eventually constant and so there is an index m for which

$$G_{m+i} \cap N = G_m \cap N \quad \text{and} \quad G_{m+i}N = G_m N$$

for all $i \ge 0$. Hence, Theorem 2.18 implies that $G_{m+i} = G_m$ for all $i \ge 0$ and so $\mathcal{F}(G) \in$ ACC. A similar argument holds for the DCC.

For part 3), let $\mathcal{F}(G_1), \mathcal{F}(G_2) \in$ ACC and let

$$P = G_1 \boxtimes G_2, \quad N_1 = G_1 \boxtimes \{1\}, \quad N_2 = \{1\} \boxplus G_2$$

Then $\mathcal{F}(G_i) \in$ ACC implies that $\mathcal{F}(N_i) \in$ ACC for $i = 1, 2$. Also, $P/N_1 \approx N_2$ and so $\mathcal{F}(P/N_1) \in$ ACC. Hence, $\mathcal{F}(P) \in$ ACC by the extension property. Conversely, if $\mathcal{F}(P) \in$ ACC, then $\mathcal{F}(N_i) \subseteq \mathcal{F}(P)$ implies that $\mathcal{F}(N_i) \in$ ACC and so $\mathcal{F}(G_i) \in$ ACC for $i = 1, 2$.□

Chain Conditions and Homomorphisms

The presence of a chain condition can have a significant impact on homomorphisms. For example, if G has the ACC on normal subgroups, then any surjective endomomorphism $\sigma: G \to G$ is also injective. To see this, consider the **kernel sequence** of σ:

$$\ker(\sigma) \le \ker(\sigma^2) \le \ker(\sigma^3) \le \cdots$$

Since this must eventually be constant, we have $\ker(\sigma^n) = \ker(\sigma^{n+k})$ for all $k \ge 0$. Now, if $a \in \ker(\sigma)$, then $a = \sigma^n b$ for some $b \in G$ and so

$$1 = \sigma a = \sigma^{n+1} b$$

and so $b \in \ker(\sigma^{n+1}) = \ker(\sigma^n)$, whence $a = \sigma^n b = 1$. Thus, $\ker(\sigma)$ is trivial.

Let us refer to the subgroups

$$\{\sigma G, \sigma^2 G, \sigma^3 G \ldots\}$$

as the **higher images** of G and the sequence

$$G \ge \sigma G \ge \sigma^2 G \ge \cdots$$

as the **image sequence** of σ. If G has the DCC on *all* subgroups, then any injective endomomorphism $\sigma\colon G \to G$ is also surjective, since the image sequence of σ must eventually be constant and so $\sigma^n G = \sigma^{n+1} G$ for some n. Hence, the injectivity of σ implies that $G = \sigma G$, whence σ is surjective. Of course, if G has the DCC on *normal* subgroups only, then we can draw the same conclusion provided that σ has normal higher images.

Theorem 4.40 *Let G be a group and let $\sigma \in \mathrm{End}(G)$.*

1) *If G has the ACC on normal subgroups, then*

$$\sigma \textit{ surjective} \quad \Rightarrow \quad \sigma \textit{ injective}$$

2) *If G has the DCC on all subgroups or if G has the DCC on normal subgroups and σ has normal higher images, then*

$$\sigma \textit{ injective} \quad \Rightarrow \quad \sigma \textit{ surjective}$$ □

Note that σ has normal higher images if σ preserves normality in general.

Definition *Let G be a group. A homomorphism $\sigma\colon G \to H$ is* **normality preserving** *if*

$$N \trianglelefteq G \quad \Rightarrow \quad \sigma N \trianglelefteq H$$ □

The composition of two normality-preserving homomorphisms is normality preserving. For endomorphisms, a stronger condition than that of preserving normality is the following.

Definition *An endomorphism $\sigma\colon G \to G$ of a group G is* **normal** *if σ commutes with all inner automorphisms γ_g of G, that is, for any $a \in G$,*

$$\sigma(a^g) = (\sigma a)^g$$

for all $g \in G$.□

Of course, the composition of normal maps is normal. Also, it is easy to see that a normal endomorphism is normality preserving. One of the advantages of normal maps over other normality-preserving maps is that if $\sigma: G \to G$ is normal and if $H \trianglelefteq G$ is σ-invariant, then $\sigma|_H: H \to H$ is also normal.

Fitting's Lemma

If G has *both* chain conditions on normal subgroups and if $\sigma \in \operatorname{End}(G)$ has normal higher images, then both the kernel and image sequences of σ must eventually be constant and so there is an $m > 0$ for which

$$K = \ker(\sigma^m) = \ker(\sigma^{m+i}) \quad \text{and} \quad H = \sigma^m G = \sigma^{m+i} G$$

for all $i \geq 0$. There is much that can be said about the subgroups H and K.

First, $\sigma H = H$ and $\sigma K \leq K$ and so H and K are σ-invariant. Also, if $a \in H \cap K$, then $a = \sigma^m b$ and $\sigma^m a = 1$. Hence, $\sigma^{2m} b = 1$ and so $b \in \ker(\sigma^{2m}) = \ker(\sigma^m)$, whence $a = 1$. Thus, $H \cap K = \{1\}$. To show that $G = H \bowtie K$, if $a \in G$, then there is a $b \in G$ for which $\sigma^m a = \sigma^{2m} b$ and so $a(\sigma^m(b^{-1})) \in \ker(\sigma^m)$, whence

$$a = [a\sigma^m(b^{-1})]\sigma^m(b) \in HK$$

Thus,

$$G = \operatorname{im}(\sigma^m) \bowtie \ker(\sigma^m)$$

Finally, since $\sigma^m(K) = \{1\}$, the map $\sigma|_K$ is nilpotent and since $\sigma(H) = H$, the map $\sigma|_H$ is surjective. We have just proved the important Fitting's lemma, which we can make a bit more general than the previous argument.

Theorem 4.41 (Fitting's lemma, 1934) *Let G be a group with the BCC on normal subgroups and let $\sigma \in \operatorname{End}(G)$ have normal higher images.*

1) There is an $m > 0$ for which

$$G = \operatorname{im}(\sigma^m) \bowtie \ker(\sigma^m)$$

where

a) $H = \operatorname{im}(\sigma^m)$ and $K = \ker(\sigma^m)$ are σ-invariant

b) $\sigma|_H$ is surjective and $\sigma|_K$ is nilpotent.

2) In particular, if G is indecomposable, then σ is either nilpotent or an automorphism of G. □

Automorphisms of Cyclic Groups

Let us examine the automorphism group of the cyclic groups. It is clear that an automorphism of a cyclic group is completely determined by its value on a generator and that this value also generates the group.

The only generators of an infinite cyclic group $C_\infty(a)$ are a and a^{-1} and so

$$\text{Aut}(C_\infty) = \{\iota, \tau\}$$

where $\tau x = x^{-1}$ for all $x \in C_\infty$. Now suppose that $C_n = \langle a \rangle$ is cyclic of order n. If $\tau \in \text{Aut}(C_n)$, then

$$\tau a = a^k$$

for some $1 \le k < n$. But $o(a) = o(a^k)$ and so we must have $k \in \mathbb{Z}_n^*$. Moreover, this condition uniquely determines an automorphism τ_k defined by

$$\tau_k(a^i) = a^{ki}$$

for $0 \le i < n$. Hence, there is precisely one automorphism τ_k for each $k \in \mathbb{Z}_n^*$.

Moreover, the map $\sigma\colon \mathbb{Z}_n^* \to \text{Aut}(C_n)$ defined by $\sigma k = \tau_k$ is an isomorphism, since it is clear that σ is bijective and if $j, k \in \mathbb{Z}_n^*$ and $jk = qn + r$ with $r \in \mathbb{Z}_n^*$, then

$$\tau_r(a^i) = a^{ir} = a^{ijk} = \tau_k(a^{ij}) = \tau_k\tau_j(a^i)$$

Thus, $\text{Aut}(C_n) \approx \mathbb{Z}_n^*$.

Theorem 4.42 *For the cyclic groups, we have*

$$\text{Aut}(C_\infty(a)) = \{\iota, \tau\colon a \mapsto a^{-1}\}$$

and

$$\text{Aut}(C_n(a)) = \{\tau_k \mid k \in \mathbb{Z}_n^*\} \approx \mathbb{Z}_n^*$$

where τ_k is defined by $\tau_k(a) = a^k$. In particular, the automorphism group of a cyclic group is abelian.□

A Closer Look at $\mathbb{Z}_n^*$

The previous theorem prompts us to take a closer look at the groups $\mathbb{Z}_n^*$.

Theorem 4.43 *If $n = p_1^{e_1} \cdots p_m^{e_m}$ where the p_i are distinct primes and $e_i \ge 1$, then*

$$\mathbb{Z}_n^* \approx \mathbb{Z}_{p_1^{e_1}}^* \boxtimes \cdots \boxtimes \mathbb{Z}_{p_k^{e_k}}^*$$

Moreover, $\mathbb{Z}_n^*$ is cyclic if and only if $n = 2, 4, p^e$ or $2p^e$ where p is an odd prime.

Proof. Let $r_i = p_i^{e_i}$. For the direct product decomposition, consider the map $\sigma\colon \mathbb{Z}_n^* \to \boxtimes\, \mathbb{Z}_{r_i}^*$ defined by

$$\sigma(u) = (u \bmod r_1, \ldots, u \bmod r_m)$$

This map is a group homomorphism and if $\sigma u = (1, \ldots, 1)$, then $u \equiv 1 \bmod r_i$, that is, $r_i \mid (u-1)$ for all i. Since the r_i's are pairwise relatively prime, it follows that $n \mid (u-1)$, that is, $u = 1$ in $\mathbb{Z}_n^*$. Thus, σ is a monomorphism and since the domain and the range of σ have the same size (see Theorem 2.30), σ is an isomorphism.

Now, Theorem 2.33 implies that $\mathbb{Z}_n^*$ is cyclic if and only if each factor $\mathbb{Z}_{p_i^{e_i}}^*$ is cyclic and the orders

$$o(\mathbb{Z}_{p_i^{e_i}}^*) = p_i^{e_i - 1}(p_i - 1)$$

are relatively prime. So let us take a look at $\mathbb{Z}_{p^e}^*$ for p prime.

Let $p > 2$. To see that $\mathbb{Z}_{p^e}^*$ is cyclic, first note that if $e = 1$, then $\mathbb{Z}_p^*$ is cyclic since $\mathbb{Z}_p$ is a field. Assume that $e > 1$. It is sufficient to find elements in $\mathbb{Z}_{p^e}^*$ of the relatively prime orders $p - 1$ and p^{e-1}.

To find an element of order $p-1$, let $a \in \mathbb{Z}_{p^e}^*$ have order $p^k m$, where $m \mid p - 1$. Then

$$a^{p^k m} = 1 \bmod p^{e-1}(p-1)$$

and $e > 1$ implies that

$$a^{p^k m} = 1 \bmod p$$

But if a is chosen so that $(a, p) = 1$, then Fermat's little theorem implies that $a^{p^k} \equiv a \bmod p$ and so $a^m \equiv 1 \bmod p$, whence $m = p - 1$. Thus, $o(a) = p^k(p-1)$ and so $o(a^{p^k}) = p - 1$, as desired.

Moreover, since $p > 2$, the expression $\beta = 1 + p$ is in p-standard form and so Theorem 1.18 implies that

$$(1+p)^{p^{e-1}} = 1 + wp^e$$

where $p \nmid w$. Hence, $1 + p$ has order p^{e-1} in $\mathbb{Z}_{p^e}^*$.

Now consider the case $p = 2$. It is easy to see that $\mathbb{Z}_2^*$ and $\mathbb{Z}_4^*$ are cyclic. For $e \geq 3$, the elements of $\mathbb{Z}_{2^e}^*$ are odd integers. For $1 + 2a \in \mathbb{Z}_{2^e}^*$, it is easy to see by induction that

$$(1+2a)^{2^k} = 1 + 2^{k+2} x_k$$

In particular,

$$(1+2a)^{2^{e-2}} = 1 + 2^e x_{e-2} \equiv 1 \bmod 2^e$$

which implies that $\mathbb{Z}_{2^e}^*$ has exponent 2^{e-2} and so cannot be cyclic.

In summary, we can say that $\mathbb{Z}_{p^e}^*$ is cyclic if and only if $p > 2$ or $p^e = 2$ or $p^e = 4$. We can now piece together our facts. As mentioned earlier, $\mathbb{Z}_n^*$ is cyclic if and only if each factor $\mathbb{Z}_{p_i^{e_i}}^*$ is cyclic and the orders

$$o(\mathbb{Z}_{p_i^{e_i}}^*) = p_i^{e_i-1}(p_i - 1)$$

are relatively prime. Since $p_i^{e_i-1}(p_i - 1)$ is even unless $p_i = 2$ and $e_i = 1$, there can be at most one factor involving an odd prime or $\mathbb{Z}_4^*$. Thus, $\mathbb{Z}_n^*$ is cyclic if and only if $n = 2, 4, p^e$ or $2p^e$ where $p > 2$ is prime.□

Exercises

1. Let $\sigma: G \to H$ be a group homomorphism.
   a) Prove that if $S \le G$, then $\sigma S \le H$.
   b) Prove that if σ is surjective and $S \trianglelefteq G$, then $\sigma S \trianglelefteq H$.
   c) Prove that if $T \le H$, then $\sigma^{-1}T \le G$.
   d) Prove that if $T \trianglelefteq H$, then $\sigma^{-1}T \trianglelefteq G$.
2. Let G and K be groups and let $H \le G$. Show that it is not always possible to extend a homomorphism $\sigma: H \to K$ to G.
3. Show that the p-quasicyclic group $\mathbb{Z}(p^\infty)$ is isomorphic to the subgroup of all complex p^nth roots of unity.
4. Show that if G has a *unique* maximal subgroup M, then G/M is cyclic of prime order.
5. Let G be a finite group with normal subgroups H and K. If $G/H \approx G/K$, does it follow that $H \approx K$?
6. a) Find a property of groups that is inherited by quotient groups but not by subgroups.
   b) Find a property of groups that is inherited by subgroups but not by quotient groups.
7. Show that a group G is abelian if and only if the map $a \mapsto a^{-1}$ is an automorphism of G.
8. Let G be a group.
   a) Determine all homomorphisms $\sigma: C_m(a) \to G$.
   b) Determine all homomorphisms $\sigma: C_\infty(a) \to G$.
9. Are all normality-preserving homomorphisms normal? *Hint*: Use the fact that S_3 is simple.
10. Let G be a group and let N be a normal cyclic subgroup of G. Prove that $G' \le C_G(N)$.
11. Let G be a group. Prove that G' is central if and only if $\operatorname{Inn}(G)$ is abelian.
12. An endomorphism $\sigma \in \operatorname{End}(G)$ is **central** if

$$a^{-1}\sigma a \in Z(G)$$

for all $a \in G$. This is equivalent to $\sigma a Z(G) = aZ(G)$, that is, σ acts like the identity on $G/Z(G)$. Prove that
   a) A normal surjective endomorphism is central.
   b) A central endomorphism is normal.

13. An abelian group A (written additively) is **divisible** if for any $a \in A$ and any positive integer n, there is a $b \in A$ for which $nb = a$. Prove that a characteristically simple abelian group A is divisible.
14. Let G be a group and let $a \in G$. Is the map $\sigma: G \to G$ defined by $\sigma b = [b, a]$ an endomorphism of G? If not, under what conditions on G is σ an endomorphism?
15. Let G be a group of order p^2 where p is a prime.
    a) Prove that if G is abelian, then G is either cyclic or isomorphic to the direct product of two cyclic groups of order p.
    b) Show that the distinct sets $\{a^G\}$ of conjugates of elements in G form a partition of G. Show that $Z(G) \neq \{1\}$. Show that G must be abelian.
16. Prove that $S_3 \approx D_6$, where S_3 is the symmetric group of order 6 and D_6 is the dihedral group of order 6.
17. Prove that if G has a periodic subgroup H of finite index, then G is periodic. (In loose terms, if "most" of the elements of G have finite order, then all elements of G have finite order.)
18. Let G be a finite group. Let $K \trianglelefteq G$. A subgroup H of G is called a **supplement** of K if $G = HK$. Let H be a minimal supplement of K. Prove that $H \cap K \leq \Phi(H)$.
19. Let $\mathcal{F}$ be a family of groups with the following properties:
    1) If $G \in \mathcal{F}$ and $H \approx G$, then $H \in \mathcal{F}$.
    2) If $G \in \mathcal{F}$ and $N \trianglelefteq G$, then $G/N \in \mathcal{F}$.
    3) If $N \trianglelefteq G$ and if $N \in \mathcal{F}$ and $G/N \in \mathcal{F}$, then $G \in \mathcal{F}$.

    Prove that if $N \in \mathcal{F}$ and $K \in \mathcal{F}$ are subgroups of G with $K \trianglelefteq G$, then $NK \in \mathcal{F}$.
20. If G is a finite group and σ is an automorphism of G that fixes only the identity element, that is, $g \neq 1$ implies $\sigma g \neq g$, show that
$$G = \{g^{-1}\sigma g \mid g \in G\}$$
21. Prove that if $H \trianglelefteq\trianglelefteq G$ and $K \trianglelefteq\trianglelefteq G$, then it does not necessarily follow that the set product HK is a subgroup of G. *Hint*: Look at the dihedral group D_8.
22. Let G be a finite group. Suppose that $A \trianglelefteq G$ is minimal among all normal subgroups of G and that A is abelian. Prove that A is an elementary abelian group.
23. Show that a subgroup H of a group G is characteristic if and only if $\sigma H = H$ for all automorphisms $\sigma \in \operatorname{Aut}(G)$.
24. Find an example of a normal subgroup H of a group G for which H is not characteristic in G.
25. Let G be a group. Prove the following:
    a) The property of being characteristic is transitive: If $A \sqsubseteq B$ and $B \sqsubseteq C$, then $A \sqsubseteq C$.
    b) If $A \sqsubseteq B$ and $B \trianglelefteq C$, then $A \trianglelefteq C$.
26. Show that $Z(G) \sqsubseteq G$.

27. Let G be a finite group and let $H \trianglelefteq G$. Show that if $(|H|,(G:H))=1$, then $H \sqsubseteq G$.
28. Let G be a group and let $H \le G$. Prove that $H \sqsubseteq G$ implies $[H,G] \sqsubseteq G$.
29. Let G be a finitely-generated group and let $H \le G$ with $(G:H) < \infty$. Show that there is a subgroup $K \le H$ that is characteristic in G and has finite index in G.
30. Let G be a group and let $N \lhd G$. Let $H < G$. Show that $N \le \Phi(H)$ implies $N \le \Phi(G)$.
31. Let G be a group. Let G' be the commutator subgroup of G. Let G'' be the commutator subgroup of G'. In general, we can continue to take commutator subgroups and $G^{(n)} = (G^{(n-1)})'$ is called the ***n*th commutator subgroup** of G. Prove that $G^{(n)} \sqsubseteq G$.
32. Let G be a group. Prove that if $\text{Inn}(G)$ is cyclic, then G is abelian.
33. a) Show that the commutator subgroup of a group is fully invariant.
    b) Show that the center of a group need not be fully invariant. *Hint*: consider $C_2(a) \boxtimes S_3$.
34. Let G be a finite abelian group of order $n = p_1^{e_1} \cdots p_m^{e_m}$ where the p_i's are distinct primes. For each prime p_i, let

$$G_{(p_i)} = \{a \in G \mid a \text{ is a } p_i\text{-element}\}$$

    a) Show that $G_{(p_i)} \sqsubseteq G$.
    b) Show that

$$G = \bigvee G_{(p_i)} \quad \text{and} \quad G_{(p_i)} \cap \bigvee_{j \neq i} G_{(p_i)} = \{1\}$$

35. (**Chinese remainder theorem**) Let $m_1, \ldots, m_k$ be pairwise relatively prime integers greater than 1. Let $f_1, \ldots, f_n$ be integers. Prove that the system of congruences

$$\begin{aligned} x &\equiv f_1 \bmod m_1 \\ &\vdots \\ x &\equiv f_n \bmod m_k \end{aligned}$$

has a unique solution modulo the product $m_1 \cdots m_k$. : *Hint*: Use $\mathbb{Z}_{m_i}$.
36. Let G be a group of order p^n where p is a prime. Show that a subgroup H of index p must be normal in G. *Hint*: Consider the map $\lambda: G \to S_{G/H}$, where $S_{G/H}$ is the symmetric group on G/H, defined by $\lambda(g)(aH) = gaH$.
37. Find the subgroup lattice of Q. Which subgroups are normal? Is Q abelian? What is the center of Q? What is Q'?
38. Prove that the quasicyclic groups $\mathbb{Z}(p^\infty)$ are the only infinite groups with the property that their proper subgroups consist entirely of a single ascending chain

$$\{1\} < S_1 < S_2 < \cdots$$

39. Let G be a group and let $n > 1$ be an integer.
    a) When is the nth power map $f_n: G \to G$ defined by $f_n(a) = a^n$ a homomorphism? Must G be abelian?
    b) When is the nth power map a homomorphism for all n?
    c) Let $G^n = \{a^n \mid a \in G\}$ and let $G^{(n)} = \{a \in G \mid a^n = 1\}$. Show that both of these sets are normal subgroups of G. What is the relationship between these subgroups?
40. Find the automorphism group $\operatorname{Aut}(V)$ of the 4-group V.
41. Let $\sigma: G \twoheadrightarrow K$ be an epimorphism. Show that if $H \trianglelefteq \trianglelefteq G$, then $\sigma H \trianglelefteq \trianglelefteq K$ and $s(\sigma H, K) \leq s(H, G)$.
42. Prove that the subgroup $\langle \sigma \rangle$ of the dihedral group

$$D_{2^{n+1}} = \langle \sigma, \rho \mid \sigma^2 = 1 = \rho^{2^n} \rangle$$

has subnormal index n.
43. If $\mathcal{F} = \{H_i \mid i \in I\}$ is a family of subnormal subgroups of a group G and if there is an integer s for which $s(H_i, G) \leq s$ for all $i \in I$, show that $\bigcap \mathcal{F}$ is subnormal in G.
44. Prove that any finitely-generated metabelian group has the generalized SJP.
45. A group G is **Hopfian** if G is not isomorphic to any proper quotient group of itself. A group G is **co-Hopfian** if G is not isomorphic to any proper subgroup of itself.
    a) Prove that G is Hopfian if and only if every endomorphism of G is an automorphism of G.
    b) Prove that G is co-Hopfian if and only if every monomorphism of G is an automorphism.
    c) Show that $\mathbb{Q}$ is both Hopfian and co-Hopfian. *Hint*: To show that $\mathbb{Q}$ is Hopfian, show that $\mathbb{Q}/H$ is not torsion free unless $H = \{1\}$. To show that $\mathbb{Q}$ is co-Hopfian, show that any proper subgroup H of $\mathbb{Q}$ is not divisible. An abelian group G is **divisible** if $a \in G$ and $n \in \mathbb{Z}$ imply that there is a $b \in G$ for which $a = nb$.
    d) Show that $\mathbb{Z}$ is Hopfian but not co-Hopfian.
    e) Show that the quasicyclic group $\mathbb{Z}(p^\infty)$ is co-Hopfian but not Hopfian.
    f) Show that the additive group of all polynomials in infinitely many variables is neither Hopfian nor co-Hopfian.
46. Find two nonisomorphic groups with isomorphic automorphism groups.
47. Prove that the multiplicative group $\mathbb{Q}^+$ of positive rational numbers is isomorphic to the additive group $\mathbb{Z}[x]$ of polynomials over the integers. *Hint*: Use the fundamental theorem of arithmetic.
48. Let G be a group. Show that if G is centerless, then

$$C_{\operatorname{Aut}(G)}(\operatorname{Inn}(G)) = \{\iota\}$$

49. Show that $\operatorname{Aut}(S_3) = \operatorname{Inn}(S_3) \approx S_3$.
50. Prove that if G is a group in which every nonidentity element is an involution, then G has a nontrivial automorphism.

51. Consider a series for G

$$\{1\} = G_0 \trianglelefteq G_1 \trianglelefteq G_2 \trianglelefteq \cdots \trianglelefteq G_n = G$$

for which each factor group G_{i+1}/G_i is abelian. Show that if H is a subgroup of G, then there is a series of subgroups

$$\{1\} = H_0 \trianglelefteq H_1 \trianglelefteq H_2 \trianglelefteq \cdots \trianglelefteq H_m = H$$

of H whose factor groups are also abelian.

52. (**Robinson**) Let $N \trianglelefteq G$ and suppose that N has the GSJP and G/N has the ACC on subnormal subgroups. Show that G also has the GSJP. *Hint*: Apply Theorem 4.35. Let $\mathcal{C} = \{H_i \mid i \in I\}$ be an arbitrary chain in $\operatorname{subn}(G)$ and let $U = \bigcup \mathcal{C}$. To show that $U \trianglelefteq\trianglelefteq G$, consider the chains

$$\mathcal{C}N = \{H_i N \mid i \in I\}$$

and

$$\mathcal{C}N/N = \{H_i N/N \mid i \in I\}$$

in G and G/N, respectively. Let $H_\beta N/N$ be a maximal member of $\mathcal{C}N/N$. Show that $U = H_\beta(U \cap N)$.

Chapter 5
Direct and Semidirect Products

In this chapter, we will explore the issue of the decomposition of groups into "products" of subgroups. To say simply that a group G can be decomposed into a set product of two proper subgroups

$$G = HK$$

leaves something to be desired. The first problem is that the representation of an element of G as a product hk for $h \in H$ and $k \in K$ need not be unique. The second problem is that, in general, we have no information about how the elements of H and K interact, for example, what is kh as a member of HK? The first problem is easily addressed. The second is not as simple.

Complements and Essentially Disjoint Products

Uniqueness in a set product decomposition is easily characterized.

Theorem 5.1 *Let $H, K \leq G$.*
1) *Every element $a \in HK$ has a* unique *representation as a product $a = hk$ for $h \in H$ and $k \in K$ if and only if H and K are essentially disjoint.*
2) *Every element $a \in G$ can be* uniquely *represented as an element of HK if and only if $G = H \bullet K$, that is, if and only if*

$$G = HK \quad \textit{and} \quad H \cap K = \{1\}$$

In this case, K is said to be a **complement** *of H in G. A subgroup H of G is* **complemented** *if it has a complement in G.*□

Note that since $G = HK$ implies $G = KH$, the concept of complement is symmetric in H and K. Also, if $G = H \bullet K$, then

$$|G| = |H \bullet K| = |H||K|$$

as cardinal numbers.

We pause for a small clarification of terminology. In the more specialized literature of group theory, devoted to the study of the subgroup lattice $\text{sub}(G)$ of a group G, such as appears in the book *Subgroup Lattices of Groups*, by Schmidt [30], the term *complement* of $H \in \text{sub}(G)$ refers to a subgroup K for which

$$G = H \vee K \quad \text{and} \quad H \cap K = \{1\}$$

This terminology is consistent with the terminology of lattice theory. Indeed, Schmidt uses the term **permutable complement** for the concept given in our definition. However, our definition follows the trend in general treatments of group theory (including most textbooks).

Even for normal subgroups, complements need not exist. For example, no nontrivial proper subgroup of $\mathbb{Z}$ has a complement. Moreover, when complements do exist, they need not be unique; for example, in S_3 every subgroup of order 2 is a complement of the alternating subgroup A_3 and so, for example,

$$S_3 = A_3 \bullet \langle(1\,2)\rangle = A_3 \bullet \langle(1\,3)\rangle$$

are two essentially disjoint product representations of S_3.

On the other hand, any complement K of a *normal* subgroup N is isomorphic to the quotient G/N, for we have

$$\frac{G}{N} = \frac{N \bullet K}{N} \approx \frac{K}{N \cap K} = \frac{K}{\{1\}} \approx K$$

Theorem 5.2 *Let G be a group. If a normal subgroup N of G is complemented, then all complements of N are isomorphic to G/N and hence to each other.*□

Complements and Transversals

Another way to characterize complements is through the notion of a transversal. Let us recall the following definition.

Definition *Let $H \le G$.*

1) *A set consisting of exactly one element from each left coset in G/H is called a* **left transversal** *for H in G (or for G/H).*
2) *A set consisting of exactly one element from each right coset in $H\backslash G$ is called a* **right transversal** *for H in G (or for $H\backslash G$).*□

It is of interest to know which *subgroups* of G are left transversals for H. The simple answer is that these are precisely the complements of H.

Theorem 5.3 *Let $H \le G$. The following are equivalent for a subgroup $K \le G$.*

1) *K is a complement of H in G.*

2) *K is a left transversal for H in G.*
3) *K is a right transversal for H in G.*

Proof. Suppose that K is a complement of H in G. If $k_1H = k_2H$, then $k_2^{-1}k_1 \in K \cap H = \{1\}$ and so $k_1 = k_2$. Also, since $G = KH$, every $a \in G$ has the form $a = kh \in kH$ for some $k \in K$. Thus, the cosets kH for $k \in K$ form a left transversal for H in G. Conversely, suppose that $K \leq G$ is a left transversal for G/H. Since H contains a single member of K, it follows that $H \cap K = \{1\}$. Also, since the cosets G/H partition G, any $a \in G$ has the form $a = kh$, for some $k \in K$ and $h \in H$ and so $G = KH$. Thus, K is a complement of H in G. A similar argument can be made for the right cosets.□

Product Decompositions

If $G = H \bullet K$, then every element of G has a *unique* representation as a product hk with $h \in H$ and $k \in K$. The problem, however, is that we have no information about how the elements of H and K interact. Without a **commutativity rule** that expresses a product kh in the form $h'k'$, where $h, h' \in H$ and $k, k' \in K$, we have no way to simply a product of the form

$$(h_1k_1)(h_2k_2)$$

The simplest commutativity rule, that is, elementwise commutativity $hk = kh$ holds if and only if both factors H and K are normal in G and this essentially reflects the fact that there is *no interaction* between H and K. Weaker forms of decomposition come by weakening the requirement that both factors be normal. Here are the relevant definitions in one place for comparison purposes.

Definition *Let G be a group and let $H, K \leq G$.*

1) *If*

$$G = H \bullet K$$

then G is called the **essentially disjoint product** *of H and K. In this case, H is called a* **complement** *of K in G.*

2) *If*

$$G = H \bullet K, \quad H \trianglelefteq G$$

then G is called the **semidirect product** *of H with K. In this case, H is called a* **normal complement** *of K in G. The semidirect product is denoted by*

$$G = H \rtimes K$$

3) *If*

$$G = H \bullet K, \quad H \trianglelefteq G, K \trianglelefteq G$$

then G is called the **direct product** *of H and K. In this case, H is called a* **direct complement** *of K in G (and vice versa). The direct product is* *denoted by*

$$G = H \bowtie K$$

Any of these products is **nontrivial** *if both factors are proper.*□

The direct product decomposition of a group is a very strong form of decomposition and so is rather specialized. However, the semidirect product is one of the most useful constructions in group theory. For example, for $n \not\equiv 2 \bmod 4$, the dihedral groups D_{2n} have no direct product decompositions, but they do have a semidirect product decomposition

$$D_{2n} = \langle \rho \rangle \rtimes \langle \sigma \rangle$$

Nevertheless, all three types of product decompositions are special. To illustrate the point, an infinite cyclic group has no nontrivial essentially disjoint product decompositions at all.

Direct Sums and Direct Products

External Direct Sums and Products

We have already defined the external direct product of a finite number of groups. The generalization of this product to arbitrary families of groups leads to two important variations. Intuitively speaking, if κ is an arbitrary (finite or infinite) cardinal number, we can consider the set of all ordered "κ-tuples" as well as the set of all ordered κ-tuples that have only a finite number of nonzero coordinates. The notion of an ordered κ-tuple is generally described by a function.

Definition Let $\mathcal{F} = \{G_i \mid i \in I\}$ be a family of groups.

1) *The* **external direct product** *of the family* $\mathcal{F}$ *is the group*

$$\boxtimes\mathcal{F} = \boxtimes G_i = \left\{ f: I \to \bigcup G_i \,\middle|\, f(i) \in G_i \right\}$$

of all functions f *from the* **index set** I *to the union of* $\mathcal{F}$ *for which the* i*th* **coordinate** $f(i)$ *of* f *belongs to* G_i *for all* i*. The group operation is componentwise product:*

$$(fg)(i) = f(i)g(i)$$

The **support** *of* $f \in \boxtimes G_i$ *is the set*

$$\operatorname{supp}(f) = \{i \in I \mid f(i) \neq 1\}$$

2) *The* **external direct sum** *of the family* $\mathcal{F}$ *is the group*

$$\boxplus\mathcal{F} = \boxplus G_i = \{f \in \boxtimes\mathcal{F} \mid f \text{ has finite support}\}$$

also under componentwise product.□

Of course, when $\mathcal{F}$ is a finite family, the direct product and direct sum coincide. Note that authors vary on their use of the terms *direct sum* and *direct product*. For example, some authors reserve the term *sum* for abelian groups and some authors use the term *cartesian product* for direct product.

Internal Direct Products

For convenience, we repeat the definition of the internal direct product.

Definition *A group G is the* **(internal) direct sum** *or* **(internal) direct product** *of a family $\mathcal{F} = \{H_i \mid i \in I\}$ of normal subgroups if $\mathcal{F}$ is strongly disjoint and $G = \bigvee\mathcal{F}$. We denote the internal direct product of $\mathcal{F}$ by*

$$\bowtie H_i \quad \textit{or} \quad \bowtie \mathcal{F}$$

or when $\mathcal{F} = \{H_1, \ldots, H_n\}$ is a finite family,

$$H_1 \bowtie \cdots \bowtie H_n$$

Each factor H_i is called a **direct summand** *or* **direct factor** *of G. We denote the family of all direct summands of G by $\mathcal{DS}(G)$.*□

We should mention that the notation for direct sums and products varies considerably among authors. For example, some authors use the notation $G = H \times K$ for both the internal and external direct sum (as well as the cartesian product), justifying this on the grounds that the two types of direct sums are isomorphic. While this may be reasonable, in an effort to avoid any ambiguity, we have adopted the following notations:

1) Set product

$$HK$$

2) Essentially disjoint set product

$$H \bullet K$$

3) Cartesian product of sets

$$H \times K$$

4) External direct product

$$H \boxtimes K \quad \text{and} \quad \boxtimes H_i$$

5) External direct sum

$$H \boxplus K \quad \text{and} \quad \boxplus H_i$$

6) Internal direct product (sum)

$$H \bowtie K \quad \text{and} \quad \bowtie H_i$$

7) Semidirect product

$$H \rtimes K$$

Note that our notation for the internal direct product emphasizes the fact that the two subgroups are normal (a juxtaposition of the symbols $\triangleright$ and $\triangleleft$ used to indicate normality) and is similar to the established notation $\rtimes$ for the semidirect product.

To see that internal and external direct sums are isomorphic, suppose that $G = \boxplus G_i$. For each k, let

$$H_k = \{f \in \boxplus G_i \mid f(i) = 1 \text{ for } i \neq k\}$$

We leave it as an exercise to show that $H_k \trianglelefteq G$ and that $G = \bowtie H_k$. On the other hand, if $G = \bowtie G_i$, then the map $\sigma: \boxplus G_i \to \bowtie G_i$ defined by

$$\sigma f = \prod\{f(i) \mid i \in I, f(i) \neq 1\}$$

is an isomorphism. For this reason, many authors drop the adjectives "internal" and "external" when discussing direct sums.

The Universal Property of Direct Products and Direct Sums

Direct products and direct sums can each be characterized by a universal property.

Projection and Injection Maps

If

$$G = \boxtimes\{H_i \mid i \in I\} \quad \text{or} \quad G = \boxplus\{H_i \mid i \in I\}$$

is an external direct product or external direct sum, we define the ith **projection map** $\rho_i: G \to H_i$ by

$$\rho_i(f) = f(i)$$

Note that ρ_i is an epimorphism and that an element $f \in G$ is uniquely determined by the values $\rho_i(f)$, which can be specified arbitrarily, as long as $\rho_i(f) \in H_i$ for all i.

The ith **injection map** $\kappa_i: H_i \to G$ is defined by

$$\kappa_i(h)(j) = \begin{cases} h & \text{if } j = i \\ 1 & \text{otherwise} \end{cases}$$

for all $h \in H_i$. These maps are injective. Note also that

$$\rho_j \circ \kappa_i = \begin{cases} \iota & \text{if } i = j \\ 0 & \text{otherwise} \end{cases}$$

For an internal direct sum

$$G = \bowtie \{H_i \mid i \in I\}$$

we have already defined the projection $\rho_i: G \to H_i$ by setting $\rho_i(a)$ to be the ith component of a. For the internal direct sum, the injection maps $\kappa_i: H_i \to G$ are just the inclusion maps, defined by $\kappa_i(h) = h$ for all $h \in H_i$.

Universality of Direct Products

Let $\{H_i \mid i \in I\}$ be a family of groups and let

$$P = \boxtimes_{i \in I} \{H_i\}$$

with projection maps ρ_i. Then the ordered pair

$$\mathcal{P} = (P, \{\rho_i\}_{i \in I})$$

has a universal property that characterizes the direct product up to isomorphism.

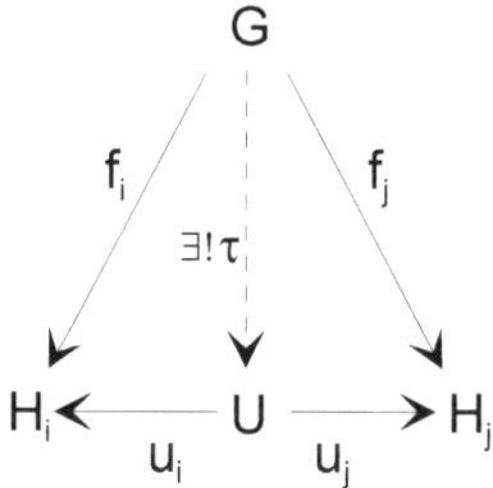

Figure 5.1

Definition *Referring to Figure 5.1, let $\mathcal{F}$ be the family of all ordered pairs*

$$\mathcal{G} = (G, \{f_i: G \to H_i\}_{i \in I})$$

where G is a group and $\{f_i: G \to H_i\}_{i \in I}$ is a family of homomorphisms. We refer to G as the **vertex** *of the pair $\mathcal{G}$. A pair*

$$\mathcal{U} = (U, \{u_i: U \to H_i\}_{i \in I})$$

in $\mathcal{F}$ is **universal** *for $\mathcal{F}$ if for any pair $\mathcal{G} \in \mathcal{F}$, there is a unique homomorphism $\tau: G \to U$ between the vertices, called the* **mediating morphism** *for $\mathcal{G}$, for which*

$$u_i \circ \tau = f_i$$

for all $i \in I$.□

Theorem 5.4 *Let $\{H_i \mid i \in I\}$ be a family of groups.*

1) The pair

$$\mathcal{P} = (P, \{\rho_i: P \to H_i\}_{i \in I})$$

where P is the direct product of the family $\{H_i\}$ and ρ_i is the ith projection map, is universal for $\mathcal{F}$.

2) *If the pair $(G,\{f_i\}_{i\in I})$ is also universal for $\mathcal{F}$, then the mediating morphism $\tau: G \to P$ is an isomorphism and so $G \approx P$.*

Proof. For part 1), for any pair $\mathcal{G} \in \mathcal{F}$, there must exist a unique mediating morphism $\tau: G \to P$ satisfying

$$\rho_i \circ \tau = f_i$$

But this is equivalent to

$$\rho_i(\tau a) = f_i(a)$$

for all $a \in G$ and this does uniquely define a homomorphism τ. The proof of part 2) is also to the proof of Theorem 4.5 and we leave the details to the reader.□

Universality of Direct Sums

Let $\{H_i \mid i \in I\}$ be a family of groups and let

$$S = \boxplus\{H_i \mid i \in I\}$$

with injections κ_i. The ordered pair

$$\mathcal{S} = (S, \{\kappa_i\}_{i\in I})$$

also has a universal property that characterizes the direct sum up to isomorphism.

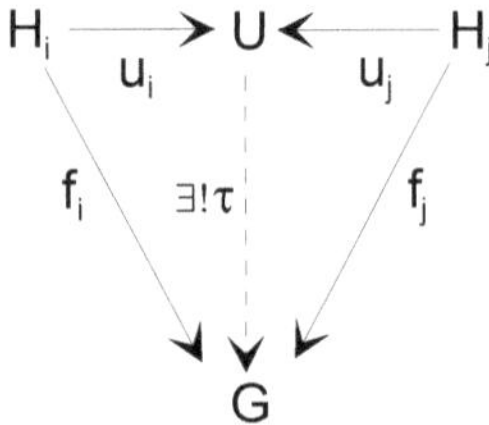

Figure 5.2

Definition *Referring to Figure 5.2, let $\mathcal{F}$ be the family of all ordered pairs*

$$\mathcal{G} = (G, \{f_i: H_i \to G\}_{i\in I})$$

where G is a group and $\{f_i: H_i \to G\}_{i\in I}$ is a family of homomorphisms. We refer to G as the **vertex** *of the pair. A pair*

$$\mathcal{U} = (U, \{u_i: H_i \to U\}_{i\in I}) \in \mathcal{F}$$

is **universal** *for $\mathcal{F}$ if for any pair $\mathcal{G} \in \mathcal{F}$, there is a unique homomorphism $\tau: U \to G$ between the vertices, called the* **mediating morphism** *for $\mathcal{G}$, for which*

$$\tau \circ u_i = f_i$$

for all $i \in I$.□

Theorem 5.5 *Let* $\{H_i \mid i \in I\}$ *be a family of groups.*

1) *The pair*

$$\mathcal{S} = (S, \{\kappa_i : H_i \to S\}_{i \in I})$$

where S *is the direct sum of the family* $\{H_i\}$ *and* κ_i *is the* i*th injection map, is universal for* $\mathcal{F}$. *That is, for any pair*

$$\mathcal{G} = (G, \{f_i : H_i \to G\}_{i \in I})$$

where G *is a group and* $\{f_i : H_i \to G\}_{i \in I}$ *is a family of homomorphisms, there is a unique homomorphism* $\tau : S \to G$ *for which*

$$\tau \circ \kappa_i = f_i$$

for all $i \in I$, *or equivalently,*

$$\tau|_{H_i} = f_i$$

for all $i \in I$. *In other words, a homomorphism* $\tau : S \to K$ *is uniquely determined by its restrictions* $\tau|_{H_i}$ *to the factors* H_i, *which may be any homomorphisms from* H_i *to* G.

2) *If the pair* $(G, \{f_i\}_{i \in I})$ *is also universal for* $\mathcal{F}$, *then the mediating morphism* $\tau : S \to G$ *is an isomorphism and so* $G \approx S$.

Proof. For part 1), for any pair $(K, \{k_i : H_i \to K\}_{i \in I})$ in $\mathcal{F}$, there must be a unique mediating morphism $\tau : S \to G$ satisfying

$$\tau \circ \kappa_i = f_i$$

But this specifies how τ is defined on any element of S that has support of size 1 and therefore on any element of S that has finite support, that is, on all of S. We leave the details of this and the proof of part 2) to the reader.□

Cancellation in Direct Sums

A group G is **cancellable in direct sums** (or **cancellable** for short) if

$$A \bowtie G \approx B \bowtie H, \quad G \approx H \quad \Rightarrow \quad A \approx B$$

We follow a line similar to Hirshon [18] to prove that any finite group is cancellable in direct sums. On the other hand, Hirshon shows that even infinite *cyclic* groups are not cancellable in direct sums.

Theorem 5.6 *Any finite group* G *is cancellable in direct sums.*

Proof. It is sufficient to show that

$$A \bowtie G = B \bowtie H, \quad G \approx H \quad \Rightarrow \quad A \approx B$$

and we prove this by induction on $o(G)$. If $o(G)=1$, the result is clear. Assume that any group of order less than $o(G)$ is cancellable and let

$$W = A \bowtie G = B \bowtie H, \quad G \approx H$$

First, we observe that if $B \cap G = \{1\}$, then $W = B \bowtie G$, for we have

$$\frac{B \bowtie G}{B} \approx G \approx H \approx \frac{W}{B}$$

and since these groups are finite, they have the same size and so they are equal, whence $B \bowtie G = W$. It follows that A and B are both direct complements of G in W and so are isomorphic. A similar argument holds if $A \cap H = \{1\}$.

Thus, we may assume that $B \cap G$ and $A \cap H$ are nontrivial. Then

$$\frac{A \bowtie G}{(A \cap H) \bowtie (B \cap G)} = \frac{H \bowtie B}{(A \cap H) \bowtie (B \cap G)}$$

and Theorem 5.19 implies that

$$\frac{A}{A \cap H} \boxtimes \frac{G}{B \cap G} \approx \frac{H}{A \cap H} \boxtimes \frac{B}{B \cap G}$$

Since $H \approx G$, we have

$$\frac{H}{\{1\}} \boxtimes \frac{A}{A \cap H} \boxtimes \frac{G}{B \cap G} \approx \frac{G}{\{1\}} \boxtimes \frac{H}{A \cap H} \boxtimes \frac{B}{B \cap G}$$

and Theorem 4.8 implies that

$$\frac{H \boxtimes A \boxtimes G}{\{1\} \boxtimes (A \cap H) \boxtimes (B \cap G)} \approx \frac{H \boxtimes G \boxtimes B}{\{1\} \boxtimes (A \cap H) \boxtimes (B \cap G)}$$

Splitting this in a different way gives

$$\frac{A}{\{1\}} \boxtimes \frac{H}{A \cap H} \boxtimes \frac{G}{B \cap G} \approx \frac{B}{\{1\}} \boxtimes \frac{H}{A \cap H} \boxtimes \frac{G}{B \cap G}$$

Since $o(G/(B \cap G)) < o(G)$, the induction hypothesis implies that $G/(B \cap G)$ is cancellable. Similarly, $H/(A \cap H)$ is cancellable and so $A \approx B$.□

The Classification of Finite Abelian Groups

We are now in a position to solve the classification problem for finite abelian groups, that is, we can identify a system of distinct representatives for the isomorphism classes of finite abelian groups.

Theorem 5.7 *Let A be a finite abelian group. If $u \in A$ has maximum order among all elements of A, then $M = \langle u \rangle$ has a direct complement, that is, there* *is a subgroup V for which*

$$A = M \bowtie V$$

Proof. The proof is by induction on the order of A. If $o(A) = 1$, the result is clear. Assume the result holds for all abelian groups of order less than that of A. We may also assume that $M < A$. If we find a subgroup $X \leq A$ for which $X \cap M = \{0\}$, since then $o(u + X) = o(u)$, the inductive hypothesis applied to the quotient G/X gives

$$\frac{A}{X} = \frac{M+X}{X} \bowtie \frac{V}{X} = \frac{M+V}{X}$$

for some $V \leq A$. Hence, $A = M + V$ and since $(M + X) \cap V = X$, it follows that $M \cap V \subseteq M \cap X = \{0\}$, whence $A = M \bowtie V$.

To this end, let X be a minimal subgroup of A that is not contained in M. If $x \in X$, then $\langle x \rangle$ is not contained in M and so $X = \langle x \rangle$. If $o(x) = ab$ with a and b relatively prime and greater than 1, then $\langle ax \rangle < \langle x \rangle$ and so $\langle ax \rangle \leq M$. Similarly, $\langle bx \rangle \leq M$ and so $x \in \langle ax \rangle + \langle bx \rangle \leq M$, a contradiction. Hence, $o(x) = p$ is prime and so $X \cap M = \{0\}$.□

Thus, if A is a finite abelian group and if u_1 has maximum order in A, then

$$A = \langle u_1 \rangle \bowtie V_1$$

for some $V_1 \leq A$. If $V_1 \neq \{0\}$ and u_2 has maximum order in V_1, then

$$A = \langle u_1 \rangle \bowtie \langle u_2 \rangle \bowtie V_2$$

where $o(u_2) \mid o(u_1)$. Since A is finite, this process must eventually result in a cyclic decomposition

$$A = \langle u_1 \rangle \bowtie \cdots \bowtie \langle u_n \rangle$$

of A, where if $\alpha_k = o(u_k)$, then

$$\alpha_n \mid \alpha_{n-1} \mid \cdots \mid \alpha_1$$

Moreover, if α_1 has the prime factorization

$$\alpha_1 = p_1^{e_{1,1}} \cdots p_m^{e_{1,m}}$$

then $\alpha_k \mid \alpha_1$ implies that

$$\alpha_k = p_1^{e_{k,1}} \cdots p_m^{e_{k,m}}$$

where $e_{k,i} \leq e_{k+1,i}$ for all $i = 1, \ldots, m$. Then Theorem 2.33 implies that

$$\langle u_k \rangle = \langle v_{k,1} \rangle \bowtie \cdots \bowtie \langle v_{k,m_k} \rangle$$

where $o(v_{k,i}) = p_i^{e_{k,i}}$ for all i. A subgroup of order a power of a prime p is called a ***p*-primary** (or just **primary**) subgroup. It is customary when writing A as a direct sum of primary cyclic subgroups to collect the terms associated with each prime.

Theorem 5.8 (The cyclic decomposition of a finite abelian group) *Let A be a finite abelian group.*

1) **(Invariant factor decomposition)** *A is the direct sum of a finite number of nontrivial cyclic subgroups*

$$A = \langle u_1 \rangle \bowtie \cdots \bowtie \langle u_n \rangle$$

where if $\alpha_k = o(u_k)$, then

$$\alpha_n \mid \alpha_{n-1} \mid \cdots \mid \alpha_1$$

The orders α_i are called the **invariant factors** *of A.*

2) **(Primary cyclic decomposition)** *If*

$$\alpha_k = p_1^{e_{k,1}} \cdots p_m^{e_{k,m}}$$

where $e_{k,i} \le e_{k+1,i}$ for all $i = 1, \ldots, m$, then A is the direct sum of primary cyclic subgroups:

$$A = [\langle u_{1,1} \rangle \bowtie \cdots \bowtie \langle u_{1,k_1} \rangle] \bowtie \cdots \bowtie [\langle u_{m,1} \rangle \bowtie \cdots \bowtie \langle u_{m,k_m} \rangle]$$

where $o(u_{i,j}) = p_i^{e_{i,j}}$. The numbers $p_i^{e_{i,j}}$ are called the **elementary divisors** *of A.*□

We note the following:

1) The product of the invariant factors is equal to the product of the elementary divisors and is the order of the group.
2) The invariant factor α_k is equal to the maximum order of the elements of the group

$$\langle u_1 \rangle \bowtie \cdots \bowtie \langle u_k \rangle$$

   In particular, the largest invariant factor α_1 is equal to the maximum order of the elements of A.
3) The *multiset* of invariant factors of A uniquely determine the *multiset* of elementary divisors of A and vice versa. In particular, the elementary divisors are determined by factoring the invariant factors and the invariant factors are determined by multiplying together appropriate elementary divisors.

Uniqueness

Although the invariant factor decomposition and the primary cyclic decomposition are *not* unique, the multiset of invariant factors and the multiset of elementary divisors are uniquely determined by A. Actually, the proof is quite easy if we use the cancellation property of Theorem 5.6. Suppose that

$$A = \langle u_1 \rangle \bowtie \cdots \bowtie \langle u_n \rangle$$

where $\alpha_k = o(u_k)$ and

$$\alpha_n \mid \alpha_{n-1} \mid \cdots \mid \alpha_1$$

and that

$$A = \langle v_1 \rangle \bowtie \cdots \bowtie \langle v_m \rangle$$

where $\beta_k = o(v_k)$ and

$$\beta_m \mid \beta_{m-1} \mid \cdots \mid \beta_1$$

Then α_1 and β_1 are both equal to the maximum order of the elements of A and so $\alpha_1 = \beta_1$. Hence $\langle u_1 \rangle \approx \langle v_1 \rangle$ and Theorem 5.6 implies that

$$\langle u_2 \rangle \bowtie \cdots \bowtie \langle u_n \rangle \approx \langle v_2 \rangle \bowtie \cdots \bowtie \langle v_m \rangle$$

Now, α_2 and β_2 are equal to the maximum orders in the two isomorphic groups above and so $\alpha_2 = \beta_2$ and we may cancel again. Hence, $m = n$ and $\alpha_k = \beta_k$ for all k.

Theorem 5.9 (Uniqueness) *Let A be a finite abelian group. The multiset of invariant factors (and elementary divisors) for A is uniquely determined by A.*□

Properties of Direct Summands

Let us explore some of the properties of direct summands. The following simple facts are worth explicit mention:

1) The direct summand property is transitive:

$$H \in \mathcal{DS}(K), \quad K \in \mathcal{DS}(G) \quad \Rightarrow \quad H \in \mathcal{DS}(G)$$

2) The property of being a direct summand in inherited by subgroups:

$$H \in \mathcal{DS}(G), \quad H \le K \le G \quad \Rightarrow \quad H \in \mathcal{DS}(K)$$

In fact,

$$G = H \bowtie J, \quad H \le K \le G \quad \Rightarrow \quad K = H \bowtie (J \cap K)$$

3) Normal subgroups of direct summands are normal in the group:

$$N \trianglelefteq H, \quad H \in \mathcal{DS}(G) \quad \Rightarrow \quad N \trianglelefteq G$$

4) The property of being characteristic is preserved by intersection with a direct summand:
$$N \sqsubseteq G, \quad H \in \mathcal{DS}(G) \quad \Rightarrow \quad N \cap H \sqsubseteq H$$

Good Order

Let G be a group and suppose that
$$G = A \bowtie A' = B \bowtie B'$$
If $A \subseteq B$, then it does not necessarily follow that $A' \supseteq B'$, as is easily seen in the Klein 4-group, for example. For convenience, let us say that an equation of the form
$$G = A \bowtie A' = B \bowtie B', \quad A \subseteq B$$
is in **good order** if $A' \supseteq B'$.

Theorem 5.10 *Let G be a group. If*
$$G = A \bowtie A' = B \bowtie B', \quad A \subseteq B$$
then we can replace either A' or B' by another subgroup to get an equation in good order. Specifically, the following are in good order:
$$G = A \bowtie [B' \bowtie (A' \cap B)] = B \bowtie B'$$
and
$$G = A \bowtie A' = B \bowtie (A' \cap AB')$$

Proof. The first equation follows directly from Dedekind's law. For the second equation,
$$\begin{aligned} B \cap (A' \cap AB') &= A' \cap (B \cap AB') \\ &= A' \cap A(B \cap B') \\ &= A' \cap A' \\ &= \{1\} \end{aligned}$$
and
$$B(A' \cap AB') = BA(A' \cap AB') = B(AA' \cap AB') = BAB' = G \qquad \square$$

Chain Conditions on Direct Summands and Remak Decompositions

Direct summands are also special with respect to chain conditions. We have seen that for the families $\operatorname{sub}(G)$ and $\operatorname{nor}(G)$, the two chain conditions (ACC and DCC) are independent. However, for the family $\mathcal{DS}(G)$ of *direct summands* of G, the two chain conditions are equivalent.

Theorem 5.11 *A group G has the ACC on $\mathcal{DS}(G)$ if and only if it has the DCC on $\mathcal{DS}(G)$.*
Proof. Suppose that G has the DCC on direct summands and let

$$D_1 \le D_2 \le \cdots$$

be an ascending sequence in $\mathcal{DS}(G)$. Theorem 5.10 implies that there is a descending sequence

$$E_1 \ge E_2 \ge \cdots$$

where $G = D_i \bowtie E_i$ for all i. This descending sequence must eventually become constant, at which time so does the original sequence of D_i's.□

A group with the the BCC on direct summands can be decomposed into a finite direct sum of indecomposable factors.

Definition *If G is a group, then any decomposition*

$$G = R_1 \bowtie \cdots \bowtie R_n$$

where each R_k is indecomposible is called a **Remak decomposition** *of G.*□

A **minimal direct summand** is a minimal member of the family $\mathcal{DS}(G) \setminus \{1\}$ of all nontrivial direct summands of G. A direct summand is minimal if and only if it is indecomposable.

Theorem 5.12 (Remak) *If a group G has either (and therefore both) chain condition on direct summands, then G has a Remak decomposition*

$$G = R_1 \bowtie \cdots \bowtie R_n$$

Proof. Let R_1 be a minimal direct summand of G. If $R_1 = G$, then G is indecomposable and we are done. Otherwise,

$$G = R_1 \bowtie E_1$$

where $E_1 \neq \{1\}$ also has BCC on direct summands. Let R_2 be a minimal direct summand of E_1, which is also a minimal direct summand of G. If $R_2 = E_1$, we are done. If not, then

$$G = R_1 \bowtie R_2 \bowtie E_2$$

Since the sequence

$$R_1 < R_1 \bowtie R_2 < \cdots$$

is a strictly increasing sequence of direct summands of G, it must become constant and so this construction must terminate after a finite number of steps, resulting in a decomposition of G into a finite direct sum of minimal direct summands.□

A Maximality Property

We paraphrase Theorem 3.15.

Theorem 5.13 *Let $\mathcal{F} = \{H_i \mid i \in I\}$ be a nonempty family of normal subgroups of a group G. For any $K \trianglelefteq G$, there is a $J \subseteq I$ that is maximal with respect to the property that the direct sum*

$$K \bowtie \left(\bowtie_{j \in J} H_j \right)$$

exists.□

One consequence of Theorem 5.13 is the following.

Theorem 5.14 *Suppose that G is the join of a family $\mathcal{F} = \{H_i \mid i \in I\}$ of minimal normal subgroups. For any $K \trianglelefteq G$, there is a $J \subseteq I$ for which*

$$G = K \bowtie \left(\bowtie_{j \in J} H_j \right)$$

and so

$$\text{nor}(G) = \mathcal{DS}(G)$$

Proof. Let $J \subseteq I$ be maximal with respect to the fact that the direct sum

$$M_J = K \bowtie \left(\bowtie_{j \in J} N_j \right)$$

exists. If $i \notin J$, then $M_J \cap N_i \trianglelefteq G$ implies that $M_J \cap N_i = \{1\}$ or $N_i \le M_J$. But if $M_J \cap N_i = \{1\}$, then the direct sum $M_{J \cup \{i\}}$ exists, contradicting the maximality of J. Hence, $N_i \le M_J$ for all $i \in I$ and so $G = M_J$.□

xY-Groups

It is natural to ask the following types of questions. Let $\mathcal{X}(G)$ be a class of subgroups of a group G, such as the class of all subgroups, all normal subgroups or all characteristic subgroups.

1) For which groups G is it true that all subgroups in $\mathcal{X}(G)$ are complemented?
2) For which groups G is it true that all subgroups in $\mathcal{X}(G)$ have normal complements?
3) For which groups G is it true that all subgroups in $\mathcal{X}(G)$ are direct summands?

There has been much research done on these and related questions and it has become customary to make the following types of definitions in this regard.

Definition

1) *An* **aC-group** *is a group in which all subgroups are complemented.*
2) *An* **nC-group** *is a group in which all normal subgroups are complemented.*
3) *An* **aD-group** *is a group in which all subgroups are direct summands.*
4) *An* **nD-group** *is a group in which all normal subgroups are direct summands.*
5) *An* **aNC-group** *is a group in which all subgroups have normal complements.* □

Note that an nNC-group is the same as an nD-group. We will characterize aD-groups, nD-groups and aNC-groups. We will also describe (without proof) the characterization of aC-groups. The theory of the least restrictive of these conditions—the nC-groups—appears to be much more complicated and we refer the reader to Christensen [8] and [9] for more details.

aD-Groups

Groups in which all subgroups are direct summands were characterized by Kertész in 1952. This is the strongest of the xY-conditions and is indeed very restrictive.

Theorem 5.15 (aD-groups: Kertész [21], 1952) *A group G is an aD-group if and only if it is the direct product of cyclic groups of prime order.*

Proof. If G is a direct sum of cyclic subgroups C_i of prime order, then since each C_i is minimal normal, Theorem 5.14 implies that

$$\operatorname{sub}(G) = \operatorname{nor}(G) = \mathcal{DS}(G)$$

and so G is an aD-group.

For the converse, suppose that G is an aD-group. Then any subgroup of G is also an aD-group. However, the only cyclic aD-groups $\langle a\rangle$ are those of square-free order (that is, order a product of *distinct* primes). For it is clear that $\langle a\rangle$ cannot have infinite order and if $p^2 \mid o(a)$ for any prime p, then $\langle a\rangle$ has an element x of order p and so

$$\langle a\rangle = \langle x\rangle \bowtie K$$

where $p \mid o(K)$. Hence, K has an element of order p as well, which is too many elements of order p for a cyclic group. Hence, every element of G has square-free order.

Now, consider the family $\mathcal{F}$ of all cyclic subgroups of prime order in G. If $a \in G$ has order $o(a) = p_1 \cdots p_m$, where the factors are distinct primes, then Theorem 2.33 implies that there are subgroups $\langle a_i\rangle$ of order p_i for which

$$\langle a\rangle = \langle a_1\rangle \bowtie \cdots \bowtie \langle a_m\rangle$$

and so $G = \bigvee \mathcal{F}$. But since each member of $\mathcal{F}$ is minimal normal, Theorem 5.14 implies that G is the direct sum of a subfamily of $\mathcal{F}$.□

nD-Groups

The nD-condition is not as strong as the aD-condition, but is still very restrictive. One reason is that normality is a finitary condition, involving individual elements and so the condition $\mathcal{DS}(G) = \text{nor}(G)$ imparts finitary properties to the otherwise global condition of being a direct summand.

In particular, the union of any ascending sequence of normal subgroups is normal and so in an nD-group, the union of any ascending sequence of direct summands is a direct summand. This condition implies a certain measure of finiteness for nD-groups. Specifically, if G is a nontrivial nD-group and

$$D_1 \le D_2 \le \cdots$$

is an ascending chain of direct summands in N, then

$$G = \left(\bigcup D_i\right) \bowtie E = \bigcup (D_i \bowtie E)$$

for some $E \le G$. Hence, if $G = \text{nc}(S, G)$ is the normal closure of a finite subset $S \subseteq G$, then $S \subseteq D_n \bowtie E$ for some $n \ge 1$ and so $G = D_n \bowtie E$, which implies that $D_{n+i} = D_n$ for all $i \ge 0$. Thus, G has the ACC on direct summands and so G has a Remak decomposition.

Note also that any normal subgroup (direct summand) D of an nD-group G is an nD-group, since

$$\mathcal{DS}(D) = \mathcal{DS}(G) \cap \text{sub}(D) = \text{nor}(G) \cap \text{sub}(D) = \text{nor}(D)$$

It follows that any nontrivial nD-group contains an indecomposable direct summand.

We can now characterize nD-groups. Suppose that G is a nontrivial nD-group and consider the family $\mathcal{F} = \{S_i \mid i \in I\}$ of all indecomposable direct summands of G, along with the trivial subgroup. Theorem 5.13 implies that there is a subset $J \subseteq I$ that is maximal with respect to the fact that the direct sum

$$S = \bowtie_{j \in J} S_j$$

exists in G and so the nD-condition implies that

$$G = S \bowtie H$$

for some $H \le G$. But if H is not trivial, then it contains an indecomposable direct summand of G, which contradicts the maximality of J. Hence, H is trivial and

$$G = \bowtie_{j \in J} S_j$$

is a direct sum of indecomposable subgroups. In particular, each S_i is minimal normal in G and so G is the direct sum of minimal normal subgroups.

Conversely, if G is the direct sum of minimal normal subgroups, then Theorem 5.14 implies that $\operatorname{nor}(G) = \mathcal{DS}(G)$, that is, G is an nD-group.

Theorem 5.16 *The following are equivalent for a group G:*
1) *G is an nD-group*
2) *G is the direct sum of minimal normal subgroups*
3) *G is the direct sum of simple subgroups.* □

aNC-Groups

We next show that aNC-groups are the same as aD-groups, by showing that an aNC-group is abelian.

Theorem 5.17 (aNC-groups: Weigold [34], 1960) *A group is an aNC-group if and only if it is an aD-group, that is, if and only if it is a direct sum of cyclic subgroups of prime order.*
Proof. An aD-group is an aNC-group. For the converse, we show that an aNC-group G has trivial commutator subgroup and so is abelian. If $A \leq G$ is abelian, then $G = N \rtimes A$ for some $N \trianglelefteq G$ and so

$$G' = [NA, NA] \leq N$$

Hence, G' is in the normal complement of any abelian subgroup of G, including all cyclic subgroups $\langle a \rangle$ of G. Hence, G' is trivial. □

aC-Groups

Finally, we state without proof the following theorem on aC-groups.

Theorem 5.18 (aC-groups)
1) **(Hall, P. [17], 1937; see also Schmidt [30], p. 122)** *A finite group G is an aC-group if and only if G is a direct sum of groups of square-free order.*
2) **(See Schmidt [30], p. 123)** *A group G is an aC-group if and only if*

$$G = \left(\bowtie_{i \in I} N_i \right) \rtimes \left(\bowtie_{j \in J} H_j \right)$$

where each N_i and each H_j is cyclic of prime order and $N_i \trianglelefteq G$ for all i. □

Behavior Under Direct Sum

The following theorem describes how some basic constructions behave with respect to direct sums and emphasizes the fact that the summands in a direct sum have a great deal of independence.

Theorem 5.19 *Let*

$$G = \bowtie_{i\in I} H_i$$

Then the following hold:

1) **(Center of *G*)**

$$Z(G) = \bowtie_{i\in I} Z(H_i)$$

2) **(Commutator of *G*)**

$$G' = \bowtie_{i\in I} H_i'$$

3) *If* $N_i \trianglelefteq H_i$ *for all* i*, then*

$$\frac{\bowtie_{i\in I} H_i}{\bowtie_{i\in I} N_i} \approx \boxplus_{i\in I} \frac{H_i}{N_i}$$

4) *If* $H_i \sqsubseteq G$*, then*

$$\operatorname{Aut}(G) \approx \boxtimes_{i\in I} \operatorname{Aut}(H_i)$$

Proof. For part 1), since the H_i's commute elementwise, it is clear that $\bowtie Z(H_i) \le Z(G)$. But if $z \in Z(G)$, then let $z = h_1 \cdots h_n$ with each h_i in a different factor H_{j_i}. Then $z \in Z(H_i)$ and $h_k \in Z(H_i)$ for $k \ne i$ and so $h_i \in Z(H_i)$, whence $z \in \bowtie Z(H_i)$.

For part 2), Theorem 3.41 implies that

$$G' = [\bowtie_i H_i, \bowtie_j H_j] = \bowtie_{i,j} [H_i, H_j] = \bowtie_i [H_i, H_i] = \bowtie_i H_i'$$

For part 3), the function

$$\sigma\colon \bowtie_{i\in I} H_i \to \boxplus_{i\in I} \frac{H_i}{N_i}$$

defined by

$$\sigma(a)(k) = \kappa_k(a) N_k = \pi_{N_k} \circ \kappa_k(a)$$

where π_{N_k} is the canonical projection map and κ_k is the kth injection map is an epimorphism. Moreover, $\sigma a = 1$ if and only if $\kappa_k(a) \in N_k$ for all k and so $\ker(\sigma) = \bowtie N_i$.

As to part 4), since $H_i \sqsubseteq G$, if $\sigma \in \operatorname{Aut}(G)$, then $\sigma|_{H_i} \in \operatorname{Aut}(H_i)$ and so we can define a map $f\colon \operatorname{Aut}(G) \to \boxtimes_{i\in I} \operatorname{Aut}(H_i)$ by

$$f(\sigma)(i) = \sigma|_{H_i}$$

This is a group homomorphism since

$$f(\sigma\tau)(i) = (\sigma \circ \tau)|_{H_i} = (\sigma|_{H_i}) \circ (\tau|_{H_i}) = f(\sigma)(i) \circ f(\tau)(i) = [f(\sigma)f(\tau)](i)$$

for all $i \in I$ and so $f(\sigma \circ \tau) = f(\sigma)f(\tau)$. Furthermore, f is injective since $\sigma|_{H_i} = \iota$ for all i implies $\sigma = \iota$.

To see that f is surjective, if $\tau \in \boxtimes_{i\in I}\mathrm{Aut}(H_i)$, then the universality of the direct sum implies that there is a unique homomorphism $\sigma: G \to G$ satisfying $\sigma|_{H_i} = \tau_i$, the ith coordinate of τ. But the bijectivity of each τ_i implies that $\sigma \in \mathrm{Aut}(G)$ and so $f(\sigma) = \tau$.□

When All Subgroups Are Normal

All subgroups of an abelian group are normal, but all subgroups of the quaternion group Q are normal and yet Q is not abelian. A **Hamiltonian group** is a *nonabelian* group all of whose subgroups are normal. In 1933, Baer [3] published a characterization of Hamiltonian groups. Baer's theorem says that the Hamiltonian groups are actually a special type of abelian group with an additional quaternion direct summand.

Theorem 5.20 (Baer [3], 1933) *A group G is Hamiltonian if and only if*

$$G = Q \bowtie A \bowtie B$$

where Q is a quaternion group, A is an elementary abelian group of exponent 2 *and B is an abelian group all of whose elements have odd order.*

Proof. First suppose that $G = Q \bowtie A \bowtie B$ and let $H \le G$. If $H \le A \bowtie B$, then $H \trianglelefteq G$. Note that if $-1 \in H$, then $G' = \{\pm 1\} \le H$ and so $H \trianglelefteq G$. Every $h \in H$ has the form

$$h = qab$$

where $q \in Q$, $a \in A$ and $b \in B$ has odd order, then

$$h^{2o(b)} = q^{2o(b)} = q^2$$

and so $q^2 \in H$. If $o(q) = 4$ for some $h \in H$, then $-1 \in H$ and so $H \trianglelefteq G$. On the other hand, if $o(q) \le 2$ for all $h \in H$, then $H \le \{\pm 1\} \bowtie A \bowtie B = Z(G)$ and so $H \trianglelefteq G$. Thus, G is Hamiltonian.

For the converse, let G be Hamiltonian. Theorem 3.38 shows that G is periodic and that any nonabelian subgroup of G contains a quaternion subgroup.

If B is the set of odd-order elements of G and M is the set of elements of order a power of 2, then $B \cap M = \{1\}$. Moreover, for any $x, y \in G$, the normality of $\langle x\rangle$ and $\langle y\rangle$ imply that $\langle x, y\rangle = \langle x\rangle\langle y\rangle$ and so $o(xy) \mid o(x)o(y)$. It follows that

B and M are (normal) subgroups of G. Finally, every $a \in G$ has order $o(a) = 2^k m$ where m is odd and so Corollary 2.11 implies that $a = bm$ for some $b \in B$ and $m \in M$. Thus,

$$G = B \bowtie M$$

Since B does not contain a quaternion subgroup, it must be abelian and since M is therefore nonabelian (since G is nonabelian), M contains a quaternion subgroup $Q = \langle x, y \rangle$.

We are left with showing that $M = Q \bowtie A$, where A is an elementary abelian group of exponent 2. The centralizer

$$C = C_M(Q)$$

of $Q = \langle x, y \rangle$ in M has exponent 2. To see this, note that if $z \in C$ has order 4, then $\langle zx \rangle \trianglelefteq M$ has order 4 and so

$$zx^3 = (zx)^y = zx \quad \text{or} \quad zx^3 = (zx)^y = z^3 x^3$$

and since the former is false, we have $z^2 = 1$. Thus, C is an elementary abelian group of exponent 2. Moreover, since C is a vector space over $\mathbb{Z}_2$, every subgroup (i.e., subspace) has a direct complement and so

$$C = \langle x^2 \rangle \bowtie A$$

where A is also an elementary abelian group of exponent 2.

Consider the quotient M/C, which contains the four distinct cosets C, xC, yC and xyC. Poincaré's theorem gives

$$\begin{aligned}(M:C) &= (M : C_M(x) \cap C_M(y)) \\ &\le (M : C_M(x))(M : C_M(y)) \\ &= |x^M||y^M| \\ &= 4\end{aligned}$$

and so

$$M/C = \{C, xC, yC, xyC\}$$

which implies that

$$M = QC = Q(\langle x^2 \rangle \bowtie A) = QA$$

But the only involution in Q is x^2, which is not in A and so $Q \cap A = \{1\}$. Thus, $M = Q \bowtie A$ and

$$G = M \bowtie B = Q \bowtie A \bowtie B \qquad \square$$

Semidirect Products

We now turn to semidirect products.

Definition *Let G be a group. If*

$$G = N \bullet H, \quad N \trianglelefteq G$$

then N is called a **normal complement** *of H in G and G is called the* **semidirect product** *of N by H, denoted by*

$$G = N \rtimes H$$

A semidirect product is **nontrivial** *if both factors are proper.*□

Example 5.21 The dihedral group D_{2n} is a nontrivial semidirect product:

$$D_{2n} = \langle \rho \rangle \rtimes \langle \sigma \rangle$$

The symmetric group is also a nontrivial semidirect product:

$$S_n = A_n \rtimes \langle (1\,2) \rangle$$

On the other hand, the quaternion group is *not* a nontrivial semidirect product, since the orders of the factors must be 2 and 4 but the only subgroup of Q of order 2 is contained in every subgroup of order 4.□

For an arbitrary semidirect product $G = N \rtimes H$, the *commutativity rule* is

$$hn = (hnh^{-1})h = n^h h$$

for $h \in H$ and $n \in N$ and this yields the *multiplication rule*

$$(n_1 h_1)(n_2 h_2) = n_1 n_2^{h_1} h_1 h_2$$

for $h_i \in H$ and $n_i \in N$. Thus, for the semidirect product, the multiplication rule shows that "cross products" are involved, in the form of conjugates $n_1^{h_1}$. This has some perhaps unexpected consequences.

For example, if $A_1 \approx A_2$ and $B_1 \approx B_2$, then $A_1 \bowtie B_1 \approx A_2 \bowtie B_2$. On the other hand, if $\langle a \rangle$ is cyclic of order 6, then

$$\langle a \rangle = \langle a^2 \rangle \rtimes \langle a^3 \rangle = \{1, a^2, a^4\} \bowtie \{1, a^3\}$$

But we also have

$$S_3 = \langle (1\,2\,3) \rangle \rtimes \langle (1\,2) \rangle = \{\iota, (1\,2\,3), (1\,3\,2)\} \rtimes \{\iota, (1\,2)\}$$

and so the nonisomorphic groups $\langle a \rangle$ and S_3 can be written as a semidirect product, where corresponding factors are isomorphic! The reason that this is not a contradiction is that the values of the conjugates are different in each group. For instance,

$$(a^2)^{a^3} = a^2$$

but

$$(1\,2\,3)^{(1\,2)} = (2\,1\,3) \neq (1\,2\,3)$$

Semidirect Products and Extensions of Epimorphisms

Let G and G_1 be groups with $H \leq G$ and let $\sigma: H \twoheadrightarrow G_1$ be an epimorphism. The key to describing the possible extensions of σ to G is to consider the possible kernels for such an extension.

Suppose that G is a group and that $H, K \leq G$ contain a normal subgroup $J \trianglelefteq G$. Then H/J and K/J are complements in G/J if and only if

$$H \cap K = J \quad \text{and} \quad G = HK$$

It will be convenient to make the following nonstandard definition.

Definition *Let G be a group and let $J \trianglelefteq G$. If $H, K \leq G$ satisfy*

$$H \cap K = J \quad \textit{and} \quad G = HK$$

we will say that H and K are **complements modulo** J. *If K is a normal complement of H modulo J, we will write*

$$G = K \rtimes H\,[\operatorname{mod} J]$$ □

Now, if $\overline{\sigma}: G \twoheadrightarrow G_1$ is an extension of $\sigma: H \twoheadrightarrow G_1$, then

$$H \cap \ker(\overline{\sigma}) = \ker(\sigma)$$

To see that $G = H\ker(\overline{\sigma})$, since σ is surjective, for any $a \in G$, there is an $h \in H$ for which $\overline{\sigma}a = \sigma h = \overline{\sigma}h$ and so $h^{-1}a \in \ker(\overline{\sigma})$, whence

$$a = h(h^{-1}a) \in H\ker(\overline{\sigma})$$

and so $\ker(\overline{\sigma})$ is a normal complement of H modulo $\ker(\sigma)$:

$$G = \ker(\overline{\sigma}) \rtimes H\,[\operatorname{mod} \ker(\sigma)] \tag{5.22}$$

It follows that every extension $\overline{\sigma}$ of σ is uniquely determined by its kernel $\ker(\overline{\sigma})$.

On the other hand, suppose that

$$G = K \rtimes H\,[\operatorname{mod} \ker(\sigma)]$$

Then the **ignore-*K* map** $\overline{\sigma}: G \to G_1$ defined by

$$\overline{\sigma}(hk) = \sigma(h)$$

for $h \in H$ and $k \in K$ is well defined since if $hk = h_1k_1$ for $h_1 \in H$ and $k_1 \in K$, then

$$h_1^{-1}h = k_1k^{-1} \in H \cap K = \ker(\sigma)$$

and so $\sigma(h_1^{-1}h) = 1$, that is, $\sigma h = \sigma h_1$. The normality of K implies that $\overline{\sigma}$ is a group homomorphism, since

$$\overline{\sigma}(khk_1h_1) = \overline{\sigma}(kk_1^h hh_1) = \sigma(hh_1) = \sigma(h)\sigma(h_1) = \overline{\sigma}(kh)\overline{\sigma}(k_1h_1)$$

The kernel of $\overline{\sigma}$ is

$$\ker(\overline{\sigma}) = \{hk \mid \sigma h = 0\} = \ker(\sigma)K = K$$

and so $\overline{\sigma}$ is the unique extension of σ with kernel K.

Theorem 5.23 *Let G and G_1 be groups, let $H \leq G$ and let $\sigma\colon H \twoheadrightarrow G_1$ be an epimorphism.*

1) *Given any normal complement K of H modulo* $\ker(\sigma)$,

$$G = K \rtimes H\,[\operatorname{mod}\ker(\sigma)]$$

the ignore-K map $\overline{\sigma}_K\colon G \to G_1$ defined by

$$\overline{\sigma}(hk) = \sigma h$$

for all $h \in H$ and $k \in K$ is the unique extension of σ with kernel K. Moreover, every extension $\overline{\sigma}$ of σ is an ignore-K map, where $K =$ $\ker(\overline{\sigma})$.

2) *If $G = K \rtimes H$, then σ has a unique extension $\overline{\sigma}\colon G \to G_1$ for which*

$$\ker(\overline{\sigma}) = K\ker(\sigma)$$

Proof. For part 2), the Dedekind law implies that

$$H \cap K\ker(\sigma) = \ker(\sigma)(H \cap K) = \ker(\sigma)$$

and so $K\ker(\sigma)$ is a normal complement of H modulo $\ker(\sigma)$.□

Semidirect Products and One-Sided Invertibility

Semidirect products are related to one-sided invertibility.

Definition *Let $\sigma\colon G \to H$ be a group homomorphism.*

1) *A* **left inverse** *of σ is a homomorphism $\sigma_L\colon H \to G$ for which $\sigma_L \circ \sigma = \iota$. If σ has a left inverse, then σ is said to be* **left invertible**.
2) *A* **right inverse** *of σ is a homomorphism $\sigma_R\colon G \to H$ for which $\sigma \circ \sigma_R = \iota$. If σ has a right inverse, then σ is said to be* **right invertible**.□

Unlike the two-sided inverse, one-sided inverses need not be unique. A left-invertible homomorphism σ is injective, since

$$\sigma a = \sigma b \quad \Rightarrow \quad \sigma_L \circ \sigma a = \sigma_L \circ \sigma b \quad \Rightarrow \quad a = b$$

and a right-invertible homomorphism σ is surjective, since if $b \in H$, then

$$b = \sigma(\sigma_R b) \in \text{im}(\sigma)$$

For *set* functions, the converses of these statements hold: σ is left-invertible if and only if it is injective and σ is right-invertible if and only if it is surjective. However, this is not the case for group homomorphisms.

Let $\sigma: G \hookrightarrow G_1$ be injective. Referring to Figure 5.3,

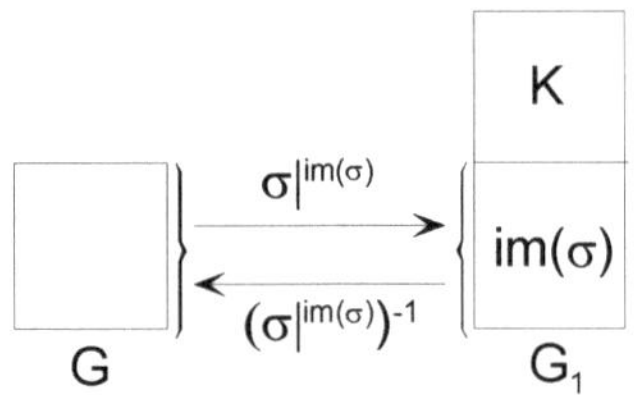

Figure 5.3

the map $\sigma|^{\text{im}(\sigma)}: G \approx \text{im}(\sigma)$ obtained from σ by restricting its range to $\text{im}(\sigma)$ is an isomorphism and the left inverses of σ are precisely the extensions of $\sigma_L = (\sigma|^{\text{im}(\sigma)})^{-1}: \text{im}(\sigma) \approx G$ to G_1. Hence, Theorem 5.23 implies that there is one left inverse σ_L for σ for each normal complement K of $\text{im}(\sigma)$ and $\ker(\sigma_L) = K$. Moreover, this accounts for all left inverses of σ. In particular, σ is left-invertible if and only if $\text{im}(\sigma)$ has a normal complement.

Now let $\sigma: G \twoheadrightarrow G_1$ be surjective. Referring to Figure 5.4,

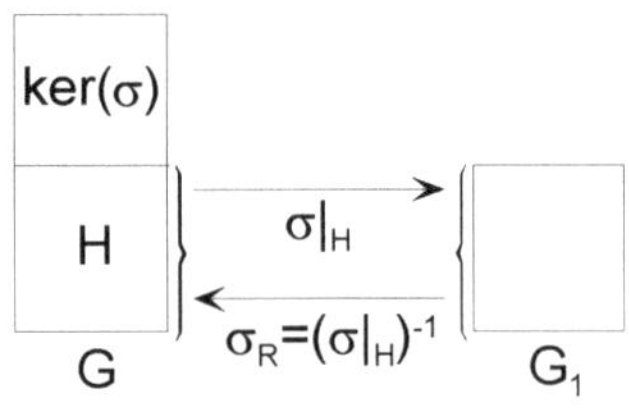

Figure 5.4

If $G = \ker(\sigma) \rtimes H$ for some $H \leq G$, then $\sigma|_H: H \approx G_1$ and $(\sigma|_H)^{-1}$ is a right inverse of σ, with image H. Conversely, if $\sigma_R: G_1 \hookrightarrow G$ is a right inverse of σ, then

$$G = \ker(\sigma) \rtimes \text{im}(\sigma_R)$$

For if $a \in \ker(\sigma) \cap \text{im}(\sigma_R)$, then $a = \sigma_R b$ and $\sigma a = 1$, whence $b = 1$ and so

$a = 1$. Also, for any $a \in G$, we have

$$a = \big(a[(\sigma_R \circ \sigma)a]^{-1}\big)((\sigma_R \circ \sigma)a) \in \ker(\sigma) \rtimes \operatorname{im}(\sigma_R)$$

Theorem 5.24 *Let $\sigma: G \to G_1$ be a group homomorphism.*

1) *If σ is injective, then σ has a unique left inverse σ_L with kernel K for each normal complement K of* $\operatorname{im}(\sigma)$. *Hence,*

$$G_1 = \ker(\sigma_L) \rtimes \operatorname{im}(\sigma)$$

This accounts for all left inverses of σ.

2) *If σ is surjective, then σ has a unique right inverse $\sigma_R = (\sigma|_H)^{-1}: G_1 \to G$ for each complement H of* $\ker(\sigma)$. *Hence, $H =$* $\operatorname{im}(\sigma_R)$ *and*

$$G = \ker(\sigma) \rtimes \operatorname{im}(\sigma_R)$$

This accounts for all of the right inverses of σ.□

Here is a nice application of this theorem.

Theorem 5.25 *Let H and K be indecomposable groups. Let $\alpha: K \to H$ and $\beta: H \to K$ be homomorphisms for which $\alpha\beta \in$* $\operatorname{Aut}(H)$ *and* $\operatorname{im}(\beta) \trianglelefteq K$. *Then α and β are isomorphisms.*

Proof. The map

$$\beta_L = (\alpha\beta)^{-1} \circ \alpha: K \to H$$

is a left inverse of β and so β is injective and

$$H = \ker(\beta_L) \bowtie \operatorname{im}(\beta)$$

But since β_L is not the zero map, β must be surjective as well. Hence, β and therefore also α are isomorphisms.□

The External Semidirect Product

To see how to externalize the semidirect product, let us review the internal version. If we write the multiplication rule for the semidirect product $N \rtimes H$ in the form

$$(n_1h_1)(n_2h_2) = n_1[\gamma_{h_1}(n_2)]h_1h_2$$

where γ_h is conjugation by h, then $N \rtimes H$ is completely described by the subgroups N and H, along with the family $\mathcal{I} = \{\gamma_h \mid h \in H\}$ of inner automorphisms. Moreover, since the map $\gamma: h \mapsto \gamma_h$ is a homomorphism, γ is a *representation* of H in $\operatorname{Aut}(N)$.

In the spirit of externalization, we separate the components by writing hk as (h, k) to get

$$(n_1, h_1)(n_2, h_2) = (n_1 \gamma_{h_1}(n_2), h_1 h_2)$$

But if the factors N and H are to be arbitrary groups, then the inner automorphisms γ_h make no sense. However, *any* representation of $\theta: H \to \operatorname{Aut}(N)$ of H in $\operatorname{Aut}(N)$ can play the role of conjugation and define a group, as we now show.

Theorem 5.26 *Let N and H be groups and let $\theta: H \to \operatorname{Aut}(N)$ be a homomorphism. We denote $\theta(h)$ by θ_h. Let $N \rtimes_\theta H$ be the cartesian product $N \times H$ together with the binary operation defined by*

$$(a, x)(b, y) = (a\theta_x(b), xy)$$

Then $G = N \rtimes_\theta H$ is a group; in fact, it is a semidirect product

$$N \rtimes_\theta H = (N \times \{1\}) \rtimes (\{1\} \times H)$$

Also,

$$(a, x) = (a, 1)(1, x)$$

and

$$(a, 1)^{(1,x)} = (\theta_x(a), 1)$$

which shows that θ does define the inner automorphisms of elements of $N \times \{1\}$ by elements of $\{1\} \times H$. The group $N \rtimes_\theta H$ is called the **external semidirect product** *of N by H* **defined by** *θ.*

Proof. To see that multiplication is associative, we have

$$\begin{aligned}[(a, x)(b, y)](c, z) &= (a\theta_x(b), xy)(c, z) \\ &= (a\theta_x(b)\theta_{xy}(c), xyz)\end{aligned}$$

and

$$\begin{aligned}(a, x)[(b, y)(c, z)] &= (a, x)(b\theta_y(c), yz) \\ &= (a\theta_x(b\theta_y(c)), xyz) \\ &= (a\theta_x(b)\theta_{xy}(c)), xyz)\end{aligned}$$

It is easy to check that $(1, 1)$ is the identity and

$$(a, x)^{-1} = (\theta_{x^{-1}}(a^{-1}), x^{-1})$$

Thus, $N \rtimes_\theta H$ is a group. For the rest, routine calculation gives

$$(a, 1)(b, 1) = (ab, 1) \quad \text{and} \quad (a, 1)^{-1} = (a^{-1}, 1)$$

and

$$(1, x)(1, y) = (1, xy) \quad \text{and} \quad (1, x)^{-1} = (1, x^{-1})$$

and so $N \times \{1\}$ and $\{1\} \times H$ are subgroups of $N \rtimes_\theta H$. To see that $N \times \{1\}$ is normal, since

$$(a, x) = (a, 1)(1, x)$$

we need only check that

$$(a, 1)^{(1,x)} = (1, x)(a, 1)(1, x^{-1}) = (1, x)(a, x^{-1}) = (\theta_x(a), 1) \in N \times \{1\}$$

and

$$(a, 1)^{(b,1)} = (b, 1)(a, 1)(b^{-1}, 1) = (bab^{-1}, 1) \in N \times \{1\}$$

Clearly, $(N \times \{1\}) \cap (\{1\} \times H) = \{(1, 1)\}$ and so $N \rtimes_\theta H$ is the (internal) semidirect product of $N \times \{1\}$ by $\{1\} \times H$.□

Note that the zero representation $\theta: H \to \mathrm{Aut}(G)$ defined by $\theta(h) = \iota_G$ defines the external *direct* product $G \boxtimes H$.

It is common to write the external version of the semidirect product using the notation of the internal version. Thus, if G and H are groups, we can specify a semidirect product $S = G \rtimes_\theta H$ by taking the set of *formal* products

$$S = \{gh \mid g \in G, h \in H\}$$

in place of ordered pairs and specifying the commutativity rule

$$hg = \theta_h(g)h$$

where $\theta: H \to \mathrm{Aut}(G)$ is a homomorphism.

Example 5.27 Given a group G, there is one rather obvious way to create an external semidirect product $G \rtimes_\theta H$, namely, by taking $H = \mathrm{Aut}(G)$ and $\theta: H \to H$ to be the identity. The group product in this case is

$$(a\sigma)(b\tau) = a\sigma(b)\sigma\tau$$

The group $G \rtimes_\theta \mathrm{Aut}(G)$ is called the **holomorph** of G.

We may generalize this by taking H to be any subgroup of $\mathrm{Aut}(G)$ and θ to be the inclusion map from H to $\mathrm{Aut}(G)$. The group $G \rtimes_\theta H$ is called the **relative holomorph** of G with respect to H.

Finally, if $\sigma \in \mathrm{Aut}(G)$, then the relative holomorph $G \rtimes_\theta \langle\sigma\rangle$ is called the **extension** of G by σ. The group product in this case is

$$(a\sigma^i)(b\sigma^j) = a\sigma^i(b)\sigma^{i+j}$$

As an example, let $G = C_{2^n}(a)$ be cyclic of order 2^n, where $n \geq 3$. If $m = 2^{n-1}$, then the power map defined by $\mu a = a^{m-1}$ is an automorphism of G

of order 2, since $m-1$ is relatively prime to 2^n and

$$(m-1)^2 = (2^{n-2}-1)2^n + 1 \equiv 1 \bmod 2^n$$

Thus, the extension of G by μ is just $SD_n = C_{2^n}(a) \rtimes_\theta C_2(x)$ and has group product

$$\begin{aligned}(a^i,1)(a^j,1) &= (a^{i+j},1)\\ (a^i,x)(a^j,1) &= (a^{i+(m-1)j},x)\\ (a^i,1)(a^j,x) &= (a^{i+j},x)\\ (a^i,x)(a^j,x) &= (a^{i+(m-1)j},1)\end{aligned}$$

for all $0 \le i,j \le 2^n - 1$. Since $(a^i,x^j) = (a,1)^i(1,x)^j$, setting $\alpha = (a,1)$ and $\xi = (1,x)$ gives

$$SD_n = \langle \alpha, \xi \rangle, \quad o(\alpha) = 2^n, \quad o(\xi) = 2, \quad \xi\alpha = \alpha^{2^{n-1}-1}\xi$$

The group SD_n is called the **semidihedral group** of order 2^{n+1}.□

Example 5.28 Recall that if C is an infinite cyclic group, then $\operatorname{Aut}(C) = \{\iota, \tau\}$ where $\tau: a \mapsto a^{-1}$ and if C is cyclic of order n, then $\operatorname{Aut}(C) \approx \mathbb{Z}_n^*$. Let $C_\infty(a)$ and $C_\infty(b)$ be infinite cyclic groups. Then a homomorphism

$$\theta: C_\infty(b) \to \operatorname{Aut}(C_\infty(a)) = \{\iota, \tau\}$$

is completely determined by the value θ_b, which can be either ι or τ. If $\theta_b = \iota$, then θ is the zero map and $C_\infty(a) \rtimes_\theta C_\infty(b)$ is direct. If $\theta_b = \tau$, then the commutativity rule in the group $C_\infty(a) \rtimes_\theta C_\infty(b)$ is

$$ba = \tau(a)b = a^{-1}b$$

The automorphisms of a finite cyclic group $C_n(a)$ are the kth power homomorphisms σ_k defined by

$$\sigma_k(a) = a^k$$

where $k \in \mathbb{Z}_n^*$. Thus, the representations $\theta_k: C_\infty(b) \to \operatorname{Aut}(C_n(a))$ are the homomorphisms defined by $\theta_k(b) = \sigma_k$ and so the possible semidirect products are

$$C_n(a) \rtimes_{\theta_k} C_\infty(b) = \{a^i b^j \mid 0 \le i \le n-1, j \in \mathbb{Z}, ba = a^k b\}$$

where $k \in \mathbb{Z}_n^*$.□

Example 5.29 To define a semidirect product $C_3(a) \rtimes_\theta C_4(b)$, we must specify a homomorphism

$$\theta: C_4(b) \to \operatorname{Aut}(C_3) = \{\iota, \sigma_2\}$$

where $\sigma_2(a) = a^2$. The zero homomorphism defines the direct product

$C_3(a) \boxtimes C_4(b)$. If $\theta_b = \sigma_2$, then the semidirect product $T = C_3(a) \rtimes_\theta C_4(b)$ defined by θ is

$$T = \langle a, b \rangle, \quad o(a) = 3, \quad o(b) = 4, \quad ba = a^2 b$$

We have not yet encountered this group of order 12. In particular, $T \not\approx A_4$ since $o(ab^2) = 6$ and $T \not\approx D_{12}$ since $o(b) = 4$. We will show later that A_4, D_{12} and T are the only nonabelian groups of order 12 (up to isomorphism).□

Example 5.30 Let us examine the possibilities for an external semidirect product of the form

$$C_{p^m}(a) \rtimes_\theta C_{p^n}(b)$$

where p is prime. The automorphisms of $C_{p^m}(a)$ are the kth power maps $\sigma_k: a \mapsto a^k$ for $k \in \mathbb{Z}_m^*$. The function $\theta: C_{p^n}(b) \to \mathrm{Aut}(C_{p^m}(a))$ satisfying $\theta_b = \sigma_k$ defines a homomorphism if and only if $o(\sigma_k) \mid o(b)$, that is, if and only if $\sigma_k^{p^n} = \iota$, or

$$a^{k^{(p^n)}} = a$$

or finally

$$k^{(p^n)} \equiv 1 \bmod p^m \tag{5.31}$$

As an example, for $m = 1$, Fermat's theorem implies that (5.31) is equivalent to

$$k \equiv 1 \bmod p$$

and since $1 \le k < p$, it follows that $k = 1$. Hence, the only semidirect product of the form

$$C_p(a) \rtimes_\theta C_{p^n}(b)$$

is the direct product.

If $n = 1$, then (5.31) is equivalent to

$$k^p \equiv 1 \bmod p^m$$

Any k of the form $k = 1 + up^{m-1}$ where $u < p$ satisfies this congruence, since

$$k^p = (1 + up^{m-1})^p = 1 + wp^m$$

Thus, for each $u < p$, there is a semidirect product

$$C_{p^m}(a) \rtimes_\theta C_p(b)$$

where

$$ba = a^{1+up^{m-1}} b$$

□

Example 5.32 Let D be a group and let $G = D^3 := D \boxtimes D \boxtimes D$. Then each permutation $\sigma \in S_3$ defines an automorphism θ_σ of G by permuting the coordinates in G. For example,

$$\theta_{(1\,3)}(x, y, z) = (z, y, x)$$

Moreover, the map $\theta\colon S_3 \to \operatorname{Aut}(G)$ sending σ to θ_σ is a homomorphism, since $\theta_\sigma\theta_\tau = \theta_{\sigma\tau}$. Thus, the semidirect product

$$G \rtimes_\theta S_3 = D^3 \rtimes_\theta S_3$$

exists. To illustrate the product, if $\sigma = (1\,3)$, then

$$\begin{aligned}((a, b, c), \sigma))((x, y, z), \tau)) &= ((a, b, c)\theta_\sigma(x, y, z), \sigma\tau)\\ &= ((a, b, c)(z, y, x), \sigma\tau)\\ &= ((az, by, cx), \sigma\tau)\end{aligned}$$

□

We will generalize this example in the next section.

*The Wreath Product

To generalize Example 5.32, let D be a group, let Ω be a nonempty set and let

$$G = \mathop{\boxtimes}_{\omega\in\Omega} D$$

be the external direct product of $|\Omega|$ copies of D, indexed by Ω. Each permutation σ of Ω defines an automorphism θ_σ of G by permuting the coordinate positions of any $f \in G$. Specifically, the ωth coordinate of f becomes the $(\sigma\omega)$-th coordinate of $\theta_\sigma(f)$, that is, $\theta_\sigma(f)(\sigma\omega) = f(\omega)$, or equivalently,

$$(\theta_\sigma f)(\omega) = f(\sigma^{-1}\omega)$$

Thus,

$$\theta_\sigma(f) = f \circ \sigma^{-1}$$

The map $\theta_\sigma\colon G \to G$ is easily seen to be an automorphism of G, since it is clearly bijective and

$$\theta_\sigma(fg) = (fg) \circ \sigma^{-1} = (f \circ \sigma^{-1})(g \circ \sigma^{-1}) = \theta_\sigma(f)\theta_\sigma(g)$$

Moreover, if $Q \le S_\Omega$, then the map $\theta\colon Q \to \operatorname{Aut}(G)$ defined by $\theta\sigma = \theta_\sigma$ is a homomorphism, since

$$\theta_{\sigma\tau} f = f \circ (\sigma\tau)^{-1} = f \circ (\tau^{-1} \circ \sigma^{-1}) = \theta_\tau(f) \circ \sigma^{-1} = \theta_\sigma\theta_\tau(f)$$

Hence, the semidirect product $G \rtimes_\theta Q$ exists. It is easy to describe the commutativity rule in words: To place the factors in the product σf in the reverse order, simply permute the coordinate positions of f using σ.

Note that it is not essential that the second coordinates in the ordered pairs $(f,\sigma) \in G \rtimes_\theta Q$ be actual permutations of Ω as long as they *act* like permutations, that is, as long as there is a homomorphism $\lambda: Q \to S_\Omega$. As is customary, we denote the permutation λq of Ω by q itself.

Thus, if Q *acts on* Ω, then for each $q \in Q$, the map $\theta_q: G \to G$ defined by

$$\theta_q f = f \circ q^{-1}$$

is an automorphism of G and the map $\theta: Q \to \operatorname{Aut}(G)$ defined by $\theta q = \theta_q$ is a homomorphism.

The semidirect product $G \rtimes_\theta Q$ of G by Q defined by θ is one version of the *wreath product* of G by Q. The other version comes by replacing the external direct product $G = \boxtimes D$ by the external direct sum $G = \boxplus D$.

Definition *Let D be a group, let Ω be a nonempty set and let Q be a group acting on Ω. We denote the action of $q \in Q$ on $\omega \in \Omega$ by $q\omega$.*

1) Let

$$G = \boxtimes_{\omega \in \Omega} D$$

Define a homomorphism $\theta: Q \to \operatorname{Aut}(G)$ by $\theta q = \theta_q$ where

$$\theta_q f = f \circ q^{-1}$$

Then the semidirect product $G \rtimes_\theta Q$ is called the **complete wreath product** *of D by Q with* **index set** *Ω and* **base** *G.*

2) If we replace G by the external direct sum

$$G = \boxplus_{\omega \in \Omega} D$$

the resulting semidirect product $G \rtimes_\theta Q$ is called the **restricted wreath product** *of G by Q with* **index set** *Ω and* **base** *G.*

A common notation for the wreath product is $D \wr Q$. To emphasize the index set, we will write $D \wr_\Omega Q$. There does not seem to be a standard notion to distinguish the two wreath products, so we use $D \wr Q$ for the restricted wreath product and $D \wr\wr Q$ for the complete wreath product. □

Note that if D and Q are finite groups and Ω is a finite set, then

$$|D \wr\wr_\Omega Q| = |D^\Omega \rtimes Q| = |D^\Omega||Q| = |D|^{|\Omega|}|Q|$$

Example 5.33 (Regular wreath product) Let D and Q be groups and let

$$G = \boxtimes_{q \in Q} D$$

be the direct product of D indexed by the group Q. Let Q act on itself by left translation, that is, $\theta_q r = qr$, that is, the action of Q is the left regular representation. In this case, the complete wreath product $D \wr Q$ is called a **regular wreath product**, which we denote by $D \wr_r Q$. Thus, the product has the form

$$(f,q)(g,r) = (f(g \circ q^{-1}), qr)$$ □

Wreath Products as Permutations

Under certain reasonable conditions, the elements of a wreath product can be thought of as permutations. Specifically, let

$$W = D \wr_\Omega Q$$

Let $G = \boxtimes_{\omega\in\Omega} D$ and assume that Q acts faithfully on Ω.

Now suppose that the group D acts faithfully on a nonempty set Λ. Then the elements of W act on the set $\Lambda \times \Omega$. In particular, for $(f,q) \in W$ define

$$(f,q)^*: \Lambda \times \Omega \to \Lambda \times \Omega$$

by

$$(f,q)^*(\lambda,\omega) = (f(q\omega)\lambda, q\omega)$$

The map $(f,q)^*$ is injective since $(f,q)^*(\lambda,\omega) = (f,q)^*(\lambda',\omega')$ implies that

$$(f(q\omega)\lambda, q\omega) = (f(q\omega')\lambda', q\omega')$$

and so $\omega' = \omega$ and $\lambda' = \lambda$. Also, $(f,q)^*$ is surjective since for any $(\lambda,\omega) \in \Lambda \times \Omega$, we have

$$(f,q)^*(f(\omega)^{-1}\lambda, q^{-1}\omega) = (\lambda,\omega)$$

Hence, $(f,q)^*$ is a permutation of $\Lambda \times \Omega$.

Moreover, the map $\sigma: W \to S_{\Lambda\times\Omega}$ defined by $\sigma(f,q) = (f,q)^*$ is a homomorphism, since

$$\begin{aligned}
[(f,q)(g,r)]^*(\lambda,\omega) &= (f(g\circ q^{-1}), qr)^*(\lambda,\omega) \\
&= ([f(g\circ q^{-1})(qr\omega)]\lambda, qr\omega) \\
&= ([f(qr\omega)(g\circ q^{-1})(qr\omega)]\lambda, qr\omega) \\
&= (f(qr\omega)g(r\omega)\lambda, qr\omega) \\
&= (f,q)^*(g(r\omega)\lambda, r\omega) \\
&= (f,q)^*(g,r)^*(\lambda,\omega)
\end{aligned}$$

As to the kernel of σ, if $(f,q)^* = \iota$, then

$$(f(q\omega)\lambda, q\omega) = (\lambda, \omega)$$

for all $\lambda \in D$ and $\omega \in \Omega$. Since the actions of Q on Ω and D on Λ are faithful, we deduce that $q = 1$ and $f(\omega) = 1$ for all ω. Thus, $\sigma\colon W \hookrightarrow S_{\Lambda\times\Omega}$ is an embedding of W into $S_{\Lambda\times\Omega}$.

When $W = D \wr_\Omega Q$ is the restricted wreath product, we can describe the image of the embedding explicitly. For each $d \in D$ and $\alpha \in \Omega$, let $d_\alpha \in G$ be defined by

$$d_\alpha(\omega) = \begin{cases} d & \text{if } \omega = \alpha \\ 1 & \text{if } \omega \neq \alpha \end{cases}$$

Let X be the subgroup of $S_{\Lambda\times\Omega}$ generated by the permutations $(d_\alpha, 1)^*$ and $(1, q)^*$, that is,

$$X = \langle (d_\alpha, 1)^*, (1, q)^* \mid d \in D, \alpha \in \Omega, q \in Q \rangle$$

Certainly, $X \subseteq \operatorname{im}(\sigma)$.

To see that the reverse inclusion holds, we observe that

$$(f, q)^* = [(f, 1)(1, q)]^* = (f, 1)^*(1, q)^*$$

and so it is sufficient to show that $(f, 1) \in X$ for all $f \in G$. Since

$$(fg, 1)^* = [(f, 1)(g, 1)]^* = (f, 1)^*(g, 1)^*$$

we have

$$(f_1 \cdots f_n, q)^* = (f_1, 1)^* \cdots (f_n, 1)^*(1, q)^*$$

Now, any $f \in G$ has finite support $\operatorname{supp}(f) = \{\omega_1, \ldots, \omega_n\}$ and so

$$f = f(\omega_1)_{\omega_1} \cdots f(\omega_n)_{\omega_n}$$

Hence,

$$(f, 1)^* = (f(\omega_1)_{\omega_1} \cdots f(\omega_n)_{\omega_n}, 1)^* = (f(\omega_1)_{\omega_1}, 1)^* \cdots (f(\omega_n)_{\omega_n}, 1)^* \in X$$

Thus, $\operatorname{im}(\sigma) = X$.

Theorem 5.34 *Let $W = D \wr\wr_\Omega Q$ be a wreath product and suppose that Q acts faithfully on Ω. Suppose also that D acts faithfully on Λ. Then the map $\sigma\colon W \to S_{\Lambda\times\Omega}$ defined by*

$$\sigma(f, q) = (f, q)^*$$

where $(f, q)^*\colon \Lambda \times \Omega \to \Lambda \times \Omega$ is defined by

$$(f, q)^*(\lambda, \omega) = (f(q\omega)\lambda, q\omega)$$

is an embedding of W into $S_{\Lambda\times\Omega}$. When $W = D \wr_\Omega Q$ is the restricted wreath product, the image of σ is

$$\operatorname{im}(\sigma) = \langle (d_\alpha, 1)^*, (1, q)^* \mid d \in D, \alpha \in \Omega, q \in Q\rangle \qquad \square$$

It is convenient to drop the $*$ notation and to think of elements of a wreath product $D \wr Q$ as permutations of $\Lambda \times \Omega$.

Example 5.35 A **permutation matrix** P is an $n \times n$ matrix with entries from the set $\{0, 1\}$ with the property that each row contains exactly one 1 and each column contains exactly one 1. Multiplication of a matrix A on the left by a permutation matrix P permutes the rows of A. Similarly, multiplication on the right by P permutes the columns of A. Let $\mathcal{P}_n$ be the multiplicative group of all $n \times n$ permutation matrices.

For $P \in \mathcal{P}_n$, let P_i denote the ith row of P and let $P^{(j)}$ denote the jth column. The rows of P also define a permutation π_P of $\Omega = \{1, \ldots, n\}$, in particular, $\pi_P(i)$ is the column number of the 1 in row P_i. In this way, P_n is isomorphic to the symmetric group S_n. Clearly, the map $f: P \mapsto \pi_P$ is bijective. If $Q \in \mathcal{P}_n$, then $(QP)_{i,j} = Q_i P^{(j)} = 1$ if and only if the column number k of the 1 in Q_i is the same as the row number k of the 1 in column $P^{(j)}$, that is, if and only if $k = \pi_Q(i)$ implies that $\pi_P(k) = j$. Hence, $\pi_{QP}(i) = j$ implies that $\pi_P(\pi_Q(i)) = j$, that is, $\pi_{QP} = \pi_P \pi_Q$. Hence, f is an anti-isomorphism from $\mathcal{P}_n$ to S_n. Since the transpose map $\tau: \mathcal{P}_n \to \mathcal{P}_n$ is an anti-automorphism of $\mathcal{P}_n$, it follows that the composite map $f \circ \tau$ is an isomorphism from $\mathcal{P}_n$ to S_n.

We can generalize the notion of a permutation matrix as follows. If H is an abelian multiplicative group, define $\mathcal{P}_n(H)$ to be the set of $n \times n$ matrices with the property that each row and each column has exactly one entry from H, all other entries being 0. Matrix multiplication is defined using the product in H along with $0 \cdot 0 = 0 \cdot a = 0$ and $0 + a = a + 0 = a$ for $a \in H$. This makes $\mathcal{P}_n(H)$ a group. Note that $\mathcal{P}_n$ is a subgroup of $\mathcal{P}_n(H)$, with the identity of H playing the role of 1.

We can describe $\mathcal{P}_n(H)$ in terms of wreath products as follows. Let S_n act on $\Omega = \{1, \ldots, n\}$ by the usual evaluation and let $\mathcal{P}_n$ act on Ω by $P(k) = \pi_P(k)$.

Let $\mathcal{D}$ be the set of all diagonal matrices in $\mathcal{P}_n(H)$. Then every matrix in $\mathcal{P}_n(H)$ is a product DP where $D \in \mathcal{D}$ and $P \in \mathcal{P}_n$. Also, $\mathcal{D}$ is a normal subgroup of $\mathcal{P}_n(H)$. Hence, $\mathcal{P}_n(H) = \mathcal{D} \rtimes \mathcal{P}_n$.

Now, any $D \in \mathcal{D}$ can be identified with the ordered n-tuple of diagonal elements of D and, in fact, $\mathcal{D}$ is isomorphic to H^n, since matrix multiplication in $\mathcal{D}$ is elementwise product in H^n. Hence,

$$\mathcal{P}_n(H) \approx H^n \rtimes \mathcal{P}_n = H \wr_\Omega \mathcal{P}_n \approx H \wr_\Omega S_n$$

When H is the group of mth roots of unity in $\mathbb{C}$, the group $\mathcal{P}_n(H)$ is called a **generalized symmetric group** or **monomial group.**□

Exercises

1. Prove that the external direct product is commutative and associative, up to isomorphism, that is,
$$H \boxtimes K \approx K \boxtimes H$$
and
$$(H \boxtimes K) \boxtimes L \approx H \boxtimes (K \boxtimes L)$$
Is there an identity for the external direct product?
2. a) Suppose that $G = \boxplus G_i$. For each k, let
$$H_k = \{f \in \boxplus G_i \mid f(i) = 1 \text{ for } i \neq k\}$$
Show that $H_k \trianglelefteq G$ and that $G = \bowtie H_k$.
b) If $G = \bowtie G_i$, show that $G \approx \boxplus G_i$.
3. Let H, H_i and K be groups and let
$$H = H_1 \boxtimes \cdots \boxtimes H_s$$
Suppose that $\sigma: H \approx K$. For each i, let
$$\overline{H}_i = \{1\} \boxtimes \cdots \boxtimes \{1\} \boxtimes H_i \boxtimes \{1\} \boxtimes \cdots \boxtimes \{1\}$$
where H_i is in the ith component. Prove that
$$K = \sigma\overline{H}_1 \bowtie \cdots \bowtie \sigma\overline{H}_s$$
4. Let $G = H_1 \cdots H_n$ be a finite group, where $H_i \trianglelefteq G$ and $(o(H_i), o(H_j)) = 1$ for $i \neq j$. Prove that the join $\bigvee H_i$ is a direct product, that is, $G = H_1 \bowtie \cdots \bowtie H_n$.
5. Suppose that $G = H \bowtie K$ and that $N \trianglelefteq G$. Prove that if $N \cap H = \{1\} = N \cap K$, then $N \leq Z(G)$.
6. Let $o(G) = mn$ where m and n are relatively prime. Let $H \leq G$ with $o(H) = m$. Show that a subgroup $K \leq G$ is a complement of H if and only if $o(K) = n$.
7. Prove that all nonabelian groups of order p^3, p prime are indecomposable. You may assume that all groups of order p^2 are abelian (which is true).
8. Prove that the group $\mathbb{Q}$ of rational numbers is indecomposible.
9. Prove that D_{2n} is indecomposable if and only if $n \not\equiv 2 \bmod 4$.
10. Let $G = \langle a \rangle$ be a cyclic group.
a) If G is infinite, show that $G = H \bullet K$ implies that $H = \{1\}$ or $K = \{1\}$.
b) If $o(G) = n$, describe precisely the conditions under which a nontrivial essentially disjoint product representation $G = H \bullet K$ exists.
11. Let $G = \bowtie_{i \in I} H_i$ and let $K \leq G$.

a) Show that it is not necessarily true that

$$K \stackrel{?}{=} \bowtie_{i\in I} (H_i \cap K)$$

even if $K \trianglelefteq G$.

b) Recall that a group that is equal to its own commutator subgroup is called **perfect**. Prove that if K is normal and perfect, then

$$K = \bowtie_{i\in I} (H_i \cap K)$$

c) Prove that if G is periodic and if for all

$$x \in \bowtie_{j\neq i} H_j$$

and $y \in H_i$, the orders $o(x)$ and $o(y)$ are relatively prime, then

$$K = \bowtie_{i\in I} (H_i \cap K)$$

This holds in particular for finite families if the factors H_i have relatively prime exponents.

12. Let G be a group and let H be a simple subgroup of G with index 2. What can you say about any other nontrivial proper normal subgroup of G? Must such a subgroup exist? (For the latter, you may assume that the alternating group A_5 is simple and that S_n is centerless for $n \geq 3$.)
13. (**Chinese remainder theorem** for groups) Let G be a group and let $H_1, \ldots, H_n$ be normal subgroups. Consider the map $\sigma: G \to \boxtimes G/H_i$ defined by

$$\sigma a = (aH_1, \ldots, aH_n)$$

a) Show that σ is a homomorphism with $\ker(\sigma) = H_1 \cap \cdots \cap H_n$.

b) Show that if the indices $(G : H_i)$ are finite and pairwise relatively prime, then σ is surjective and so for any $a_1, \ldots, a_n \in G$, there is a $g \in G$ for which $g \in \bigcap a_i H_i$. This can also be written

$$\begin{gathered} g \equiv a_1 \bmod H_1 \\ \vdots \\ g \equiv a_n \bmod H_n \end{gathered}$$

c) Discuss the uniqueness of the solution g in part b).

14. Let G be a group and let N be minimal normal in G.

a) Prove that if G is characteristically simple, then there is a $A \subseteq \operatorname{Aut}(G)$ such that

$$G = N \bowtie \left(\bowtie_{\sigma\in A} \sigma N \right)$$

Also, N is simple, as is every term in the sum above and so G is the direct sum of isomorphic simple subgroups.

b) Prove that if G has the DCC on normal subgroups, then N is the direct sum of isomorphic simple groups.

15. To see that infinite groups are not, in general, cancellable in direct products, prove that $\mathbb{Z}[x] \boxtimes \mathbb{Z} \boxtimes \mathbb{Z} \approx \mathbb{Z}[x] \boxtimes \mathbb{Z}$, where $\mathbb{Z}[x]$ is the abelian group of polynomials over $\mathbb{Z}$ but $\mathbb{Z} \boxtimes \mathbb{Z} \not\approx \mathbb{Z}$.
16. Let $G = \bowtie S_i$, where S_i is simple for all i. Prove that the center $Z(G)$ is the direct sum of those factors S_i that are abelian. Hence, G is centerless if and only if all of the factors S_i are nonabelian.
17. Let
$$G = \bowtie_{i \in I} S_i$$
be centerless, where each S_i is simple. Prove that the only minimal normal subgroups of G are the summands S_i.
18. Let
$$G = \bowtie_{i \in I} S_i$$
be centerless, where each S_i is simple. Prove that the normal subgroups of G are precisely the direct sums of the S_i's taken over the subsets of I.
19. Let $G = \bowtie_{i \in I} H_i$ where H_i are simple subgroups and $H_i \approx H_j$ for all $i, j \in I$. Prove that G is characteristically simple. *Hint*: Consider the centers of the H_i.

Minimal Normal Subgroups

20. Let G be a group. Show that if H and K are distinct minimal normal subgroups of G, then H and K commute elementwse.
21. Let A be an abelian minimal normal subgroup of a group G. Show that if $G = AH$ where $H < G$, then $A \cap H = \{1\}$.
22. Let $N_1, \ldots, N_n$ be minimal normal subgroups of G, let $M = N_1 \cdots N_n$ and let $K \trianglelefteq G$. Prove that there is a subset of $N_1, \ldots, N_n$, say after reindexing $N_1, \ldots, N_m$, such that
$$KM = K \bowtie N_1 \bowtie \cdots \bowtie N_m$$
23. Let $A \le B \le G$ and assume that A is a minimal normal subgroup of B and B is a minimal normal subgroup of G. Assume further that the set $\{A\} = \{gAg^{-1} \mid g \in G\}$ of conjugates of A is finite. Show that A is simple and that B is the direct product of conjugates of A.
24. Prove that the epimorphic image of a minimal normal subgroup is either trivial or minimal normal.

Semidirect Products

25. Let $G = H \rtimes K$. Show that if $o(H) = 2$, then $G = H \bowtie K$.
26. Let $G = H \rtimes_\theta K$. Show that if K is simple, then $o(K) \mid o(\mathrm{Aut}(H))$.

27. Show that for every positive integer of the form $n = 2(2k+1)$ there is a centerless group of order n.
28. Prove that if $G = H \rtimes K$, then $G = H \rtimes K^a$ for any $a \in G$. Hence, if K is a complement of a normal subgroup, then so is any conjugate of K.
29. Show that it is not always possible to extend a homomorphism $\sigma: H \to G'$ from a subgroup $H \leq G$ to G.
30. Show that if $G = H \rtimes K$ is an internal semidirect product, then G is isomorphic to an external semidirect product $G \approx H \rtimes_\theta K$ for some $\theta: K \to \mathrm{Aut}(H)$.
31. Prove that

$$\mathbb{Z}_6 = A \rtimes B \quad \text{and} \quad S_3 = C \rtimes D$$

where $A \approx \mathbb{Z}_3 \approx C$ and $B \approx \mathbb{Z}_2 \approx D$, but clearly $\mathbb{Z}_6 \not\approx S_3$.
32. Prove that $S_n = A_n \rtimes \{\iota, (1\,2)\}$.
33. a) Prove that $D_8 \approx A \rtimes B_1$, where $A \approx \mathbb{Z}_4$ and $B_1 \approx \mathbb{Z}_2$.
    b) Prove that $D_8 \approx C \rtimes B_2$, where $C \approx V$ and $B_2 \approx \mathbb{Z}_2$.
    What does this say about semidirect product decompositions?
34. Prove that $D_{2n} \approx A \rtimes B$, where $A \approx \mathbb{Z}_n$ and $B \approx \mathbb{Z}_2$.
35. What is wrong, if anything, with the following argument? Let $G = H \rtimes K$. Then the projection maps $\rho_A: G \to H$ and $\rho_K: G \to K$ defined by $\rho_H(hk) = h$ and $\rho_K(hk) = k$ have kernel K and H, respectively, whence both H and K are normal subgroups of G and so $G = H \bowtie K$.
36. Recall that $G = GL(n, F)$ is the general linear group of all invertible $n \times n$ matrices over the field F and $S = SL(n, K)$ is the subgroup of matrices with determinant equal to 1.
    a) Prove that $SL(n, K) \trianglelefteq GL(n, F)$.
    b) Find a complement of $SL(n, F)$ in $GL(n, F)$. *Hint*: How does one get a special matrix from a general one?
37. Let G be a group. For any $a \in G$, denote left translation by a by ℓ_a. Thus, $\ell_a: G \to G$ is defined by $\ell_a(x) = ax$. Let $\mathcal{L} = \{\ell_a \mid a \in G\}$. Let $\mathcal{A} = \mathrm{Aut}(G)$. Finally, let $H = \langle \mathcal{L}, \mathcal{A} \rangle$ be the subgroup of the symmetric group S_G generated by $\mathcal{L}$ and $\mathcal{A}$.
    a) Prove that $H = \mathcal{L} \rtimes \mathcal{A}$.
    b) Prove that $C_H(\mathcal{L}) = \mathcal{R}$, the subgroup of all right translations $r_a: x \mapsto xr$.
38. In this exercise, we describe all groups of order pq where $p < q$ are primes. We will assume a fact to be proved later in the book: If G is a finite group and if p is the smallest prime dividing $o(G)$, then any subgroup of index p is normal in G.
    a) Describe the automorphism group $\mathrm{Aut}(C_p)$, where $C_p = \langle a \rangle$ is cyclic of order a prime p.
    b) Show that any group of order pq is a semidirect product of a group of order q by a group of order p.
    c) Describe the possible external direct products of C_q by C_p, where p and q are distinct primes.

d) "Internalize" the external semidirect products in the previous part to show that up to isomorphism, the groups of order pq are the direct product $C_q \bowtie C_p$ and for each $m \neq 1$ satisfying $m^p \equiv 1 \bmod q$, a group described by

$$G = \langle \alpha, \beta \rangle, o(\alpha) = p, o(\beta) = q, \alpha\beta = \beta^m \alpha$$

39. Let $A = \{m2^n \mid m, n \in \mathbb{Z}\}$ be the additive subgroup of $\mathbb{Q}$ and let $B = \mathbb{Z}x$ be an additive infinite cyclic group. Prove that the group $G = A \rtimes_\theta B$, where $\theta_x(a) = 2a$, has the ACC on normal subgroups but that the normal subgroup A of G does not have the ACC on normal subgroups.

Wreath Products

40. One must be cautious in working with the action of $\sigma \in S_n$ on the product D^n. Recall that σ permutes the *coordinates* of an element of D^n. Suppose that $(d_1, \ldots, d_n) \in D^n$. Then

$$\sigma(d_1, \ldots, d_n) = (d_{\sigma 1}, \ldots, d_{\sigma n})$$

For $\tau \in S_n$, compute $(\tau\sigma)(d_1, \ldots, d_n)$. Are you sure?

41. In this exercise, we take a combinatorial look at the wreath product $W = S_2 \wr S_5$, with the help of Figure 5.5.

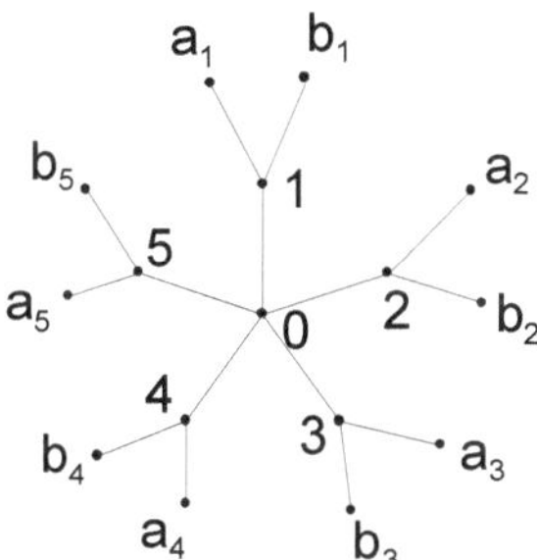

Figure 5.5

Informally speaking, a **graph** is a set of **vertices** (or **nodes**) together with a set of **edges** connecting pairs of vertices. Figure 5.6 is an example of a graph $\mathcal{G}$. Two vertices are said to be **adjacent** if there is an edge between them. A bijection of the vertices of a graph that preserves adjacency is called an **isomorphism**. Show that the automorphisms of $\mathcal{G}$ are isomorphic to the wreath product $W = S_2 \wr S_5$.

42. Show that the wreath product is associative. Specifically, let G act on Δ, H act on Ω and K act on Λ, all actions being faithful. Show that $(G \wr\wr_\Omega H) \wr\wr_\Lambda K \approx G \wr\wr_\Omega (H \wr\wr_\Lambda K)$.

43. Show that the regular wreath product is not associative. Specifically, if G, H and K are nontrivial finite groups, explain why $(G \wr_r H) \wr_r K$ cannot possibly be isomorphic to $G \wr_r (H \wr_r K)$.

44. Show that the restricted wreath product $C_2 \wr C_2$ is isomorphic to the dihedral group D_8.
45. Let X be a nonempty set of size nk and let $\mathcal{P} = \{B_1, \ldots, B_n\}$ be a partition of X into equal-sized blocks of size k. Call a permutation $\sigma \in S_X$ *nice* if it also permutes the blocks, that is, for all i, there is a j such that $\sigma B_i = B_j$. Show that the set N of nice permutations is a group isomorphic to $S_k \wr S_n$.
46. Referring to Example 5.35, let $H = \{\pm 1\}$ be the multiplicative group of square roots of unity. The group $\mathcal{P}_n(H)$ is called the **hyperoctahedral group**. Show that $\mathcal{P}_n(H)$ is isomorphic to the subgroup $G \leq S_{2n}$ of all permutations of $X = \{-n, \ldots, -1, 1, \ldots, n\}$ with the property that $\pi(-k) = -\pi(k)$.
47. Let $W = D \wr\wr_{\Omega} Q$ be a wreath product where D act faithfully on Λ. Suppose that both actions are **transitive**, that is, for any $\omega, \omega' \in \Omega$ there is a $q \in Q$ for which $q\omega = \omega'$ and similarly for the other action. Prove that the permutation representation of W is also transitive, that is, for any pair $(\lambda, \omega), (\lambda', \omega') \in \Lambda \times \Omega$ there is an $(f, q) \in W$ for which
$$(f, q)^*(\lambda, \omega) = (\lambda', \omega')$$

Chapter 6
Permutation Groups

Permutations are fundamental to many branches of mathematics. In this chapter, we examine permutations from a group-theoretic perspective.

The Definition and Cycle Representation

A **permutation** of a nonempty set X is a bijective function on X. The set of all permutations of X is denoted by S_X. As is customary, when

$$X = I_n := \{1, \ldots, n\}$$

we write S_X as S_n. As we have seen, the set S_X of permutations of a nonempty set X forms a group under composition of functions. For the record, we have:

Definition *Let X be a nonempty set. The* **symmetric group** *S_X on the set X is the group of all permutations of X, under composition of functions.*□

There are various notations for permutations. The notation

$$\sigma = \begin{pmatrix} a_1 & a_2 & \cdots & a_n \\ a_{i_1} & a_{i_2} & \cdots & a_{i_n} \end{pmatrix}$$

is sometimes used to denote the permutation that sends a_j in the top row to a_{i_j} in the bottom row. This notation is a bit combersome and we prefer the *cycle notation* described in an earlier chapter. In particular, if $a_i \neq a_j \in X$ for $i \neq j$, then

$$\sigma = (a_1 \cdots a_k)$$

denotes the permutation $\sigma \in S_X$ defined, for $0 \leq i \leq n$, by

$$\sigma a_i = \begin{cases} a_{i+1} & \text{for } 1 \leq i \leq k-1 \\ a_1 & \text{for } i = k \end{cases}$$

and sending all other elements of X to themselves. Such a permutation σ is called a k-**cycle** in S_X. A 2-cycle

$$\sigma = (a\,b)$$

is called a **transposition**. Two cycles $(a_1 \cdots a_k)$ and $(b_1 \cdots b_m)$ are **disjoint** if $a_i \neq b_j$ for all i, j.

When X is an infinite set containing the elements $\{a_i \mid i \in \mathbb{Z}\}$, then we can define the **infinite cycle**

$$\sigma = (\ldots, a_{-1}, a_0, a_1, \ldots)$$

where $a_i \neq a_j$ for $i \neq j$ as the permutation σ sending a_k to a_{k+1} for all $k \in \mathbb{Z}$ and leaving all other elements of X fixed.

Theorem 6.1 *For a permutation group S_X, the following hold.*
1) *Disjoint cycles in S_X commute.*
2) *Every permutation in S_X is a product of disjoint cycles, the product being unique except for the order of factors. A representation of σ as a product of disjoint cycles (with or without 1-cycles) is called a* **cycle representation** *or* **cycle decomposition** *of σ. The* **cycle structure** *of a permutation σ is the sequence of cycle lengths in a cycle decomposition of σ, or equivalently, the number of cycles of each length in a cycle decomposition of σ.*

Proof. Part 1) we leave to the reader. For part 2), let $\sigma \in S_X$ and define an equivalence relation on X by $x \equiv y$ if $y = \sigma^k x$ for some integer k. For $x \in X$, the equivalence class containing x is

$$[x] = \{\sigma^i x \mid i \in \mathbb{Z}\}$$

Note that the restriction $\sigma|_{[x]}$ is a permutation of $[x]$. In fact, if the elements $\sigma^i x$ are distinct for all $i \in \mathbb{Z}$, then $\sigma|_{[x]}$ is the infinite cycle

$$\sigma|_{[x]} = (\ldots, \sigma^{-2}x, \sigma^{-1}x, x, \sigma x, \sigma^2 x, \ldots)$$

On the other hand, if $\sigma^i x = \sigma^j x$ for $i < j$, then $\sigma^{j-i}x = x$ and if m is the smallest positive integer for which $\sigma^m x = x$, then $\sigma|_{[x]}$ is the m-cycle

$$\sigma|_{[x]} = (x, \sigma x, \sigma^2 x, \ldots, \sigma^{m-1}x)$$

The distinct equivalence classes form a partition of X and σ is a product of the disjoint cycles $\sigma|_B$ as B varies over these equivalence classes.□

A cycle of length k has order k in S_X. Thus, a transposition is an involution, as is any product of disjoint transpositions. A power of a cycle need not be a cycle; for example

$$(1\,2\,3\,4)^2 = (1\,3)(2\,4)$$

Since disjoint cycles commute, if $\sigma = c_1 \cdots c_m$ is a cycle representation of σ, then for any integer k,

$$\sigma^k = c_1^k \cdots c_m^k$$

Moreover, $\sigma^k = \iota$ if and only if $c_i^k = \iota$ for each i.

Theorem 6.2 *The order of $\sigma \in S_X$ is the least common multiple of the lengths (orders) of the disjoint cycles in a cycle decomposition of σ.*$\square$

The group properties of S_X do not depend upon the nature of the set X, but only upon its cardinality. More formally put, if $|X| = |Y|$, then the groups S_X and S_Y are isomorphic. Accordingly, we will feel free to state theorems in terms of either S_X or S_n (when X is finite).

A Fundamental Formula Involving Conjugation

When a permutation $\tau \in S_X$ is written as a product of disjoint cycles, it is very easy to describe the conjugates τ^σ of τ. It is also remarkably easy to tell when two permutations are conjugate.

Theorem 6.3

1) *Let $\sigma \in S_n$. For any k-cycle $(a_1 \cdots a_k)$,*

$$(a_1 \cdots a_k)^\sigma = (\sigma a_1 \cdots \sigma a_k)$$

Hence, if $\tau = c_1 \cdots c_k$ is a cycle decomposition of τ, then

$$\tau^\sigma = c_1^\sigma \cdots c_k^\sigma$$

is a cycle decomposition of τ^σ.

2) *Two permutations are conjugate if and only if they have the same cycle structure.*

Proof. For part 1), we have

$$(a_1 \cdots a_k)^\sigma (\sigma a_i) = \begin{cases} \sigma a_{i+1} & i < k \\ \sigma a_1 & i = k \end{cases}$$

Also, if $b \neq \sigma a_i$ for any i, then $\sigma^{-1} b \neq a_i$ and so

$$(a_1 \cdots a_k)^\sigma b = \sigma (a_1 \cdots a_k)(\sigma^{-1} b) = \sigma(\sigma^{-1} b) = b$$

Hence, $(a_1 \cdots a_k)^\sigma$ is the cycle $(\sigma a_1 \cdots \sigma a_k)$. For part 2), if $\tau = c_1 \cdots c_m$ is a cycle decomposition of τ, then

$$\tau^\sigma = c_1^\sigma \cdots c_m^\sigma$$

and since c_i^σ is a cycle of the same length as c_i, the cycle structure of τ^σ is the same as that of τ.

For the converse, suppose that σ and τ have the same cycle structure. If σ and τ are cycles, say

$$\sigma = (a_1 \cdots a_n) \quad \text{and} \quad \tau = (b_1 \cdots b_n)$$

then any permutation λ that sends a_i to b_i satisfies $\sigma^\lambda = \tau$. More generally, if

$$\sigma = c_1 \cdots c_m \quad \text{and} \quad \tau = d_1 \cdots d_m$$

are the cycle decompositions of σ and τ, ordered so that c_k has the same length as d_k, then we can define a permutation λ that sends the element in the ith position of c_k to the element in the ith position of d_k. Then $\sigma^\lambda = \tau$.□

The previous theorem implies that it is easy to tell when a subgroup H of the symmetric group S_X is normal.

Theorem 6.4 *A subgroup $H \le S_X$ is normal if and only if whenever $\sigma \in H$, then so are all permutations in S_X with the same cycle structure as σ.*□

Parity

Every cycle is a product of transpositions, to wit

$$(a_1 a_2 \cdots a_m) = (a_1 a_m)(a_1 a_{m-1}) \cdots (a_1 a_3)(a_1 a_2)$$

and so every permutation is a product of transpositions. While this representation is far from unique, the *parity* of the number of transpositions is unique.

Theorem 6.5 *Let $\sigma \in S_n$.*

1) *Exactly one of the following holds:*
   a) *σ can be written as a product of an even number of transpositions, in which case we say that σ is* **even** *(or has* **even parity***).*
   b) *σ can be written as a product of an odd number of transpositions, in which case we say that σ is* **odd** *(or has* **odd parity***).*
2) *The symbol $(-1)^\sigma$ or $\operatorname{sg}(\sigma)$, called the* **sign** *or* **signum** *of σ, is equal to 1 if σ is even and -1 if σ is odd. We have*

$$(-1)^\sigma = (-1)^{n-k}$$

where k is the number of cycles in the cycle decomposition of σ (including 1-cycles).

Proof. For part 1), if σ can be written as a product of an even number of transpositions and an odd number of transpositions, say

$$\sigma = \rho_1 \cdots \rho_{2v} = \tau_1 \cdots \tau_{2u+1}$$

then the identity can be written as a product of an odd number of transpositions

$$\iota = \tau_1 \cdots \tau_{2u+1} \rho_{2v} \cdots \rho_1$$

To show that this is not possible, we choose a representation of ι as a product of an odd number of transpositions as follows:

1) Find the smallest odd integer $m > 1$ for which ι is the product of m transpositions.
2) Choose an integer k that appears in at least one such representation of ι.
3) Among all representations of ι as a product of m transpositions that contain k, let τ be the one whose rightmost appearence of k is as far to the left as possible, say

$$\tau = \theta_1 \cdots \theta_{t-1}(x\,k)\theta_{t+1} \cdots \theta_m$$

where $\theta_{t+1} \cdots \theta_m$ does not involve k.

Note that $t \geq 2$, since otherwise the only appearance of k is in θ_1 and so $\tau \neq \iota$. However, we can easily move this rightmost occurrence of k one transposition to the left by using the following substitutions. Suppose that $\theta_{t-1} = (a\,b)$. Note that $(a\,b) \neq (x\,k)$ since otherwise the two transpositions would cancel, contradicting the definition of m.

1) If $(a\,b)$ and $(x\,k)$ are disjoint, then they commute

$$(a\,b)(x\,k) = (x\,k)(a\,b)$$

2) If $(a\,b) = (x\,b)$ for $b \neq k$, then

$$(x\,b)(x\,k) = (x\,k\,b) = (b\,x\,k) = (b\,k)(b\,x)$$

3) If $(a\,b) = (k\,b)$ where $a \neq k$, then write

$$(k\,b)(x\,k) = (k\,x\,b) = (x\,b\,k) = (x\,k)(x\,b)$$

Thus, we can move the rightmost occurrence of k to the left, contradicting the construction of τ and proving part 1).

For part 2), a cycle of length $m \geq 2$ can be written as a product of $m - 1$ transpositions as above. Now suppose that the cycle decomposition of σ is

$$\sigma = c_1 \cdots c_r d_1 \cdots d_s$$

where $\text{len}(c_i) = m_i \geq 2$ and $\text{len}(d_i) = 1$. Then σ can be written as a product of the following number of transpositions:

$$\sum_{i=1}^{r} (m_i - 1) = (n - s) - r = n - k \qquad \square$$

Generating Sets for S_n and A_n

There are a variety of useful generating sets for the symmetric group S_n. For example, we have seen that the set of transpositions generates S_n.

Theorem 6.6 *The following sets generate S_n.*
1) The set of transpositions.
2) The "transpositions of 1"

$$(1\,2), (1\,3), \ldots, (1\,n)$$

3) The "adjacent transpositions"

$$(1\,2), (2\,3), (3\,4), \ldots, (n-1\,n)$$

4) The n-cycle $(1\cdots n)$ and the transposition $(1\,2)$.
5) The cycle $(2\cdots n)$ and a single transposition $(1\,k)$ for $2 \le k \le n$.
Proof. For part 2) we have

$$(a\,b) = (1\,b)^{(1\,a)}$$

and so the subgroup generated by the transpositions of 1 contains all transpositions. For part 3), if A is the subgroup generated by the adjacent transpositions, then $(1\,2) \in A$ and if $(1\,k) \in A$, then so is

$$(1\,k+1) = (1\,k)^{(k\,k+1)}$$

Hence, part 2) implies that $A = S_n$. For part 4), if $P = \langle (2\cdots n), (1\,2) \rangle$, then $(1\,2) \in P$ and if $(k-1\,k) \in P$, then so is

$$(k, k+1) = (k-1\,k)^{\sigma}$$

for all $k \le n-1$. Hence, $P = S_n$. For part 5), if $\sigma = (2\cdots n)$, then

$$\begin{aligned}(1\,k)^{\sigma} &= (1\,k+1)\\ (1\,k+1)^{\sigma} &= (1\,k+2)\\ &\vdots\\ (1\,n-1)^{\sigma} &= (1\,n)\end{aligned}$$

and since $\mu = \sigma^{-1} = (n\cdots 2)$,

$$\begin{aligned}(1\,k)^{\mu} &= (1\,k-1)\\ (1\,k-1)^{\mu} &= (1\,k-2)\\ &\vdots\\ (1\,3)^{\mu} &= (1\,2)\end{aligned}$$

and so we get all transpositions of 1.□

As to the alternating group, we have the following.

Theorem 6.7 *For $n \ge 3$, A_n is generated by the 3-cycles.*
Proof. Any even permutation is a product of pairs of distinct transpositions. For "overlapping" transpositions, we have

$$(a\,b)(a\,c) = (a\,c\,b)$$

and for disjoint transpositions, we can arrange for overlaps as follows:

$$(a\,b)(c\,d) = (a\,b)(b\,c)(b\,c)(c\,d) = (b\,c\,a)(c\,d\,b)$$

Hence, A_n is generated by 3-cycles.□

Subgroups of S_n and A_n

The alternating group A_n sits very comfortably inside S_n.

Theorem 6.8

1) *The signum map $\phi\colon \sigma \mapsto (-1)^\sigma$ is a group homomorphism from S_n onto the multiplicative group $\{-1, 1\}$, with kernel A_n. Hence, $A_n \trianglelefteq S_n$ has index 2.*
2) *A_n is the only subgroup of S_n of index 2.*
3) **(Subgroups of S_n)** *If $H \le S_n$, then either $H \le A_n$ or H contains an equal number of even and odd permutations and so has even order.*
4) **(Subgroups of A_n)** *For $n \ge 3$, A_n has no subgroup of index 2.*

Proof. For part 2), let H be a subgroup of S_n of index 2. Theorem 3.17 implies that $\sigma^2 \in H$ for any $\sigma \in S_n$ and so the squares of all 3-cycles are in H. But any 3-cycle is the square of another 3-cycle:

$$(a\,b\,c) = (a\,c\,b)^2$$

and so H contains all 3-cycles, which generate A_n, whence $H = A_n$. Part 3) also follows from Theorem 3.17. For part 4), if $(A_n : H) = 2$, then $\sigma^2 \in H$ for all $\sigma \in A_n$ and so H contains all 3-cycles, a contradiction.□

The Alternating Group Is Simple

We wish to show that the alternating group A_n is simple for $n \ne 4$, but not for $n = 4$. We will leave proof of the cases $n \le 4$ to the reader and assume that $n \ge 5$.

Suppose that $N \trianglelefteq S_n$ is nontrivial. We have seen that any two 3-cycles are conjugate in S_n. However, it is also true that any two 3-cycles are conjugate in A_n, for $n \ge 5$. To see this let $\alpha = (a_1 a_2\, a_3)$ and $\beta = (b_1\, b_2\, b_3)$ be distinct 3-cycles with $a_1 \ne b_i$ for all i. If

$$\sigma = (a_1\, b_1)(a_2\, b_2)(a_3\, b_3)$$

where $(a_i\, b_i)$ is the identity when $a_i = b_i$, then $\alpha^\sigma = \beta$. Finally, if σ is not even, then we can take $x \notin \{a_1, b_1, b_2, b_3\}$, in which case $\alpha^{(a_1\, x)\sigma} = \beta$.

Thus, since A_n is generated by 3-cycles, we can prove that A_n is simple by showing that any nontrivial normal subgroup of A_n contains at least one 3-cycle.

A_5 is Simple

To see that A_5 is simple, note that the possible cycle representations of elements of A_5, excluding 1-cycles, are

$$(\cdot\cdot\cdot\cdot\cdot),\quad (\cdot\cdot\cdot)\quad \text{and}\quad (\cdot\cdot)(\cdot\cdot)$$

If N contains a 5-cycle, we may assume that $\sigma = (1\,2\,3\,4\,5) \in N$. To shorten the 5-cycle to a 3-cycle, we reverse the effects of σ on 1 and 3 by taking the product

$$\tau = (2\,1\,4\,3\,5)(1\,2\,3\,4\,5) = (2\,5\,4)$$

and since

$$(2\,1\,4\,3\,5) = (1\,2\,3\,4\,5)^{(1\,2)(3\,4)} \in N$$

we deduce that N contains the 3-cycle τ.

If N contains the product of two disjoint transpositions, we may assume that

$$(1\,2)(3\,4) \in N$$

To get a 3-cycle, recall that the product of two distinct *nondisjoint* transpositions is a 3-cycle. Thus,

$$(1\,2)(3\,4)(1\,2)(4\,5) = (3\,4)(4\,5) = (4\,5\,3)$$

and since

$$(1\,2)(4\,5) = [(1\,2)(3\,4)]^{(3\,4\,5)} \in N$$

we see that N contains a 3-cycle in this case as well.

A_n is Simple for $n \geq 5$

We now examine the general case $n \geq 5$. Let $N \trianglelefteq A_n$ and let $\sigma \in N$.

Case 1

Suppose that the cycle decomposition of σ contains a cycle of length $k \geq 4$, say

$$\sigma = \pi(a_1 \cdots a_m\, x\, y\, z)$$

where $m \geq 1$. If

$$\lambda = (x\, z\, y\, a_m \cdots a_1)$$

then λ and π are disjoint and

$$\tau := \lambda\pi^{-1}\sigma = (x\, z\, y\, a_m \cdots a_1)(a_1 \cdots a_m\, x\, y\, z) = (x\, a_m\, z)$$

is a 3-cycle. Moreover,

$$\pi\lambda^{-1} = \pi(a_1 \cdots a_m\, y\, z\, x) = [\pi(a_1 \cdots a_m\, x\, y\, z)]^{(x\, y\, z)} = \sigma^{(x\, y\, z)} \in N$$

and so $\lambda\pi^{-1} \in N$, which implies that the 3-cycle τ is in N.

Case 2

Suppose that the cycle decomposition of σ contains two or more 3-cycles, say

$$\sigma = (a\, b\, c)(x\, y\, z)\pi$$

Then N contains the permutation

$$\tau := \sigma^{(a\, y\, z)} = (y\, b\, c)(x\, z\, a)\pi$$

and therefore also the 5-cycle

$$\sigma\tau^{-1} = (a\, b\, c)(x\, y\, z)(a\, z\, x)(c\, b\, y) = (a\, x\, b\, z\, y)$$

Hence, case 1) completes the proof.

Case 3

If the cycle decomposition of σ consists of a single 3-cycle, with possibly some additional transpositions,

$$\sigma = (a\, b\, c)\pi_1 \cdots \pi_m$$

then N contains the 3-cycle

$$\sigma^2 = (a\, b\, c)^2 = (a\, c\, b)$$

Case 4

If the cycle decomposition of σ is a product of at least three disjoint transpositions,

$$\sigma = (a\, b)(x\, y)(z\, w)\pi_1 \cdots \pi_m$$

then N also contains

$$\tau := \sigma^{(b\, x)(y\, z)} = (a\, x)(b\, z)(y\, w)\pi_1 \cdots \pi_m$$

and therefore also

$$\sigma\tau = (a\, b)(x\, y)(z\, w)(a\, x)(b\, z)(y\, w) = (a\, y\, z)(b\, w\, x)$$

and so case 2) completes the proof.

Case 5

Suppose that the cycle decomposition of σ has the form

$$\sigma = (a\, b)(c\, d)$$

If $x \notin \{a, b, c, d\}$, then N contains

$$\tau := \sigma^{(a\,b\,x)} = (b\,x)(c\,d)$$

and therefore also the 3-cycle

$$\sigma\tau = (a\,b)(c\,d)(b\,x)(c\,d) = (a\,b)(b\,x) = (b\,x\,a)$$

Thus, in all cases N contains a 3-cycle and so $N = A_n$.

Theorem 6.9 *The alternating group A_n is simple if and only if $n \neq 4$.*□

As to the normal subgroups of S_n, we have the following.

Theorem 6.10 *If $n \neq 4$, then S_n has no normal subgroups other than $\{1\}, A_n$ and S_n.*

Proof. This is easily checked for $n \leq 2$ so assume that $n \geq 3$. If $N \trianglelefteq S_n$, then $N \cap A_n \trianglelefteq A_n$. Hence, $N \cap A_n = \{\iota\}$ or $N \cap A_n = A_n$. But if $N \cap A_n = \{\iota\}$, then $N = \{\iota, \sigma\}$ where σ is an odd involution and so has cycle decomposition

$$\sigma = (a_1\,b_1)\cdots(a_k\,b_k)$$

where the transpositions $(a_i\,b_i)$ are disjoint. Since N is normal, it must contain all permutations with the same cycle structure as σ, which is not the case if $n \geq 3$. Hence, $A_n \leq N$ and so $N = A_n$ or $N = S_n$.□

Example 6.11 (An infinite simple group) Using the fact that A_n is simple for $n \neq 4$, we can construct an example of an infinite simple group. Let X be an infinite set and let S_X be the symmetric group on X. Thus, $\sigma|_{\text{supp}(\sigma)}$ is a permutation of $\text{supp}(\sigma)$ with no fixed points and σ is the identity on $X \setminus \text{supp}(\sigma)$. Let $S_{(X)}$ be the subgroup of S_X consisting of all elements of S_X that have finite support. Let $A_{(X)}$ be the subgroup of $S_{(X)}$ consisting of those permutations $\sigma \in S_{(X)}$ for which $\sigma|_{\text{supp}(\sigma)}$ is an even permutation. We leave it as an exercise to show that $A_{(X)}$ is an infinite simple group. Moreover, $A_{(X)}$ is the smallest nontrivial normal subgroup in S_X.□

Some Counting

It is sometimes useful to know how many permutations there are with a particular cycle structure.

Theorem 6.12

1) *The number of cycles of length k in S_n is*

$$\binom{n}{k}(k-1)! = \frac{n!}{k(n-k)!}$$

In particular, the number of n-cycles in S_n is $(n-1)!$.

2) *The number of permutations in S_n whose cycle structure consists of r_i cycles of length k_i, for $i = 1, \ldots, m$ is*

$$\frac{n!}{r_1! \cdots r_m! k_1^{r_1} \cdots k_m^{r_m}}$$

3) *Let $s(n,k)$ be the number of permutations in S_n whose cycle structure has exactly k cycles (including 1-cycles). Then*

$$s(n,1) = (n-1)!$$

and for $k \geq 2$

$$s(n,k) = s(n-1,k-1) + (n-1)s(n-1,k)$$

Also,

$$\sum_{k=1}^{n} s(n,k)x^k = x^{(n)} := x(x+1)\cdots(x+n-1)$$

The numbers $s(n,k)$ are known as the **Stirling numbers of the first kind**. *The expression $x^{(n)}$ is known as the nth* **upper factorial**.

Proof. We leave proof of part 1) to the reader. For part 2), we write down a template consisting of r_i cycles of length k_i, with dots in place of the elements of $I_n = \{1, \ldots, n\}$. For example, if the cycle lengths are $3, 3, 2, 1$, then the template is

$$(\cdot\cdot\cdot)(\cdot\cdot\cdot)(\cdot\cdot)(\cdot)$$

Now, the dots can be replaced by the elements of I_n in $n!$ different ways. However, there are two ways in which the number $n!$ is an overcount of the desired number. First, for each of the r_i cycle templates of length k_i, a cycle can start at any of its k_i elements, so we must divide by $n!$ by $k_i^{r_i}$, giving

$$\frac{n!}{k_1^{r_1} \cdots k_m^{r_m}}$$

Second, for each cycle length k_i, the $r_i!$ arrangements of the r_i cycles of length k_i are counted separately in the number above, but give the same permutation, so we must also divide by $r_1! \cdots r_m!$.

For part 3), it is easy to see that

$$s(n,1) = (n-1)!$$

In general, we group the permutations in S_n with exactly k cycles into two groups, depending on whether the permutation contains the 1-cycle (n). There are $s(n-1,k-1)$ permutations containing the 1-cycle (n). The other permutations are formed by inserting n after any of the $n-1$ elements in the $s(n-1,k)$ permutations of $n-1$ elements into k cycles. Thus, for $k \geq 2$

$$s(n,k) = s(n-1,k-1) + (n-1)s(n-1,k)$$

Now we can verify the formula for $s(n,k)$ by induction. It is easy to see that the formula holds for $n=1$. Assume the formula is true for $s(m,k)$ where $m<n$. Then

$$\begin{aligned}
&\sum_{k=1}^{n} s(n,k)x^k \\
&\quad = (n-1)!x + \sum_{k=2}^{n} s(n-1,k-1)x^k + (n-1)\sum_{k=2}^{n} s(n-1,k)x^k \\
&\quad = (n-1)!x + x\sum_{k=1}^{n-1} s(n-1,k)x^k \\
&\qquad\qquad + (n-1)\left[\sum_{k=1}^{n} s(n-1,k)x^k - (n-2)!x\right] \\
&\quad = (n-1)!x + x(x)^{(n-1)} + (n-1)[x^{(n-1)} - (n-2)!x] \\
&\quad = (x+n-1)x^{(n-1)} \\
&\quad = x^{(n)}
\end{aligned}$$

as desired.□

Exercises

1. Let
$$\sigma = (1\,2\,3)(4\,5)(6)$$
and
$$\tau = (4\,5\,6)(1\,3)(2)$$
be elements of S_6. Find a permutation $\rho \in S_6$ for which $\sigma^\rho = \tau$. Is ρ unique?
2. Find the smallest normal subgroup of S_4 containing $\sigma = (1\,2)(3\,4)$. Is this the smallest subgroup of S_4 containing σ?
3. Let $\sigma \in S_n$ have prime order p. Prove that σ is a product of disjoint p-cycles. If σ moves all elements of I_n, show that $p \mid n$.
4. Show that two even permutations may be conjugate in S_5 but not in A_5.
5. Prove that A_n has no subgroup of index 2 without using the fact that A_n is simple for $n \neq 4$. How does this relate to Lagrange's theorem?
6. a) Find all subgroups of A_4.
   b) Find all normal subgroups of A_4. Is A_4 the join of all of its proper normal subgroups?
7. Prove that the 3-cycles $(123), (124), \ldots, (12n)$ generate A_n.
8. Let $n \geq 3$ and $k \geq 1$. Prove that A_n is generated by the cycles of length $2k+1$.

9. Prove that for $n \geq 5$, A_n is generated by the set of all products of pairs of *disjoint* transpositions. Does this hold for $n < 5$?
10. Let $n \geq 5$. Prove that the only proper subgroup H of S_n with $(S_n : H) < n$ is A_n.
11. Prove that S_n is centerless for $n \geq 3$. Prove that A_4 is centerless. What about A_n in general?
12. A subgroup H of S_n is **transitive** if for any $j, k \in I_n$, there is a $\sigma \in H$ for which $\sigma j = k$. Prove that the order of a transitive subgroup H of S_n is divisible by n.
13. A subgroup H of S_n is k-**ply transitive** if for any distinct $i_1, \ldots, i_k \in I_n$ and distinct $j_1, \ldots, j_k \in I_n$ there is a permutation $\sigma \in H$ for which $\sigma i_u = j_u$. Prove that A_n is $(n-2)$-ply transitive. Is it $(n-1)$-ply or n-ply transitive?
14. Let X be an infinite set. Define the **alternating group** A_X is defined to be the set of all permutations in S_n that can be written as the product of an even number of transpositions. Let H_X be the set of all permutations in S_X that fix all but a finite number of elements of X.
    a) What is the relationship between A_X, H_X and S_X (including normality and index)?
    b) Prove that A_X is simple.
15. In S_n, for each $k = 3, \ldots, n$, let

$$\sigma_k = \prod_{i=1}^{\lfloor \frac{k}{2} \rfloor} (i\, k - i)$$

for example,

$$\begin{aligned} \sigma_3 &= (1\,2) \\ \sigma_4 &= (1\,3) \\ \sigma_5 &= (1\,4)(2\,3) \\ \sigma_6 &= (1\,5)(2\,4) \\ \sigma_7 &= (1\,6)(2\,5)(3\,4) \\ \sigma_8 &= (1\,7)(2\,6)(3\,5) \end{aligned}$$

Show that the permutations $\sigma_1, \ldots, \sigma_n$ generate S_{n-1}.
16. What is the largest order of the elements of S_{10}?
17. (Determining the parity of a permutation) Let $X = \{x_1, \ldots, x_n\}$. Let P be the set of all polynomials in the x_i's with rational coefficients. Then we can apply a $\sigma \in S_n$ to the elements of P by applying σ to the variables individually. For example, if

$$p = x_1 x_3 - \frac{1}{2} x_2^3$$

then

$$\sigma p = \sigma(x_1)\sigma(x_3) - \frac{1}{2}\sigma(x_2)^3 \in P$$

a) Show that for $p, q \in P$,

$$\sigma(p+q) = \sigma(p) + \sigma(q), \sigma(pq) = \sigma(p)\sigma(q)$$

and for $\sigma, \tau \in S_X$,

$$\sigma(\tau p) = (\sigma\tau)p$$

b) Consider the polynomial

$$p = p(x_1, \ldots, x_n) = \prod_{i>j} \frac{x_i - x_j}{i - j} \in P$$

Show that if $\sigma = (x_1\, x_a)$ is a transposition, then

$$\sigma p = \prod_{i>j} \frac{\sigma(x_i) - \sigma(x_j)}{i - j} = -p$$

c) Show that for any $\sigma \in S_n$,

$$\sigma p = (-1)^\sigma p$$

Hence, if $\sigma \in S_n$, then since $p(1, 2, \ldots, n) = 1$, we have

$$\sigma p(1, 2, \ldots, n) = \prod_{i>j} \frac{\sigma(i) - \sigma(j)}{i - j} = (-1)^\sigma$$

d) Let $\sigma \in S_n$. An **inversion** in σ is a pair (j, i) of indices with $1 \le i, j \le n$ satisfying $j < i$ but $\sigma(j) > \sigma(i)$. Prove that the sign of σ is the parity of the number of inversions in σ.

e) Determine whether the permutation $\sigma \in S_9$ is even or odd, where

$$\sigma = (7\,2\,4\,9\,3\,8\,6\,5\,1)$$

18. For each $\tau \in S_X$, we may associate the conjugation map $\lambda_\tau \in S_X$ defined by $\lambda_\tau(\sigma) = \sigma^\tau$.
    a) Show that the map $\lambda: G \to S_X$ defined by $\lambda(\tau) = \lambda_\tau$ is a homomorphism.
    b) Find the kernel of λ.
    c) Suppose that $\emptyset \ne A \subseteq X$ is **invariant** under conjugation, that is, $a \in A$ implies that $a^\sigma \in A$. Then we can restrict λ_σ to a permutation of A. Find the kernel of the map $\lambda': D \to S_A$.
19. Let G be any normal subgroup of S_n (such as A_n or $S_n$itself).

a) Show that if $\sigma, \tau, \rho \in G$, then

$$\sigma^\tau = \sigma^\rho \Leftrightarrow \tau \in \rho C_G(\sigma)$$

b) Find a one-to-one correspondence between the set σ^G of conjugates of σ and the set of cosets of $C_G(\sigma)$ in G. The set σ^G is referred to as the **conjugacy class** of σ or the **orbit** of σ under conjugation. The centralizer $C_G(\sigma)$ is also referred to as the **stabilizer** of σ under conjugation.

c) Prove the **orbit-stabilizer relationship** for conjugation in G: For any $\sigma \in G$,

$$|\sigma^G| = (G : C_G(\sigma)) = \frac{|G|}{|C_G(\sigma)|}$$

d) Let $\alpha \in S_n$ be an n-cycle. Prove that $C_{S_n}(\alpha) = \langle \alpha \rangle$. Put another way, a permutation $\sigma \in S_n$ commutes with α if and only if it is a power of α.

20. If $\sigma, \tau \in A_n$ are conjugate in S_n, it does not necessarily follow that σ and τ are conjugate in A_n. Suppose that $\sigma \in A_n$ commutes with an *odd* permutation $\lambda \in S_n$
    a) Prove that if σ and τ are conjugate in S_n, then they are also conjugate in A_n.
    b) Prove that the centralizers of σ are related as follows:

$$C_{S_n}(\sigma) = C_{A_n}(\sigma) \cup \lambda C_{A_n}(\sigma)$$

and, in particular,

$$|C_{S_n}(\sigma)| = 2|C_{A_n}(\sigma)|$$

Note: It is not hard to find permutations $\sigma \in A_n$ that commute with an odd permutation. For instance, if $\sigma \in A_n$ does not move either a or b, then σ commutes with the transposition $(a\,b)$. Thus, for instance, an $(n-2)$-cycle in A_n commutes with a transposition. As another example, if $\sigma \in A_n$ interchanges a and b, then $\sigma = \mu(a\,b)\mu'$ where μ and μ' fix a and b. It follows that σ commutes with $(a\,b)$.

21. Let X be a nonempty set. Define the **support** of a permutation $\sigma \in S_X$ to be the set of elements of X that are moved by σ:

$$\text{supp}(\sigma) = \{x \in X \mid \sigma x \neq x\}$$

Let $S_{(X)}$ be the set of all permutations in S_X with finite support, that is, for which $\text{supp}(\sigma)$ is a finite set. This set is called the **restricted symmetric group** on the set X and is used in defining the determinant of an infinite matrix.
    a) Show that $\text{supp}(\sigma) = \text{supp}(\sigma^{-1})$.
    b) Show that $\text{supp}(\sigma\tau) \subseteq \text{supp}(\sigma) \cup \text{supp}(\tau)$.
    c) Show that $\text{supp}(\tau\sigma\tau^{-1}) = \tau(\text{supp}(\sigma))$.
    d) Show that $\text{supp}(\sigma) \cap \text{supp}(\tau) = \emptyset$ implies that $\sigma\tau = \tau\sigma$.

e) Show that $S_{(X)} \trianglelefteq S_X$ and that $S_{(X)} = S_X$ if and only if X is finite.
f) Show that if X is infinite, then $S_{(X)}$ is an infinite group in which every element has finite order and that $S_X/S_{(X)}$ is infinite.
g) Let X be infinite and let $A_{(X)}$ be the subgroup of $S_{(X)}$ consisting of those permutations $\sigma \in S_{(X)}$ for which $\sigma|_{\text{supp}(\sigma)}$ is an even permutation. Show that $A_{(X)}$ is an infinite simple group. Moreover, $A_{(X)}$ is the smallest normal subgroup in S_X.

22. For the Stirling numbers $s(n,k)$ of the first kind, show that

$$\sum_{k=1}^{n} s(n,k) = n!$$

23. (Stirling numbers of the second kind) For the curious, we present a brief discussion of the "other" Stirling numbers. We have seen that the Stirling numbers $s(n,k)$ of the first kind count the number of ways to partition a set of size n into k disjoint nonempty "cycles." The **Stirling numbers of the second kind**, denoted by $S(n,k)$, count the number of ways to partition a set of size n into k nonempty disjoint subsets. It is clear that

$$S(n,1) = 1$$

a) Find $S(n,2)$.
b) Show that for $n > 0$,

$$S(n,k) = kS(n-1,k) + S(n-1,k-1)$$

c) Show that

$$x^n = \sum_{k=1}^{n} S(n,k)x^{(k)}$$

Chapter 7
Group Actions; The Structure of p-Groups

Group Actions

We have had a few occasions to use group actions $\lambda: G \to S_X$ in the past and we now wish to make a more systematic study of group actions, beginning with the definition.

Definition *An* **action** *of a group G on a nonempty set X is a group homomorphism $\lambda: G \to S_X$, called the* **representation map** *for the action. The permutation λa is often denoted by λ_a, or simply by a itself when no confusion can arise. Thus,*

$$1x = x \quad \textit{and} \quad (ab)x = a(bx)$$

for all $x \in X$. When λ is a group action, we say that G **acts** *on X by λ or that X is a* **G-set** *under λ. An action is* **faithful** *if it is an embedding.*

1) *An element $a \in G$ is said to* **fix** *$x \in X$ if $ax = x$ and* **move** *x if $ax \neq x$.*
2) *An element $x \in X$ is* **stable** *if every element of G fixes x. We will denote the set of all stable elements by* $\mathrm{Fix}_X(G)$ *or just* $\mathrm{Fix}(G)$.□

Let X^X denote the set of all functions from X to X. If X is a finite set, then any *function* $\lambda: G \to X^X$ that satisfies $\lambda_1 = \iota$ and $\lambda_{ab} = \lambda_a \lambda_b$ is an action of G on X, since

$$\begin{aligned} \lambda_a x = \lambda_a y &\Rightarrow \lambda_{a^{-1}} \lambda_a x = \lambda_{a^{-1}} \lambda_a y \\ &\Rightarrow \lambda_{a^{-1}a} x = \lambda_{a^{-1}a} y \\ &\Rightarrow \lambda_1 x = \lambda_1 y \\ &\Rightarrow x = y \end{aligned}$$

and so λ_a is injective and therefore a permutation of X. Thus, for *finite* sets, we do not need to check *explicitly* that λ_a is a permutation of X.

Orbits and Stabilizers

Two elements $x, y \in X$ are **G-equivalent** if there is an $a \in G$ for which $ax = y$. Since λ is a homomorphism, G-equivalence is an equivalence relation on X. The equivalence classes

$$Gx = \{ax \mid a \in G\}$$

are called the **orbits** of the action. We will use the notations Gx, $\text{orb}(x)$ and $\text{orb}_G(x)$ for the orbit of x under G. The distinct orbits form a partition of the set X.

On the group side of an action, we can associate to each element $x \in X$ the set of all element of G that fix x:

$$\text{stab}(x) = \text{stab}_G(x) = \{a \in G \mid ax = x\}$$

This subgroup of G is called the **stabilizer** of x.

Definition *Let G be a group and X a nonempty set.*

1) *An action of G on X is* **transitive** *if every pair of elements of X are G-equivalent, that is, if there is only one orbit in X. In this case, we also say that G is* **transitive** *on X.*
2) *An action of G on X is* **regular** *if it is transitive and if the stabilizer* $\text{stab}(x)$ *is trivial for every $x \in X$. In this case, we also say that G is* **regular** *on X.*$\square$

The Kernel of the Representation Map

The kernel of the representation map $\lambda\colon G \to S_X$ is

$$\ker(\lambda) = \{a \in G \mid \lambda_a = \iota\}$$

As we have remarked, the action λ is **faithful** if $\ker(\lambda)$ is trivial.

Theorem 7.1 *The kernel of an action $\lambda\colon G \to S_X$ is the intersection of the stabilizers of all elements of X,*

$$\ker(\lambda) = \bigcap_{x \in X} \text{stab}(x) \qquad \square$$

The importance of the kernel $K = \ker(\lambda)$ of the representation map $\lambda\colon G \to S_X$ stems from two facts: K is a normal subgroup of G and there is an embedding of G/K into the symmetric group S_X. In particular, if $|X| = n$ is finite, then $(G : K) \mid n!$ and if λ is faithful, then $o(G) \mid n!$.

The Key Relationships

A set consisting of precisely one element from each orbit of G in X is a **system of distinct representatives**, or **SDR** for the orbits in X.

For a given SDR, we will have occasion to separate the representatives of the 1-element orbits from the representatives of the orbits of size greater than 1. Accordingly, we denote a system of distinct representatives for the orbits of size greater than 1 by $\text{SDR}_{>1}$.

Our immediate goal is to establish some key facts concerning group actions. First, as the various elements of a group G act on an element $x \in X$, the orbit of x is described. Of course, different elements of G may have the same effect on x. In fact,

$$ax = bx \quad \Leftrightarrow \quad b^{-1}a \in \text{stab}(x) \quad \Leftrightarrow \quad a\text{stab}(x) = b\text{stab}(x)$$

Thus, $ax = bx$ if and only if a and b are in the same coset of $\text{stab}(x)$ in G and so we can think of the cosets themselves as acting on the elements $x \in X$, describing each element of the orbit of x *with no duplications*, that is, distinct cosets send $x \in X$ to distinct elements of X.

Put another way, there is a bijection between $G/\text{stab}(x)$ and Gx. This gives the **orbit-stabilizer relationship**

$$|Gx| = (G : \text{stab}(x))$$

It follows that

$$|X| = \sum_{x \in \text{SDR}} |Gx| = \sum_{x \in \text{SDR}} (G : \text{stab}(x))$$

We will refer to this equation as the **class equation** for the action of G on X. (Actually, this term is traditionally reserved for a specific case of this equation, arising from the specific action of G on itself by conjugation. We will encounter this specific case a bit later in the chapter.)

Another key property of a group action is the following. If $x, y \in X$ are G-equivalent, say $y = ax$ for $a \in G$, then the stabilizers of x and y are related as follows:

$$\begin{aligned} \text{stab}(ax) &= \{b \in G \mid bax = ax\} \\ &= \{b \in G \mid a^{-1}bax = x\} \\ &= \{b \in G \mid a^{-1}ba \in \text{stab}(x)\} \\ &= \text{stab}(x)^a \end{aligned}$$

Finally, suppose that G acts on X and that the restriction of this action to a subgroup $H \leq G$ is transitive on X. Then the action of G is duplicated by the action of H, that is, for any $g \in G$ and $x \in X$, there exists an $h \in H$ for which $hx = gx$ or, equivalently, $g \in h\text{stab}_G(x)$. Hence,

$$G = H\text{stab}_G(x)$$

Moreover, if H is regular on X, then

$$H \cap \mathrm{stab}_G(x) = \mathrm{stab}_H(x) = \{1\}$$

and so

$$G = H \bullet \mathrm{stab}_G(x)$$

Now we summarize.

Theorem 7.2 *Let the group G act on the set X.*

1) **(Orbit-stabilizer relationship)** *For any $x \in X$,*

$$|Gx| = (G : \mathrm{stab}(x))$$

Thus, for a finite group G,

$$|Gx| = \frac{|G|}{|\mathrm{stab}(x)|}$$

and $|Gx|$ divides $|G|$.

2) **(The class equation)**

$$|X| = \sum_{x \in \mathrm{SDR}} (G : \mathrm{stab}(x))$$

where the sum is taken over a system of distinct representatives SDR *for the orbits in X. We can also write this as*

$$|X| = |\mathrm{Fix}_X(G)| + \sum_{x \in \mathrm{SDR}_{>1}} (G : \mathrm{stab}(x))$$

3) **(The stabilizer relationship)** *For any $x \in X$ and $a \in G$,*

$$\mathrm{stab}(ax) = \mathrm{stab}(x)^a$$

Thus stabilizers of an orbit in X form a conjugacy class of G and therefore the stabilizers of G-equivalent elements have the same cardinality.

4) **(The Frattini argument)** *If the action of $H \leq G$ is transitive on X, then*

$$G = H\mathrm{stab}_G(x)$$

and if H is regular on X, then

$$G = H \bullet \mathrm{stab}_G(x)$$

and so $\mathrm{stab}_G(x)$ is a complement of H in G.□

When the group action is transitive, the class equation and orbit-stabilizer relationship become quite simple.

Theorem 7.3 *If a group G acts transitively on a set X, then all stabilizers are conjugate and the orbit-stabilizer relationship (and the class equation) is simply*

$$|X| = (G : \text{stab}(x))$$

for any $x \in X$. Hence, if G is finite, then $|X|$ divides $|G|$.□

Congruence Relations on a G-Set

If G acts on the elements of a set X, then there is a natural way in which G also acts on the power set $\wp(X)$ of X, namely,

$$aS = \{as \mid s \in S\}$$

for all $S \subseteq X$ and $a \in G$. Let us refer to this action as the **induced action** of G on $\wp(X)$.

Now, a G-set is a set X with some structure, namely, the group action of G on X and an equivalence relation $\equiv$ on X is compatible with this action if it satisfies the following definition.

Definition *An equivalence relation $\equiv$ on a G-set X is called a* **G-congruence relation** *on X if it preserves the group action, that is, if*

$$x \equiv y \quad \Rightarrow \quad ax \equiv ay \quad \textit{for all } a \in G$$

We denote the set of all congruence classes under $\equiv$ by $X/\equiv$ and the congruence class containing $x \in X$ by $[x]$.□

Thus, if $\equiv$ is a G-congruence relation on X, then the induced action is an action on the partition $X/\equiv$ and

$$a[x] = [ax]$$

for all $x \in X$. Conversely, suppose that the induced action of G on $\wp(X)$ is an action on a partition $\mathcal{P} = \{B_i \mid i \in I\}$ of X, that is, $aB_i \in \mathcal{P}$ for all $a \in G$ and $B_i \in \mathcal{P}$. Then the equivalence relation $\equiv$ associated to $\mathcal{P}$ is a G-congruence relation on X.

In other words, the partitions of X that correspond to the G-congruence relations are the partitions $\mathcal{P} = \{B_i \mid i \in I\}$ that are closed under the induced action of G on $\wp(X)$.

Moreover, if G acts transitively on X and if $\equiv$ is a G-congruence relation on X, then G also acts transitively on $X/\equiv$ and so

$$\frac{X}{\equiv} = GS := \{aS \mid a \in G\}$$

is the orbit of any given conjugacy class $S \in X/\equiv$. Hence, the partitions of X that correspond to the G-congruence relations of a transitive group G are the partitions of the form GS, where $S \subseteq X$ is nonempty.

But if GS is a partition of X, then $aS = S$ or $aS \cap S = \emptyset$ for all $a \in G$. Conversely, if $aS = S$ or $aS \cap S = \emptyset$ for all $a \in G$, then the distinct members of GS form a partition of X. To see this, suppose that $x \in aS \cap bS$. Then $a^{-1}x \in S \cap a^{-1}bS$ and so $a^{-1}bS = S$, whence $bS = aS$. Moreover, the transitivity of G implies that $GS = X$.

Hence, if $S \subseteq X$ is nonempty, then GS is a partition of X if and only if

$$aS = S \quad \text{or} \quad aS \cap S = \emptyset$$

for all $a \in G$.

Theorem 7.4 *Let a group G act on a nonempty set X.*

1) *The partitions of X that correspond to the G-congruence relations on X are the partitions of X that are closed under the induced action of G on $\wp(X)$.*
2) *Suppose that G acts transitively on X.*
   a) *The partitions of X that correspond to the G-congruence relations are the partitions of the form*

   $$GS = \{aS \mid a \in G\}$$

   where $S \subseteq X$ is nonempty.
   b) *A nonempty subset S of X is a congruence class under some G-congruence if and only if S has the property that*

   $$aS = S \quad \text{or} \quad aS \cap S = \emptyset$$

   for all $a \in G$. Such a subset S of X is called a **block** *of G.*□

Thus, $S \subseteq X$ is a block of G if and only if GS is a partition of X. It follows that if S is a block of G, then so is aS for all $a \in G$.

Now that we have established the basic properties of group actions, we can examine a few of the most important examples of group actions. Then we will use group actions to explore the structure of p-groups. In the next chapter, we will use group actions to explore the structure of finite groups and to prove the famous Sylow theorems.

Translation by G

Earlier in the book, we discussed two fruitful examples of the action of translation by elements of a group G, namely, on the elements of G and on the elements of a quotient set G/H. The action of G on itself is the **left regular representation** of G as a subgroup of the symmetric group S_G, as described by Cayley's theorem.

Let us review the action of translation by G on G/H:

$$\lambda_g(aH) = gaH$$

This action is transitive and so all stabilizers are conjugate. Since

$$\text{stab}(aH) = \text{stab}(H)^a = H^a$$

the kernel of the action is the normal closure of H,

$$\ker(\lambda) = \bigcap_{a\in G} H^a = H^\circ$$

which is the largest normal subgroup of G contained in H. If $(G:H) = m$, then the embedding

$$\frac{G}{H^\circ} \hookrightarrow S_{(G:H)} \approx S_m$$

implies that

$$(G:H^\circ) \mid (G:H)!$$

The consequences of this action were recorded earlier in Theorem 4.20, but we repeat them here for easy reference.

Theorem 7.5 *Let G be a group and let $H < G$ have finite index. Then*

$$G/H^\circ \hookrightarrow S_{G/H}$$

and so

$$(G:H^\circ) \mid (G:H)!$$

In particular, $(G:H^\circ)$ is also finite and

$$(H:H^\circ) \mid ((G:H)-1)!$$

1) *Any of the following imply that $H \trianglelefteq G$:*
   a) *H is periodic and $(G:H) = p$ is equal to the smallest order among the nonidentity elements of H.*
   b) *G is finite and $o(H)$ and $((G:H)-1)!$ are relatively prime, that is, for all primes p,*

   $$p \mid o(H) \quad \Rightarrow \quad p \geq (G:H)$$

   This happens, in particular, if $(G:H)$ is the smallest prime dividing $o(G)$.
2) *If G is finitely generated, then G has at most a finite number of subgroups of any finite index m.*

3) *If G is simple, then*

$$o(G) \mid (G:H)!$$

   a) *If G is infinite, then G has no proper subgroups of finite index.*
   b) *If G is finite and $o(G) \nmid m!$ for some integer m, then G has no subgroups of index m or less.* $\square$

Conjugation by G on the Conjugates of a Subgroup

The elements of G act by conjugation on $\text{sub}(G)$,

$$\lambda_a(H) = H^a$$

for all $H \le G$. The orbit of a subgroup H is the conjugacy class $\text{conj}_G(H)$ and the stable elements are

$$\text{Fix}(G) = \text{nor}(G)$$

The stabilizer of a subgroup H is its normalizer and so

$$N_G(H^a) = N_G(H)^a$$

Also, the orbit-stabilizer relationship is

$$|\text{conj}_G(H)| = (G : N_G(H))$$

which we discussed earlier in the book (Theorem 3.27).

Conjugation by G on a Normal Subgroup

Let $N \trianglelefteq G$ and let G act on the elements of N by conjugation:

$$\lambda_g(a) = a^g$$

for all $a \in N$. The orbits of this action

$$Ga = a^G := \{a^x \mid x \in G\}$$

are called the **conjugacy classes** of N **under** G. The stabilizer of $a \in N$ is its centralizer $C_G(a)$ and the kernel of the representation λ is

$$\ker(\lambda) = \bigcap_{a \in N} C_G(a) = C_G(N)$$

The orbit-stabilizer relationship is

$$\left|a^G\right| = (G : C_G(a))$$

as we saw in Theorem 3.23.

The stable elements of N are the elements of N that commute with every element of G and so

$$\operatorname{Fix}_N(G) = Z(G) \cap N$$

Hence, the class equation is

$$|N| = |Z(G) \cap N| + \sum_{a \in \mathrm{SDR}_{>1}} (G : C_G(a))$$

When G acts on itself by conjugation, that is, when $N = G$, the class equation is

$$|G| = |Z(G)| + \sum_{a \in \mathrm{SDR}_{>1}} (G : C_G(a))$$

This is the equation to which the name *class equation* is traditionally applied and is one of the most useful tools in finite group theory.

The Structure of Finite p-Groups

We now wish to study the structure of a very special type of finite group.

Definition *Let G be a nontrivial group and let p be a prime.*

1) *An element $a \in G$ is called a* **p-element** *if $o(a) = p^k$ for some $k \geq 0$.*
2) *G is a* **p-group** *if every element of G is a p-element.*
3) *A nontrivial subgroup S of G is called a* **p-subgroup** *of G if S is a p-group.* □

As we saw earlier in the book, Lagrange's theorem and Cauchy's theorem conspire to give the following result.

Theorem 7.6 *A finite group G is a p-group if and only if the order of G is a power of p.* □

When a p-group G acts on a set X, the class equation has the property that all of the terms $(G : \operatorname{stab}(x))$ that are greater than 1 are divisible by p. This gives the following simple but useful result.

Theorem 7.7 *If a p-group G acts on a set X, then*

$$|X| \equiv |\operatorname{Fix}_X(G)| \bmod p$$ □

We now turn to the key properties of finite p-groups.

The Center-Intersection Property

It will be convenient to make the following nonstandard definition.

Definition *A group G has the* **center-intersection property** *if every nontrivial normal subgroup of G intersects the center of G nontrivially.* □

Note that a finite group G has the center-intersection property if and only if every nontrivial normal subgroup of G contains a central subgroup of prime order. Any finite p-group G has the center-intersection property, for if $N \trianglelefteq G$ is nontrivial, then G acts on the elements of N by conjugation and Theorem 7.7 implies that

$$|N| \equiv |Z(G) \cap N| \bmod p$$

which shows that $|Z(G) \cap N| > 1$.

Theorem 7.8 *A finite p-group G has the center-intersection property.*
1) $Z(G)$ *is nontrivial.*
2) G *is simple if and only if* $o(G) = p$.□

The fact that the center of a p-group is nontrivial tells us something very significant about groups of order p^2.

Corollary 7.9 *If* $o(G) = p^2$, *then G is abelian. In fact, G is either cyclic or is the direct product of two cyclic subgroups of order p.*
Proof. We must have $|Z(G)| = p$ or p^2. But if $|Z(G)| = p$, then $G/Z(G)$ is cyclic and so G is abelian, which is a contradiction. Hence, $|Z(G)| = p^2$ and G is abelian. If G is not cyclic, then G is elementary abelian of exponent p and so is the direct product of two cyclic subgroups of order p.□

p-Series and Nilpotence

We next show that p-groups have normal subgroups of all possible orders. But first a couple of definitions.

Definition (Central Series and p-Series) *Let G be a group.*
1) A normal series

$$H_0 \trianglelefteq H_1 \trianglelefteq \cdots \trianglelefteq H_n$$

in G is **central** *in G if each factor group H_{k+1}/H_k is central in G/H_k, that is,*

$$H_{k+1}/H_k \le Z(G/H_k)$$

A group is **nilpotent** *if it has a central series starting at the trivial subgroup $\{1\}$ and ending at G.*
2) If p is a prime, then a **p-series** *from H to G is a series*

$$H = H_0 \triangleleft H_1 \triangleleft \cdots \triangleleft H_n = G$$

whose steps $H_k < H_{k+1}$ have index p.□

Definition *Let p be a prime. If $H < K$ is an extension in G of index p, we refer to K as a* **p-cover** *of H. (K is a cover of H in the lattice* sub(G).)□

Theorem 7.5 implies that if K is a p-cover of H, then $H \lhd K$.

Theorem 7.10 *Let G be a finite p-group. Then every $H < G$ has a p-cover K and if $H \lhd G$, then K can be chosen so that $H \lhd K$ is central in G.*
1) *There is a p-series from H to G.*
2) *If $H \unlhd G$, then there is a central p-series from H to G. In particular, G has a normal subgroup containing H of every order between $o(H)$ and $o(G)$ (under division).*
3) *G is nilpotent.*

Proof. The proof is by induction on $o(G)$. The theorem is true if $o(G) = p$. Assume $o(G) > p$ and that the theorem is true for all groups smaller than G. Let $H < G$ and let N be central in G of order p. If $H = \{1\}$, then N is a p-cover of H and $H \lhd N$ is central in G. Assume that $H \neq \{1\}$. We may also assume that $N \leq H$ if $H \lhd G$. In any case, $N \leq H$ or $N \cap H = \{1\}$.

If $N \cap H = \{1\}$, then NH is a p-cover of H and if $N \leq H$, then the induction hypothesis implies that H/N has a p-cover K/N and so K is a p-cover of H. Also, if $H \lhd G$, the inductive hypothesis implies that $H/N \lhd K/N$ is central in G/N for some $K \leq G$ and so Theorem 4.11 implies that $H \lhd K$ is central in G.□

The Normalizer Condition

Theorem 7.10 implies that a p-group has the normalizer condition, that is,

$$H < G \quad \Rightarrow \quad H < N_G(H)$$

and therefore several other nice properties (see the discussion following Theorem 4.35).

Corollary 7.11 *The following hold in a finite p-group G:*
1) *G has the normalizer condition*
2) *Every subgroup of G is subnormal*
3) *Every maximal subgroup of G is normal*
4) *$G/\Phi(G)$ is abelian.*□

Maximal and Minimal Subgroups

Maximal and minimal subgroups play a key role in the study of finite p-groups. For a finite p-group G, Cauchy's theorem implies that a subgroup H is minimal if and only if $o(H) = p$ and the center-intersection property implies that H is minimal normal if and only if it is central of order p.

As to the maximal subgroups of G, Theorem 7.10 implies that a subgroup H is maximal in G if and only H has index p and that H is maximal normal in G if and only if H has index p.

Theorem 7.12 (Maximal and minimal subgroups) *Let G be a finite p-group and let $H \le G$.*

1) *H is minimal if and only if it has order p.*
2) *H is minimal normal if and only if it is central of order p.*
3) *The following are equivalent:*
   a) *H is maximal*
   b) *H is maximal normal*
   c) *$(G : H) = p$.*□

The Frattini Subgroup of a p-Group; The Burnside Basis Theorem

The fact that any maximal subgroup M of a p-group G is normal and has index p implies that if $a \in G$, then

$$(aM)^p = M$$

and so $a^p \in M$. Thus, $G^p \subseteq \Phi(G)$, the Frattini subgroup of G. It follows that $G/\Phi(G)$ is an elementary abelian group of exponent p.

Conversely, if G/K is elementary abelian, then it is characteristically simple and so $\Phi(G/K) = \{K\}$. Hence,

$$\Phi(G) \le \bigcap_{\substack{K \le M < G \\ M \text{ maximal}}} M \le K$$

We have shown that $\Phi(G)$ is the smallest normal subgroup K of G for which the quotient G/K is elementary abelian.

Theorem 7.13 *Let p be a prime. Let G be a group of order p^n, with Frattini subgroup $\Phi(G)$ of order p^m.*

1) *$\Phi(G)$ is the smallest normal subgroup of G for which $G/\Phi(G)$ is an elementary abelian group. Moreover, $G/\Phi(G)$ has exponent p and so is a vector space over $\mathbb{Z}_p$, of dimension $n - m$.*
2) *$\Phi(G) = G'G^p$*
3) **(The Burnside Basis Theorem)** *Any generating set for G contains a generating set of size $n - m$.*

Proof. Part 1) has been proved. For part 2), since $\{G'a^p \mid a \in G\}$ is a subgroup of G/G', it follows that $G'G^p$ is a normal subgroup of G. In fact, $G/G'G^p$ is elementary abelian of exponent p and so part 1) implies that $\Phi(G) \le G'G^p \le \Phi(G)$. Hence, $\Phi(G) = G'G^p$.

For part 3), write $\Phi = \Phi(G)$. We show that

$$G = \langle g_1, \ldots, g_k \rangle \quad \Leftrightarrow \quad G/\Phi = \langle g_1\Phi, \ldots, g_k\Phi \rangle$$

One direction is clear and since Φ is the set of nongenerators of G, we have

$$G/\Phi = \langle g_1\Phi, \ldots, g_k\Phi \rangle \quad \Rightarrow \quad G = \langle \{g_1, \ldots, g_k\} \cup \Phi \rangle = \langle g_1, \ldots, g_k \rangle$$

Thus, since G/Φ is a $\mathbb{Z}_p$-space of dimension $n-m$, any generating set for G/Φ contains a generating set of size $n-m$. Hence, the same holds true for G.□

Number of Subgroups of a p-Group

We now wish to inquire about the number of subgroups of a given size p^d in a p-group G of order p^n. Let $\text{sub}_d(G)$ and $\text{nor}_d(G)$ denote the families of subgroups and normal subgroups, respectively, of G of size p^d. Then G acts by conjugation on $\text{sub}_d(G)$ and the stable set is $\text{nor}_d(G)$, whence

$$|\text{sub}_d(G)| \equiv |\text{nor}_d(G)| \bmod p$$

Our plan is to show that $|\text{nor}_d(G)| \equiv 1 \bmod p$.

Theorem 7.14 *Let G be a nontrivial p-group of order* p^n.
1) The number of maximal subgroups of G is

$$|\text{sub}_{n-1}(G)| = |\text{nor}_{n-1}(G)| = \frac{p^{n-m}-1}{p-1} \equiv 1 \bmod p$$

2) For any $0 \le d \le n$,

$$|\text{sub}_d(G)| \equiv |\text{nor}_d(G)| \equiv 1 \bmod p$$

Proof. For part 1), if $o(\Phi(G)) = p^m$, then Theorem 7.13 implies that $A = G/\Phi(G)$ is a vector space over $\mathbb{Z}_p$ of dimension $n-m$. In general, if V is a vector space over $\mathbb{Z}_p$ of dimension k, then the number of subspaces of V of dimension d is

$$V(k,d) = \frac{(p^k-1)(p^k-p)\cdots(p^k-p^{d-1})}{(p^d-1)(p^d-p)\cdots(p^d-p^{d-1})}$$

(We ask the reader to supply a proof in the exercises.) Hence, the number of subgroups (subspaces) of V of order p^{k-1} is

$$V(k,k-1) = \frac{(p^k-1)(p^k-p)\cdots(p^k-p^{k-2})}{(p^{k-1}-1)(p^{k-1}-p)\cdots(p^{k-1}-p^{k-2})} = \frac{p^k-1}{p-1}$$

In particular, the number of maximal subgroups of $G/\Phi(G)$ is

$$V(n-m,n-m-1) = \frac{p^{n-m}-1}{p-1}$$

and this is also the number of maximal subgroups of G.

For part 2), let $u_d(G) = |\text{nor}_d(G)|$ and let $\equiv$ stand for congruence modulo p. Then part 1) implies that $u_{u-1}(G) \equiv 1$. We show that $u_d(G) \equiv 1$ by induction on $o(G)$. If $o(G) = p$, the result is clear. Assume it is true for p-groups smaller than G.

If $M \in \mathrm{nor}_{n-1}(G)$, then G acts on $\mathrm{sub}_d(M)$ by conjugation. The stable set is $\mathrm{nor}_d(G) \cap \mathrm{sub}_d(M)$ and so the inductive hypothesis implies that

$$|\mathrm{nor}_d(G) \cap \mathrm{sub}_d(M)| \equiv |\mathrm{sub}_d(M)| \equiv 1$$

Hence, the set

$$\mathcal{S} = \{(N, M) \mid N < M, N \in \mathrm{nor}_d(G), M \in \mathrm{nor}_{n-1}(G)\}$$

has size $|\mathcal{S}| \equiv u_{n-1}(G) \equiv 1$.

On the other hand, for each $N \in \mathrm{nor}_d(G)$, there is one $M \in \mathrm{nor}_{n-1}(G)$ containing N for each maximal subgroup of G/N and since $u_{n-d-1}(G/N) \equiv 1$, we have $|\mathcal{S}| \equiv u_d(G)$. Thus, $u_d(G) \equiv 1$.□

*Conjugates in a p-Group

In the study of finite p-groups, it can be useful to examine the conjugates of an element a by the powers of another element b.

Theorem 7.15 *Let G be a finite p-group and let $a \in G$ have order $o(a) = p^m$. Let $b \in G$ and suppose that*

$$a^b = a^\alpha$$

for some integer $\alpha \not\equiv 1 \bmod p^m$. Let $a^{\langle b \rangle}$ be the set of conjugates of a by the elements of $\langle b \rangle$. Then

$$\left|a^{\langle b \rangle}\right| = p^d$$

where $d \geq 1$ and p^d is the smallest power of p for which b^{p^d} commutes with a.

1) *If $p > 2$ or if $\alpha \not\equiv 3 \bmod 4$, then*

$$a^{\langle b \rangle} = \{a^{1+kp^{m-d}} \mid k = 1, \ldots, p^d\}$$

where $m - d \geq 1$ and $a^{-1} \notin a^{\langle b \rangle}$.

2) *If $p = 2$ and $\alpha \equiv 3 \bmod 4$, then one of the following holds:*
   a) *If $d \geq 1$, then*

$$a^{\langle b \rangle} = \{a^{e_k + kp^{m-d}} \mid k = 1, \ldots, p^d\}$$

   where $m - d \geq 1$, $a^{-1} \notin a^{\langle b \rangle}$ and half of the e_k's are 1 and half are -1.

   b) *If $d = 0$, then*

$$a^{\langle b \rangle} = \{a, a^{-1+2^{m-1}}\} \quad or \quad a^{\langle b \rangle} = \{a, a^{-1}\}$$

3) a) *If $p > 2$, then no element of G of order p can be conjugate to one of its own powers other than the first power.*

b) *If $p=2$, then no element $a \in G$ of order 2 can be conjugate to one of its own powers other than a or a^{-1}.*

Proof. The conjugates of a by $\langle b \rangle$ are

$$b^k a b^{-k} = a^{\alpha^k}$$

for $k = 1, \ldots, r$, where $r = \left| a^{\langle b \rangle} \right|$. In fact, r is the smallest positive integer for which b^r commutes with a. Also, $a^{\langle b \rangle}$ is the orbit of a under conjugation by $\langle b \rangle$ and so $r = p^d$ for some $d \geq 1$.

Note that r is also the smallest positive integer for which $a^{\alpha^r} = a$, that is, the smallest positive integer for which $\alpha^r \equiv 1 \bmod p^m$ and so

3) $p^m \mid \alpha^{p^d} - 1$
4) $p^m \nmid \alpha^{p^{d-1}} - 1$

Furthermore, since $\alpha^{p^d} \equiv 1 \bmod p$, Fermat's little theorem implies that $\alpha \equiv 1 \bmod p$. Thus, if $\alpha = e + cp^t$ is in p-standard form, then Lemma 1.18 implies that for any $u \geq 0$,

$$\alpha^{p^u} = e^{p^u} + wp^{u+t}$$

where $p \nmid w$. In particular,

5) $p^{d+t-1} \mid \alpha^{p^{d-1}} - e^{p^{d-1}}$
6) $p^{d+t+1} \nmid \alpha^{p^d} - e^{p^d} = \alpha^{p^d} - 1$

From 3) and 6), we see that $m \leq d+t$.

Now, if $e^{p^{d-1}} = 1$ then 4) and 5) imply that $d+t \leq m$ and so $m = t+d$. This implies that $m-d = t \geq 1$. Also,

$$\alpha = e + cp^{m-d}$$

and so for $1 \leq k \leq p^d$,

$$\alpha^k = (e + cp^{m-d})^k = e^k + p^{m-d} w_k$$

Hence,

$$a^{\alpha^k} = a^{e^k + p^{m-d} w_k}$$

where no two distinct w_k's are congruent modulo p^d, since otherwise we would not get p^d distinct conjugates. Thus, we can assume that w_k ranges over the set $\{1, \ldots, p^d\}$ and so

$$a^{\langle b \rangle} = \{ a^{e_k + kp^{m-d}} \mid k = 1, \ldots, p^d \}$$

Now, if $p > 2$ or $\alpha \not\equiv 3 \bmod 4$, then $e = 1$ and so $e_k = 1$ for all k. If $p = 2$ and

$\alpha \equiv 3 \bmod 4$ but $d > 1$, then we still have $e^{p^d} = 1$ but since $e = -1$, as k ranges from 1 to p^d, the term e^k alternates between -1 and 1 and so half of the terms e_k are 1 and half are -1.

Also, as k ranges from 1 to p^d, the exponents $e_k + kp^{m-d}$ range from $\pm 1 + p$ to $\pm 1 + p^m$. But a is conjugate to itself and so one of these exponents must be conjugate to 1 modulo p^m. Therefore, the last exponent is $1 + p^m$ and since no other exponent is conjugate to -1 modulo p^m, it follows that $a^{-1} \notin a^{\langle b \rangle}$.

The case $e^{p^{d-1}} = -1$ occurs precisely when $p = 2$, $\alpha \equiv 3 \bmod 4$ and $d = 1$, in which case a has exactly two conjugates and $\alpha = -1 + c2^t$ with c odd and $t \geq 2$.

If $t \geq m$, then $a^\alpha = a^{-1+c2^t} = a^{-1}$ and so $a^{\langle b \rangle} = \{a, a^{-1}\}$. If $t < m$, then $m \leq 1 + t$ implies that $t = m - 1$ and so $\alpha = -1 + c2^{m-1}$ where c is odd, that is, $c = 1$. Hence, the two conjugates of a are a itself and

$$a^\alpha = a^{-1+2^{m-1}}$$

Note finally that the case $a^{\langle b \rangle} = \{a, a^{-1}\}$ does occur in the dihedral group $D_{2^{m+1}} = \langle \rho, \sigma \rangle$ where $o(\rho) = 2^m$ and $o(\sigma) = 2$ and $\rho^\sigma = \rho^{-1}$. Also, the case

$$a^{\langle b \rangle} = \{a, a^{-1+2^{m-1}}\}$$

occurs in the semidihedral group

$$SD_m = \langle \alpha, \xi \rangle, \quad o(\alpha) = 2^m, \quad o(\xi) = 2, \quad \xi\alpha = \alpha^{2^{m-1}-1}\xi$$

For part 3), if $o(a) = p$, then the number of conjugates of a is both a power of p and at most $p - 1$, whence it must equal 1.□

***Unique Subgroups in a *p*-Group**

A cyclic p-group has a unique subgroup of order p. If $p > 2$, then the converse of this is true: A p-group that has a unique subgroup of order p is cyclic. We begin with a definition.

Definition *A* **generalized quaternion group** *of order* 2^n, $n \geq 2$ *is a group* Q_n *with the following properties:*

$$Q_n = \langle a, b \rangle, o(a) = 2^{n-1}, o(b) = 4, b^2 = a^{2^{n-2}}, bab^{-1} = a^{-1}$$

If $n = 3$, *then* Q_n *is a quaternion group.*□

We will show later in the book that such a group exists: It is a special case of the dicyclic group. We leave it as an exercise to show that $\langle b^2 \rangle$ is the only subgroup of order 2 in Q_m but that for any $2 < 2^s < 2^n$, the group Q_n has at least two subgroups of order 2^s. Also, any $x \in Q_n \setminus \langle a \rangle$ has the form $x = a^k b$, where

$$(a^k b)^2 = a^k (ba^k) b = b^2 = a^{2^{m-1}}$$

and so $o(a^k b) = 4$. Thus, any element of $Q_n \setminus \langle a \rangle$ has order 4. It follows that if $n \geq 4$, then $\langle a \rangle$ is the unique cyclic subgroup of Q_n of order 2^{n-1} and so $\langle a \rangle \sqsubseteq Q_n$.

We will prove that if a p-group G has a unique subgroup of order p, then G is cyclic if $p > 2$ and G is either cyclic or generalized quaternion if $p = 2$. First, let us show that if G has a unique subgroup H_s of *any* order p^s, where $p \leq p^s < o(G)$, then G must have a unique subgroup of order p.

Since a p-group has subgroups of all orders dividing the order of the group, we have for any subgroup $K \leq G$,

$$o(H_s) \leq o(K) \quad \Rightarrow \quad H_s \leq K$$

Also, since any subgroup $K \leq G$ of order less than p^s is contained in some subgroup of order p^s, we have

$$o(K) \leq o(H_s) \quad \Rightarrow \quad K \leq H_s$$

Thus, all subgroups of G either contain H_s or are contained in H_s. In this sense, H_s forms a *bottleneck* in the lattice of subgroups of G. It follows that H_s is cyclic, for if $a \notin H_s$, then $\langle a \rangle \not\leq H_s$ and so $H_s \leq \langle a \rangle$, whence H_s is cyclic. Since H_s is cyclic, the subgroup lattice of G has the form shown in Figure 7.1 and so G contains exactly one subgroup of each order p^d with $0 \leq d \leq s$. In particular, G has a unique subgroup of order p. Thus, we have shown that G has a unique subgroup of some order p^s, where $p \leq p^s < o(G)$ if and only if G has a unique subgroup of order p.

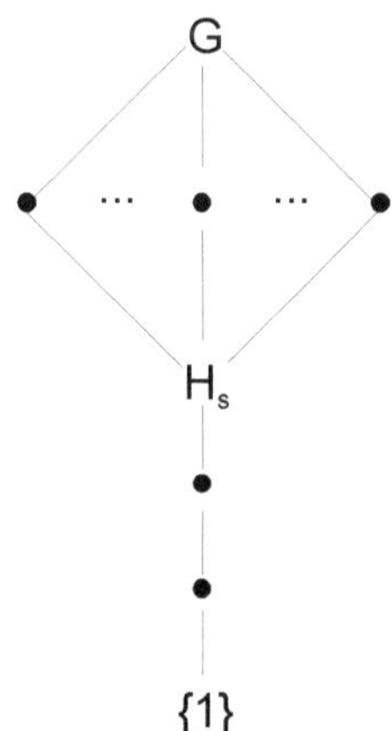

Figure 7.1

Let G be a *noncyclic* group of order $p^n > 1$. To show that G has more than one subgroup of order p, it suffices to show G contains a nontrivial subgroup A as well as an element of order p that is *not* in A. We first consider the case $p > 2$.

Theorem 7.16 *Let $p > 2$ be prime and let $o(G) = p^n > 1$.*

1) *If G is noncyclic and if $a \in G$ is an element of maximum order, then G has an element of order p that is not contained in $\langle a \rangle$.*
2) *If G has a unique subgroup of order p^s for some $1 \leq s < n$, then G is cyclic.*

Proof. We have already seen that part 2) follows from part 1). To prove part 1), assume that G is noncyclic. Let $A = \langle a \rangle$, where $a \in G$ has maximum order p^m. If $m = 1$, then all nonidentity elements of G have order p and so we may assume that $2 \leq m < n$.

Let $A \lhd B$ where $o(B) = p^{m+1}$. If $b \in B \setminus A$, then $b^p \in A$ and so $b^p = a^t$ for some $t \geq 0$. If $t = 0$, then $o(b) = p$ and we are done, so let us assume that $t > 0$. Since $a^t = b^p$ does not have maximum order in G, it follows that $p \mid t$ and so

$$b^p = a^{up}$$

for some $0 < u < p^{m-1}$.

Now, if b commutes with a, then $ba^{-u} \notin A$ has order p. Assume now that no $b \in B \setminus A$ commutes with a. We wish to show that there is still an integer x for which $o(ba^x) = p$. If $b \in B \setminus A$, then to get a formula for $(ba^x)^p$, we need a commutativity rule for a and b. But since $A \lhd B$, there is an $\alpha \neq 1$ for which

$$a^b = a^\alpha$$

and since b^p commutes with a, Theorem 7.15 implies that

$$a^{\langle b \rangle} = \{a^{1+kp^{m-1}} \mid k = 1, \ldots, p\}$$

Moreover, $b \notin A$ implies that $b^i \in B \setminus A$ for all $1 \leq i < p$ and so we may replace b by an appropriate power of b so that $\alpha = 1 + p^{m-1}$.

Now, $ba = a^\alpha b$ and so

$$ba^x = a^{\alpha x} b$$

Then an easy induction shows that for $k \geq 1$,

$$(ba^x)^k = a^{x(\alpha + \alpha^2 + \cdots + \alpha^k)} b^k$$

and so

$$(ba^x)^p = a^{x(\alpha + \alpha^2 + \cdots + \alpha^p)} b^p = a^{up + x(\alpha + \alpha^2 + \cdots + \alpha^p)}$$

Hence, we want an integer x for which

$$x\alpha\frac{1-\alpha^p}{1-\alpha} \equiv -up \pmod{p^m}$$

Since $\alpha = 1 + p^{m-1}$, Theorem 1.18 implies that

$$\alpha^p = (1 + p^{m-1})^p = 1 + wp^m$$

where $p \nmid w$ and so

$$x\alpha\frac{1-\alpha^p}{1-\alpha} = x(1 + p^{m-1})wp \equiv xwp \pmod{p^m}$$

Hence, we want an integer x for which

$$xw \equiv -u \pmod{p^{m-1}}$$

But w is invertible in $\mathbb{Z}^*_{p^{m-1}}$ and so we may take $x = -uw^{-1}$.□

Now we turn to the case $p = 2$. (The reader may wish to skip the proof upon first reading.)

Theorem 7.17 *Let G be a nontrivial group of order 2^n.*
1) *G has a unique subgroup of order 2 if and only if G is cyclic or a generalized quaternion group.*
2) *If G has a unique subgroup of order 2^s for some $1 < s < n$, then G is cyclic.*

Proof. We have already seen that part 2) follows from part 1). Let G be a noncyclic group of order 2^n with a unique involution. We will show that G is a generalized quaternion group. Let $a \in G$ be an element of maximum order 2^m and let $A = \langle a \rangle$. Clearly, we may assume that $m \geq 2$.

In one case we will need to be a bit more specific about the choice of the cyclic subgroup A. Namely, if $m = 2$, then since the unique subgroup of order 2 is normal in G, it has a 2-cover N of order 4, which is cyclic since the 4-group has two involutions. In this case, we let $A = N$ and so $A \trianglelefteq G$. Thus, if $m = 2$, we may assume that $A \triangleleft G$.

If $B = \langle b, A \rangle$ is a 2-cover of A, then $o(B/A) = 2$ and so $b^2 \in A$, which implies that $b^2 = a^k$ for some $k > 0$. But $o(a^k) = o(b^2) < o(b) \leq o(a)$ and so $2 \mid k$, that is,

$$b^2 = a^{2^t u}$$

for $t \geq 1$ and u odd. Since $\langle a^u \rangle = A$, we can rename a^u to a to get

$$b^2 = a^{2^t}$$

for $t \geq 1$. Since $o(b^2) = o(a^{2^t}) = 2^{m-t}$, we also have

$$o(b) = 2^{m-t+1}$$

where $t \le m - 1$ since $o(b) > 2$.

If b commutes with a, then $ba^{-2^{t-1}} \notin A$ is an involution, contrary to assumption. Thus, b does not commute with a; in fact, A is not properly contained in any abelian subgroup of G.

However, since $A \lhd B$ and $b^2 \in A$, Theorem 7.15 implies that

$$a^{\langle b \rangle} = \{a, a^{-1+2^{m-1}}\} \quad \text{or} \quad a^{\langle b \rangle} = \{a, a^{-1}\}$$

and so for any 2-cover $B = \langle b, A \rangle$ of A, there is an $a \in A$ for which

$$bab^{-1} = a^{\alpha} \quad \text{and} \quad b^2 = a^{2^t}$$

where either $\alpha = -1 + 2^{m-1}$ or $\alpha = -1$.

Conjugating the second equation by b and using the first equation gives

$$b^2 = ba^{2^t}b^{-1} = a^{\alpha 2^t} = a^{-2^t}$$

and so $a^{2^t} = a^{-2^t}$, that is, $a^{2^{t+1}} = 1$. Hence, $2^m \mid 2^{t+1}$, which implies that $m \le t + 1 \le m$, that is, $t = m - 1$.

Now, if for any 2-cover $B = \langle b, A \rangle$, we have $\alpha = -1 + 2^{m-1}$, then $\alpha = -1 + 2^t$ and so

$$bab^{-1} = a^{-1+2^t} = b^2 a^{-1}$$

which implies that $ab^{-1} \notin A$ is an involution, a contradiction. Hence, for all 2-covers $B = \langle b, A \rangle$ of A, we have $\alpha = -1$ and

$$bab^{-1} = a^{-1} \quad \text{and} \quad b^2 = a^{2^{m-1}}$$

It follows that $o(b) = 4$ and so B can be described as follows:

$$B = \langle a, b \rangle, o(a) = 2^m, o(b) = 4, b^2 = a^{2^{m-1}}, bab^{-1} = a^{-1}$$

that is, $B = Q_{m+1}$ is generalized quaternion and so every element of $B \setminus A$ has order 4 and if $m \ge 3$, then $A \sqsubseteq B$.

We want to show that $B = G$. If not, then B has a 2-cover C, that is,

$$A \lhd B \lhd C \le G$$

where $o(C) = p^{m+2}$. Recall that if $m = 2$, then we have chosen A so that $A \unlhd G$ and if $m \ge 3$, then $A \sqsubseteq B$ and so in either case, $A \lhd C$.

The quotient group C/A is either $C_4(cA)$ or $C_2(cA) \bowtie C_2(dA)$. In the latter case, the subgroups $\langle c, A\rangle$ and $\langle d, A\rangle$ are 2-covers of A and so

$$a^c = a^{-1} = a^d$$

Hence,

$$a^{cd} = a$$

which implies that $A < \langle A, cd\rangle$ is abelian, a contradiction. Hence, $C/A = C_4(cA)$. But then $\langle c^2, A\rangle$ is a 2-cover of A and so

$$a^{c^2} = a^{-1}$$

However, since A is not properly contained in an abelian subgroup of G, the smallest power of c that commutes with a is $c^4 \in A$ and so Theorem 7.0 (where $d = 2$) implies that $a^{-1} \notin a^{\langle c\rangle}$, a contradiction.

Thus $G = B = Q_{m-1}$ is generalized quaternion of order 2^m.□

*Groups of Order p^n With an Element of Order p^{n-1}

We can use the previous result to take a close look at nonabelian groups G of order p^n that have an element of order p^{n-1}. We will restrict attention to the case $p > 2$.

Let $o(a) = p^{n-1}$ and $A = \langle a\rangle$. Theorem 7.16 implies that there is a $b \in G \setminus A$ with $o(b) = p$. Then

$$G = \langle a\rangle \rtimes \langle b\rangle$$

and it remains to see how a and b interact. Since $A \trianglelefteq G$, we have $b^{-1}ab = a^k$ for some $k > 1$ and Theorem 7.15 implies that the conjugates of a by $\langle b\rangle$ are

$$a^{\langle b\rangle} = \{a^{1+kp^{n-2}} \mid k = 1, \ldots, p\}$$

Since any nonidentity element of $\langle b\rangle$ generates $\langle b\rangle$, we can take $k = 1$ and write

$$bab^{-1} = a^{1+p^{n-2}}$$

Theorem 7.18 *Let $p > 2$ be a prime. Let G be a nonabelian group of order p^n with an element a of order p^{n-1}. Then*

$$G = \langle a\rangle \rtimes \langle b\rangle$$

where $o(b) = p$ and

$$bab^{-1} = a^{1+p^{n-2}}$$

□

To see that such a group exists, recall from Example 5.30 that there is a semidirect product

$$C_{p^{n-1}}(a) \rtimes_\theta C_p(b)$$

where

$$\theta_b(a) = a^{1+p^{n-2}}$$

□

*Groups of Order p^3

We have seen that groups of order p are cyclic and that groups of order p^2 are either cyclic or the direct product of two cyclic subgroups of order p (Corollary 7.9). Theorem 7.18 gives us insight into groups of order p^3.

If $p = 2$, we have seen that, up to isomorphism, the groups of order $p^3 = 8$ are

1) C_8
2) $C_4 \boxtimes C_2$
3) $C_2 \boxtimes C_2 \boxtimes C_2$
4) Q, the (nonabelian) quaternion group
5) D_8, the (nonabelian) dihedral group

More generally, we will show that for any prime p, the groups of order p^3 are (up to isomorphism):

1) C_{p^3}
2) $C_{p^2} \boxtimes C_p$
3) $C_p \boxtimes C_p \boxtimes C_p$
4) $UT(3, \mathbb{Z}_p)$, the unitriangular matrix group (described below)
5) The group $G = \langle a, b \rangle$ where

$$G = \langle a, b \rangle, o(a) = p^2, o(b) = p, bab^{-1} = a^{1+p}$$

Thus, there are only two nonabelian groups of order p^3 (up to isomorphism). We will leave analysis of the abelian groups of order p^3 to a later chapter, where we will prove that any finite abelian group is the direct product of cyclic groups.

So let $p > 2$ be prime and let G be a nonabelian group of order p^3. If G has an element a of order p^2, then Theorem 7.18 implies that

$$G = \langle a \rangle \rtimes \langle b \rangle$$

where $o(b) = p$ and

$$b^{-1}ab = a^{1+p}$$

It remains to consider the case where G has exponent p. The center $Z = Z(G)$ is nontrivial but cannot have order p^2, since then G/Z is cyclic and so G is abelian. Hence, $o(Z) = p$ and so $o(G/Z) = p^2$. Hence, G/Z is abelian with exponent p and so

$$G/Z = \langle aZ \rangle \bowtie \langle bZ \rangle$$

Moreover, since $z := [b, a] \in Z$, we have $Z = \langle z \rangle$. Hence,

$$G = \langle a, b, z \rangle, o(a) = o(b) = o(z) = 1, z \in Z(G), ba = zab$$

To see that this does describe a group, we have the following.

Definition *Let R be a commutative ring with identity. A matrix $M \in GL(n, R)$ is* **unitriangular** *if it is upper triangular (has 0's below the main diagonal) and has 1's on the main diagonal. We denote the set of all unitriangular matrices by $UT(n, R)$.*□

We will leave it as an exercise to show that

$$|UT(n, \mathbb{Z}_p)| = p^{(n^2-n)/2}$$

and that for $p > 2$, the group $UT(3, \mathbb{Z}_p)$ has order p^3 and exponent p. Also, $UT(3, \mathbb{Z}_2) \approx Q$.

Exercises

1. A **left action** of G on X is sometimes defined as a map from the cartesian product $G \times X$ to X, sending (a, x) to an element $ax \in X$, satisfying
   a) $1x = x$ for all $x \in X$
   b) $(ab)x = a(bx)$ for all $x \in X$, $a, b \in G$.
   A **right action** of G on X is a map from the cartesian product $X \times G$ to X, sending (x, a) to an element $xa \in X$, satisfying
   c) $x1 = x$ for all $x \in X$
   d) $x(ab) = (xa)b$ for all $x \in X$, a, $b \in G$. Given a left action, show that the map $(x, g) = g^{-1}x$ is a right action. What about $(x, g) = gx$?
2. Let $\lambda: G \to S_X$ be an action of G on X.
   a) Prove that λ is regular if and only if it is transitive and $\text{stab}(x) = \{1\}$ for some $x \in X$.
   b) Prove that λ is regular if and only if it is transitive and for all distinct $g, h \in G$, we have $gx \neq hx$ for all $x \in X$.
   c) Prove that if λ is faithful and transitive and if G is abelian, then the action is regular.
3. Let G be a finite group and let p be the smallest prime dividing $o(G)$. Prove that any normal subgroup of order p is central.
4. Let G be an infinite group. Use normal interiors (not Poincaré's theorem) to prove that if H and K have finite index in G, then so does $H \cap K$.
5. Show that the condition that G be finitely generated cannot be removed from the hypotheses of Theorem 7.5.
6. Let G be a finite simple group and let $H \leq G$ have prime index $(G : H) = p$. Prove that p must be the largest prime dividing $o(G)$ and that p^2 does not divide $o(G)$.

7. Let G be a finite group. Prove that a transitive action of G on X is regular if and only if $|G| = |X|$.
8. Let $o(G) = 2n$ where $n \geq 1$ is odd. Let $a \in G$ have order 2. Show that under the left regular representation of G on itself, the element a corresponds to an odd permutation. Show that G is not simple.
9. a) Prove that if G is a finitely generated infinite group and H is a subgroup of finite index in G, then G has a characteristic subgroup K of finite index for which $K \leq H$.
   b) Show that the condition that G be finitely generated is necessary.
10. The action of a group G on a set X is **2-transitive** if for any pairs $(x, y), (u, v) \in X \times X$ where $x \neq y$ and $u \neq v$, there is an $a \in G$ for which $ax = u$ and $ay = v$. Prove that for a 2-transitive action, the stabilizer $\operatorname{stab}(x)$ is a maximal subgroup of G for all $x \in X$.

Equivalence of Actions

Two group actions $\lambda: G \to S_X$ and $\mu: H \to S_Y$ are **equivalent** if there is a pair (α, f) where $\alpha: G \to H$ is a group isomorphism and $f: X \to Y$ is a bijection satisfying the condition

$$f(gx) = (\alpha g)(fx)$$

In this case, we refer to (α, f) as an **equivalence** from λ to μ.

11. a) Show that the inverse of an equivalence is an equivalence.
    b) Show that the (coordinatewise) composition of two "compatible" equivalences is an equivalence.
12. Let $\lambda: G \to S_X$ be a transitive action and let $x \in X$. Show that λ is equivalent to the action of left-translation by G on $G/\operatorname{stab}(x)$.
13. Suppose that $\lambda: G \to S_X$ and $\mu: H \to S_Y$ are equivalent transitive actions, under the equivalence (α, f). Prove that $\operatorname{stab}(x) \approx H_y$ for any $x \in X$, where $y = fx$.

Conjugacy

14. Let G be a group and let $g \in G$. Show that $\langle g^G \rangle$ is a normal subgroup of G.
15. Let G be a finite group and let $g \in G$. Show that $|C_G(g)| \geq |G/G'|$ where G' is the commutator subgroup of G.
16. Let G be a finite group and let $H \leq G$ with $[G : H] = 2$. Suppose that $C_G(h) \leq H$ for all $h \in H$. Prove that $G \setminus H$ is a conjugacy class of G.
17. Let G be a p-group and let $H \leq G$ be a nonnormal subgroup of G and let $a \in G$. Show that the number of conjugates of H that are fixed by every element of H^a is positive and divisible by p.
18. a) Let G be a finite group and let $H \trianglelefteq G$. Show that

$$k(G/H) = k(G) - k_G(H) + 1$$

where $k_G(H)$ is the number of G-conjugacy classes of H.

b) Let G be a finite nonabelian group such that $G/Z(G)$ is abelian. Show that

$$k(G) \geq |G/Z(G)| + |Z(G)| - 1$$

19. a) Find all finite groups (up to isomorphism) that have exactly one conjugacy class.
b) Find all finite groups (up to isomorphism) that have exactly two conjugacy classes.
c) Find all finite groups (up to isomorphism) that have exactly three conjugacy classes.
20. a) If $q \in \mathbb{Q}$ and $n > 0$, show that there are only finitely many solutions $k_1, \ldots, k_n$ in positive integers to the equation

$$q = \frac{1}{k_1} + \cdots + \frac{1}{k_n}$$

Hint: Use induction on n. Look at the smallest denominator first.
b) Show that for any integer $n > 0$, there are only finitely many finite groups (up to isomorphism) that have exactly n conjugacy classes. *Hint*: Use the class equation.
21. a) Let H be a proper subgroup of a finite group G. Show that the set

$$S = \bigcup_{g \in G} H^g$$

is a proper subset of G.
b) If H is a proper subgroup of a group G and $(G : H) < \infty$, then the set

$$S = \bigcup_{g \in G} H^g$$

is a proper subset of G.
22. Let X be a conjugacy class of G and let $X^{-1} = \{x^{-1} \mid x \in X\}$.
a) Show that X^{-1} is also a conjugacy class of G.
b) Show that if G has odd order, then $X = \{1\}$ is the only conjugacy class for which $X = X^{-1}$.
c) Show that if G has even order, then there is a conjugacy class X other than $\{1\}$ for which $X = X^{-1}$.
d) Show that if G is finite and $k(G)$ is even, then $o(G)$ is even. Show by example that the converse does not hold.
23. Let G be a group of order $2m$. Suppose that G has a conjugacy class of size m. Prove that m is odd, and that G has an abelian normal subgroup of size m.
24. Let H be normal in G and suppose that $(G : H) = p$ is a prime. Let $x \in H$ have the property that there is a $g \in G \setminus H$ such that $gx = xg$.
a) Show that $|C_G(x)| = p|C_H(x)|$.
b) Show that $x^H = x^G$.

p-Groups and p-Subgroups

25. a) Let $f: G \to H$ be a group homomorphism. If G is a p-group, under what conditions, if any, is H a p-group?
    b) Let $f: G \to H$ be a group homomorphism. If H is a p-group, under what conditions, if any, is H a p-group?
    c) Let $H \trianglelefteq G$. If H and G/H are both p-groups, under what conditions, if any, is G a p-group?
26. Let G be a finite p-group.
    a) Prove that any cover of $H < G$ has index p.
    b) Prove that a cover of the center $Z(G)$ is abelian.
27. Let $H \leq G$. Prove that G is a p-group if and only if H and G/H are p-groups.
28. Let G be a finite simple nonabelian group. Show that $o(G)$ is divisible by at least two distinct prime numbers.
29. Prove that the derived group G' of a p-group G is a proper subgroup of G.
30. Let G be a p-group. Show that if $H \leq G$ and $(G:H) < \infty$, then $(G:H)$ is a power of p.
31. Let $G = \bowtie G_p$ be a direct product of p-subgroups for distinct primes p. Show that if $H \leq G$, then $H = \bowtie(H \cap G_p)$. What if the primes are not distinct?
32. Show that the generalized quaternion group

$$Q_m = \langle a, b \rangle, o(a) = 2^{m-1}, o(b) = 4, b^2 = a^{2^{m-2}}, bab^{-1} = a^{-1}$$

    has only is single involution.
33. Prove that a finite p-group has the normalizer condition using the action of $N_G(H)$ on the conjugates $\operatorname{conj}_G(H)$ of H by conjugation.
34. Let G be a nonabelian group of order p^3, where p is a prime. Determine the number $k(G)$ of conjugacy classes of G.

Additional Problems

35. Let p be a prime.
    a) Show that

$$|GL(n, \mathbb{Z}_p)| = (p^n - 1)(p^n - p)\cdots(p^n - p^{n-1})$$

    b) Show that $UT(n, \mathbb{Z}_p)$ is a p-group, in fact,

$$|UT(n, \mathbb{Z}_p)| = p^{(n^2-n)/2}$$

    c) Show that $UT(n, \mathbb{Z}_p)$ is a Sylow p-subgroup of $GL(n, \mathbb{Z}_p)$.
    d) For $n = 2$ or $n - 1 \leq p$, show that $UT(n, \mathbb{Z}_p)$ has exponent p.
    e) Show that $UT(3, \mathbb{Z}_2) \approx Q$.
36. Let F be a finite field of size q and let V be an n-dimensional vector space over F. Show that the number of subspaces of V of dimension k is

$$\binom{n}{k}_q := \frac{(q^n - 1)(q^n - q)\cdots(q^n - q^{k-1})}{(q^k - 1)(q^k - q)\cdots(q^k - q^{k-1})}$$

The expressions $\binom{n}{k}_q$ are called **Gaussian coefficients**. *Hint*: Show that the number of k-tuples of linearly independent vectors in V is

$$(q^n - 1)(q^n - q)\cdots(q^n - q^{k-1})$$

37. Let $H, K \leq G$.
   a) Show that the distinct **double cosets** AgB, where $a \in G$, form a partition of G.
   b) Show that $|AgB| = (A : B^g \cap A)$.

Chapter 8
Sylow Theory

In 1872, the Norwegian mathematician Peter Ludwig Mejdell Sylow [32] published a set of theorems which are now known as the *Sylow theorems*. These important theorems describe the nature of maximal p-subgroups of a finite group, which are now called *Sylow p-subgroups*. (For convenience, we will collect the Sylow theorems into a single theorem.)

Sylow Subgroups

We begin with the definition of a Sylow subgroup.

Definition *Let G be a group and let p be a prime. A* **Sylow p-subgroup** *of G is a maximal p-subgroup of G (under set inclusion). The set of all Sylow p-subgroups of G is denoted by $\mathrm{Syl}_p(G)$. The number of Sylow p-subgroups of a group G is denoted by $n_p(G)$, or just n_p when the context is clear.* $\square$

Of course, if a prime p divides $o(G)$, then G contains a Sylow p-subgroup; in fact, every p-subgroup of G is contained in a Sylow p-subgroup. Also, if G is an infinite group and if H is a p-subgroup of G, then an appeal to Zorn's lemma shows that G has a Sylow p-subgroup containing H.

Since conjugation is an order isomorphism and also preserves the group order of elements, it follows that if S is a Sylow p-subgroup of G, then so is every conjugate S^a of S.

Note also that if G is finite and $o(G) = p^n m$ where $(p, m) = 1$, then any subgroup of order p^n is a Sylow p-subgroup. We will prove the converse of this a bit later: Any Sylow subgroup of G has order p^n.

The Normalizer of a Sylow Subgroup

Let G be a finite group. If a Sylow p-subgroup S of G happens to be normal in G, then G/S has no nonidentity p-elements. Hence, $p \nmid (G : S)$ and so S is the set of all p-elements of G. It also follows that $S \sqsubseteq G$, since automorphisms preserve order.

Of course, S is always normal in its normalizer $N_G(S)$.

Theorem 8.1 *Let G be a finite group and let $S \in \mathrm{Syl}_p(G)$.*
1) *S is the set of all p-elements of $N_G(S)$.*
2) *Any p-element $a \in G \setminus S$ moves S by conjugation, that is, $S^a \neq S$.*
3) *S is the only Sylow p-subgroup of $N_G(S)$.*
4) *$p \nmid (N_G(S) : S)$.*
5) *$S \sqsubseteq N_G(S)$.*□

If $S \in \mathrm{Syl}_p(G)$, then

$$S^a \leq N_G(S)^a$$

for any $a \in G$. Hence, if a normalizes $N_G(S)$, then

$$S^a \leq N_G(S)$$

and since S^a is also a Sylow p-subgroup of $N_G(S)$, Theorem 8.1 implies that $S^a = S$. In other words, if a normalizes $N_G(S)$, then a also normalizes S and so

$$N_G(N_G(S)) = N_G(S)$$

Theorem 8.2 *The normalizer $N_G(S)$ of a Sylow subgroup of G is self-normalizing, that is,*

$$N_G(N_G(S)) = N_G(S)$$ □

Soon we will be able to prove that not only is $N_G(S)$ self-normalizing, but so is any subgroup of G *containing* $N_G(S)$.

The Sylow Theorems

Let G be a finite group and let $S \in \mathrm{Syl}_p(G)$. The fact that any p-element $a \notin S$ moves S by conjugation prompts us to look at the action of a p-subgroup K of G by conjugation on the set

$$\mathrm{conj}_G(S) = \{S^a \mid a \in G\}$$

of conjugates of S in G. As to the stabilizer of S^a, we have

$$\mathrm{stab}(S^a) = N_G(S^a) \cap K = S^a \cap K$$

and so

$$|\mathrm{orb}_K(S^a)| = (K : S^a \cap K)$$

which is divisible by p unless $K \leq S^a$, in which case the orbit has size 1.

Hence,

$$\mathrm{Fix}_{\mathrm{conj}_G(S)}(K) = \{S^a \mid K \leq S^a\}$$

and so

$$|\mathrm{conj}_G(S)| \equiv |\{S^a \mid K \leq S^a\}| \bmod p$$

Now if K is a Sylow p-subgroup of G, then $K \leq S^a$ if and only if $K = S^a$ and so

$$|\mathrm{conj}_G(S)| \equiv \begin{cases} 1 \bmod p & \text{if } K \in \mathrm{conj}_G(S) \\ 0 \bmod p & \text{if } K \notin \mathrm{conj}_G(S) \end{cases}$$

It follows that $K \notin \mathrm{conj}_G(S)$ is impossible and so $\mathrm{Syl}_p(G) = \mathrm{conj}_G(S)$ is a conjugacy class and

$$n_p \equiv 1 \bmod p$$

Note also that

$$n_p = |\mathrm{conj}_G(S)| = (G : N_G(S)) \mid o(G)$$

Finally, we can determine the order of a Sylow p-subgroup S, since

$$(G : S) = (G : N_G(S))(N_G(S) : S)$$

and neither of the factors on the right is divisible by p. Hence, the order of S is the *largest* power of p dividing $o(G)$. We have proved the *Sylow theorems.*

Theorem 8.3 (The Sylow theorems [32], 1872) *Let G be a finite group and let $o(G) = p^n m$, where p is a prime and $p \nmid m$.*

1) *The Sylow p-subgroups of G are the subgroups of G of order p^n.*
2) $\mathrm{Syl}_p(G)$ *is a conjugacy class in* $\mathrm{sub}(G)$.
3) *The number n_p of Sylow p-subgroups satisfies*

$$n_p \equiv 1 \bmod p \quad \textit{and} \quad n_p = (G : N_G(S)) \mid o(G)$$

where $S \in \mathrm{Syl}_p(G)$.

4) *Let $S \in \mathrm{Syl}_p(G)$.*
   a) *S is normal if and only if $n_p = 1$.*
   b) *S is self-normalizing if and only if $n_p = (G : S) = m$, in which case all Sylow p-subgroups of G are self-normalizing.*
5) *If K is a p-subgroup of G, then*

$$|\{S \in \mathrm{Syl}_p(G) \mid K \leq S\}| \equiv 1 \bmod p$$

□

We will prove later in the chapter that every *normal* Sylow p-subgroup of a finite group is complemented. This is implied by the famous Schur–Zassenhaus theorem. However, we also have the following simple consequence of Theorem 3.1 concerning supplements of Sylow subgroups.

Theorem 8.4 *Let G be a finite group. Then any Sylow p-subgroup of G and any subgroup whose index is a power of p are supplements.* $\square$

Sylow Subgroups of Subgroups

Let G be a finite group and let $H \leq G$. We wish to explore the relationship between $\mathrm{Syl}_p(G)$ and $\mathrm{Syl}_p(H)$. On the one hand, every $S \in \mathrm{Syl}_p(H)$ is contained in a $T \in \mathrm{Syl}_p(G)$ and so the set

$$\mathrm{Syl}_p(S;G) = \{T \in \mathrm{Syl}_p(G) \mid S \leq T\}$$

is nonempty. Moreover, since $A \in \mathrm{Syl}_p(S;G)$ implies that $A \cap H$ is a p-subgroup of H containing S, we have

$$A \in \mathrm{Syl}_p(S;G) \quad \Rightarrow \quad A \cap H = S$$

In particular, the families $\mathrm{Syl}_p(S;G)$ are disjoint, that is,

$$S \neq T \in \mathrm{Syl}_p(H) \quad \Rightarrow \quad \mathrm{Syl}_p(S;G) \cap \mathrm{Syl}_p(T;G) = \emptyset$$

and so

$$n_p(H) \leq n_p(G)$$

On the other hand, if $S \in \mathrm{Syl}_p(G)$, then the intersection $S \cap H$ need not be a Sylow p-subgroup of H, as can be seen by taking H and S to be distinct Sylow p-subgroups of G. However, if HS is a subgroup of G, then $o(S) \mid o(HS)$ and so

$$|H \cap S|\frac{|HS|}{|S|} = |H|$$

where $|HS|/|S|$ is not divisible by p. Hence, $|H \cap S|$ and $|H|$ are divisible by the same powers of p and so $H \cap S \in \mathrm{Syl}_p(H)$.

Theorem 8.5 *Let G be a finite group and let $H \leq G$.*

1) *If $S \in \mathrm{Syl}_p(H)$, then*

$$A \in \mathrm{Syl}_p(S;G) \quad \Rightarrow \quad A \cap H = S$$

and

$$S \neq T \in \mathrm{Syl}_p(H) \quad \Rightarrow \quad \mathrm{Syl}_p(S;G) \cap \mathrm{Syl}_p(T;G) = \emptyset$$

and so

$$n_p(H) \leq n_p(G)$$

2) *If* $S \in \mathrm{Syl}_p(G)$ *and* $HS \leq G$, *then*

$$S \cap H \in \mathrm{Syl}_p(H) \qquad \square$$

Some Consequences of the Sylow Theorems

Let us consider some of the more-or-less direct consequences of the Sylow theorems.

A Partial Converse of Lagrange's Theorem

A Sylow p-subgroup S of a group G has subgroups of all orders dividing $o(S)$. This gives a partial converse to Lagrange's theorem.

Theorem 8.6 *Let* G *be a finite group and let* p *be a prime. If* $p^k \mid o(G)$, *then* G *has a subgroup of order* p^k.$\square$

More on the Normalizer of a Sylow Subgroup

Recall that the normalizer $N_G(S)$ of a Sylow subgroup S of G is self-normalizing. Now we can say more.

Theorem 8.7 *Let* G *be a finite group and let* $S \in \mathrm{Syl}_p(G)$. *If*

$$S \leq N_G(S) \leq H \leq G$$

then H *is self-normalizing. In particular, if* $H < G$, *then* H *is not normal in* G.
Proof. Conjugating by any $a \in N_G(H)$ gives

$$S^a \leq N_G(S)^a \leq H \leq G$$

and so both S and S^a are Sylow p-subgroups of H. It follows that S and S^a are conjugate *in* H. Hence, there is an $h \in H$ for which $S^{ha} = S$, that is, $ha \in N_G(S) \leq H$. Thus, $a \in H$ and so $N_G(H) = H$.$\square$

The normalizer of a Sylow p-subgroup has a somewhat stronger property than is expressed in Theorem 8.7. In the exercises, we ask the reader to prove that $N_G(S)$ is abnormal.

Counting Subgroups in a Finite Group

In an earlier chapter, we proved that if G is a p-group and $p^k \mid o(G)$, then the number $n_{p,k}(G)$ of subgroups of G of order p^k satisfies

$$n_{p,k}(G) \equiv 1 \bmod p \tag{8.8}$$

We have just proved that for *any* finite group G for which $p \mid o(G)$,

$$n_p(G) \equiv 1 \bmod p$$

To see that (8.8) holds for all finite groups, we count the size of the set

$$\mathcal{F}_k = \{(H,S) \mid H \le S, S \in \operatorname{Syl}_p(G), o(H) = p^k\}$$

modulo p. On the one hand, for each $S \in \operatorname{Syl}_p(G)$, there are $n_{p,k}(S) \equiv 1$ subgroups of S of order p^k and so

$$|\mathcal{F}_k| \equiv n_p(G) \cdot 1 \equiv 1$$

On the other hand, for each $H \le G$ of order p^k, Theorem 8.3 implies that

$$|\{S \in \operatorname{Syl}_p(G) \mid H \le S\}| \equiv 1$$

and so

$$|\mathcal{F}_k| \equiv n_{p,k}(G) \cdot 1 \equiv n_{p,k}(G)$$

Hence, $n_{p,k}(G) \equiv 1$ and we have proved an important theorem of Frobenius.

Theorem 8.9 (Frobenius [13], 1895) *Let G be a group with $o(G) = p^n m$ where $(m,p) = 1$. Then for each $1 \le k \le n$, the number $n_{p,k}(G)$ of subgroups of G of order p^k satisfies*

$$n_{p,k}(G) \equiv 1 \bmod p \qquad \square$$

When All Sylow Subgroups Are Normal

Several good things happen when all of the Sylow subgroups of a group G are normal. In particular, let G be a finite group. In an earlier chapter (see Theorem 4.22 and Theorem 4.35), we showed that among the conditions:

1) Every subgroup of G is subnormal
2) G has the normalizer condition
3) Every maximal subgroup of G is normal
4) $G/\Phi(G)$ is abelian

the following implications hold:

$$1) \Leftrightarrow 2) \Rightarrow 3) \Leftrightarrow 4)$$

We also promised to show that these four conditions are equivalent, which we can do now, adding several additional equivalent conditions into the bargin.

First, let us speak about arbitrary (possibly infinite) groups. If G is a group, let G_{tor} denote the set of all torsion (periodic) elements of G. If G is abelian, then G_{tor} is a subgroup of G. However, in the nonabelian general linear group $GL(2,\mathbb{C})$, the elements

$$A = \begin{pmatrix} 0 & -1 \\ 1 & 0 \end{pmatrix} \quad \text{and} \quad B = \begin{pmatrix} 0 & 1 \\ -1 & -1 \end{pmatrix}$$

are torsion but their product is not. Hence, G_{tor} is not always a subgroup of G. Note that when G_{tor} is a subgroup of G, then $G_{\text{tor}} \sqsubseteq G$ since automorphisms preserve order.

Theorem 8.10 *Let G be a group in which every Sylow subgroup is normal. Let the Sylow subgroups of G be $\{Y_p \mid p \in \mathcal{P}\}$. Then*

$$G_{\text{tor}} = \bowtie_{p\in\mathcal{P}} Y_p$$

and so $G_{\text{tor}} \leq G$. Thus, the product of two elements of finite order has finite order.

Proof. Since the Sylow p-subgroups are normal and pairwise essentially disjoint, they commute elementwise. In particular, if $a_1, \ldots, a_n$ come from distinct Sylow subgroups, then

$$o(a_1 \cdots a_n) = o(a_1) \cdots o(a_n)$$

and so the family of Sylow subgroups is strongly disjoint and

$$Y := \bowtie_{p\in\mathcal{P}} Y_p \subseteq G_{\text{tor}}$$

For the reverse inclusion, if $a \in G_{\text{tor}}$ has order $n = p_1^{e_1} \cdots p_m^{e_m}$ where the primes p_i are distinct, then Corollary 2.11 implies that $a = a_1 \cdots a_m$, where $o(a_k) = p_k^{e_k}$ and so $a_k \in Y_{p_k}$, whence $a \in Y$.□

Now we turn to finite groups in which all Sylow subgroups are normal.

Theorem 8.11 *Let G be a finite group, with Sylow subgroups $\{Y_p \mid p \in \mathcal{P}\}$. The following are equivalent:*

1) *Every Sylow subgroup of G is normal.*
2) *G is the direct product of its Sylow p-subgroups*

$$G = \bowtie_{p\in\mathcal{P}} Y_p$$

3) *If $H \leq G$, then*

$$H = \bowtie_{p\in\mathcal{P}} (H \cap Y_p)$$

4) *G is the direct product of p-subgroups.*
5) **(Strong converse of Lagrange's theorem)** *If $n \mid o(G)$, then G has a normal subgroup of order n.*
6) *Every subgroup of G is subnormal.*
7) *G has the normalizer condition.*
8) *Every maximal subgroup of G is normal.*
9) *$G/\Phi(G)$ is abelian.*

If these conditions hold, then G has the center-intersection property. In particular, $Z(G)$ is nontrivial.

Proof. Theorem 8.10 shows that 1) implies 2) and the converse is clear. If 1) holds, then $HY_p \le G$ and so the Sylow p-subgroups of H are $\{H \cap Y_p \mid p \in \mathcal{P}\}$. Moreover, since $H \cap Y_p \trianglelefteq H$, it follows that H is the direct product of its Sylow p-subgroups and so 3) holds. It is clear that 3) implies 2) and so 1)–3) are equivalent. Also, it is clear that 2) $\Rightarrow$ 4). If 4) holds and p is a prime dividing $o(G)$, then we can isolate the factors in the direct product of G that have exponent p, say

$$G = P \bowtie Q$$

where P is a direct product of p-subgroups and Q is a direct product of q-subgroups for various primes $q \neq p$. Then P is the set of all p-elements of G and so P a Sylow p-subgroup of G. But $P \trianglelefteq G$ and so 1) holds. Thus, 1)–4) are equivalent.

It is clear that 5) $\Rightarrow$ 1). To see that 2) $\Rightarrow$ 5), any divisor n of $o(G)$ is a product $n = \prod d_p$ where $d_p \mid o(Y_p)$ and since Y_p has a normal subgroup of order d_p, the direct product of these subgroups is a normal subgroup of G of order n. Thus, 1)–5) are equivalent and we have already remarked that

$$6) \Leftrightarrow 7) \Rightarrow 8) \Leftrightarrow 9)$$

To see that 3) $\Rightarrow$ 6), if $H < G$, then one of the factors $H \cap Y_p$ is proper in Y_p and so there is a subgroup N_p for which

$$H \cap Y_p \triangleleft N_p \le Y_p$$

Hence,

$$H \triangleleft \bowtie_{q \neq p} (H \cap Y_q) \bowtie N_p$$

Since H is an arbitrary proper subgroup of G, it follows that all subgroup of G are subnormal. To see that 6) $\Rightarrow$ 1), if Y_p is not normal in G, then since $N_G(Y_p) < G$ is subnormal, there is a subgroup H_p for which $N_G(Y_p) \triangleleft H_p \le G$. But this contradicts the fact that $N_G(Y_p)$ is self-normalizing. Hence, $Y_p \triangleleft G$. Thus, 1)–7) are equivalent and imply 8).

Similarly, if 8) holds but Y_p is not normal in G, then there is a maximal subgroup $M \le G$ for which

$$Y_p \le N_G(Y_p) \le M \triangleleft G$$

which contradicts Theorem 8.7. Hence, $Y_p \trianglelefteq G$ and 1) holds.

Finally, if 3) holds and $N \le G$ is nontrivial, then $N \cap Y_q \trianglelefteq Y_q$ is nontrivial for some $q \in \mathcal{P}$ and

$$N \cap Z(G) = \bowtie_{p \in \mathcal{P}} (N \cap Z(G) \cap Y_p)$$

But $N \cap Z(G) \cap Y_q = N \cap Z(Y_q)$ is nontrivial and therefore so is $N \cap Z(G)$.□

The hypotheses of the previous theorems hold for all abelian groups.

Corollary 8.12 *Let G be an abelian group.*

1) **(Primary decomposition)** *Then G_{tor} is the direct product of its Sylow p-subgroups.*
2) **(Converse of Lagrange's theorem)** *If G is finite and $n \mid o(G)$, then G has a subgroup of order n.*□

We will add one additional characterization (nilpotence) to Theorem 8.11 in a later chapter (see Theorem 11.8).

When a Subgroup Acts Transitively; The Frattini Argument

The Frattini argument (Theorem 7.2) shows that if a group G acts on a nonempty set X and if $H \leq G$ is transitive on X, then

$$G = H\text{stab}_G(x)$$

and if H is regular on X, then

$$G = H \bullet \text{stab}_G(x)$$

and so $\text{stab}_G(x)$ is a complement of H in G. To apply this idea, let G be a finite group and let

$$S \leq H \trianglelefteq G$$

where $S \in \text{Syl}_p(H)$. Let G act on $\text{conj}_G(S)$ by conjugation. Since $S^a \in \text{Syl}_p(H)$ for any $a \in G$, it follows that H acts transitively on $\text{conj}_G(S)$. Hence,

$$G = H\text{stab}_G(S) = HN_G(S)$$

This specific argument is also referred to as the *Frattini argument.*

Theorem 8.13 *Let G be a finite group and let $H \leq G$. If $S \in \text{Syl}_p(H)$, then*

$$G = HN_G(S)$$

and if the action of H by conjugation on $\text{conj}_G(S)$ is regular, then

$$G = H \bullet N_G(S)$$ □

This theorem can be used to show that the Frattini subgroup of a finite group G has the property that all of its Sylow subgroups are normal in G.

Theorem 8.14 (Frattini [12], 1885) *If G is a finite group, then the Frattini subgroup $\Phi(G)$ has the property that all of its Sylow subgroups are normal in G.*

Proof. If $S \in \mathrm{Syl}_p(\Phi)$, then $S \leq \Phi \trianglelefteq G$ and the Frattini argument shows that

$$G = \Phi N_G(S)$$

But if $N_G(S) < G$, then there is a maximal subgroup M of G for which $N_G(S) \leq M$ and so $G \leq M$, a contradiction. Hence, $N_G(S) = G$ and $S \trianglelefteq G$.□

The Search for Simplicity

The Sylow theorems, along with group actions and counting arguments, provide powerful tools for the analysis of finite groups. A key issue with respect to finite groups is the question of simplicity. As we will discuss in a later chapter, the issue of which finite groups (up to isomorphism) are simple appears to be resolved, but the resolution is so complex that some mathematicians may still have questions regarding its completeness and its accuracy.

We have seen that a group of prime-power order p^n has a normal subgroup of each order $p^k \mid p^n$. Accordingly, we will do no further direct analysis of p-groups in this chapter.

Throughout our discussion, p will denote a prime, Y_p will denote an arbitrary Sylow p-subgroup and, as always, n_p denotes the number of Sylow p-subgroups of G. Recall that

1) $n_p = 1 + kp$ for some integer $k \geq 0$.
2) $n_p = (G : N_G(Y_p)) \mid o(G)$.

Note that if $o(G) = p^n m$, where $p \nmid m$, then $n_p \mid o(G)$ if and only if $n_p \mid m$.

The following facts (among others) are useful in showing that a group is not simple:

3) Y_p is normal if and only if $n_p = 1$.
4) If $(G : H)$ is equal to the smallest prime dividing $o(G)$, then $H \lhd G$.
5) The kernel of any action $\lambda\colon G \to S_X$ is a normal subgroup of G.
6) If $H < G$, then $(G : H^\circ) \mid (G : H)!$. Hence, if $o(G) \nmid (G : H)!$, then H° is a nontrivial proper normal subgroup of G.

We will also have use for the fact that if p is prime and $1 \leq e < p$, then ep is the *smallest* integer for which $p^e \mid (ep)!$.

The n_p-Argument

It happens quite often that for some odd prime $p \mid o(G)$, the integers $1 + kp$ do not divide $o(G)$ unless $k = 0$, in which case $n_p = 1$ and $Y_p \trianglelefteq G$. Let us refer to the argument

$$n_p = 1 + kp \mid o(G) \quad \Rightarrow \quad k = 0$$

as the **n_p-argument**. Note that the n_p-argument does not hold if $p = 2$, unless $o(G)$ is a power of 2, since $1 + 2k \mid o(G)$ for some $k > 1$.

Example 8.15 If $o(G) = 9982 = 2 \cdot 7 \cdot 23 \cdot 31$, then routine calculation shows that the n_7 argument holds:

$$1 + 7k \mid o(G) \quad \Rightarrow \quad k = 0$$

and so $Y_7 \lhd G$ and G is not simple.□

Example 8.16 If $o(G) = p^n m$ for $n \geq 1$, $m > 1$ and $p \nmid m$, then $n_p = (1 + kp) \mid m$ and so if $m < p$, then $k = 0$, whence $Y_p \lhd G$. Thus, groups of order

$$p^n, 2p^n, \ldots, (p-1)p^n$$

for p prime and $n \geq 1$ have $Y_p \lhd G$ and so are not simple.□

A little programming shows that among the orders up to 10000 (not including prime powers) there are only 569 orders (less than 6%) that are *not* succeptible to the n_p-argument for some p. Thus, the vast majority of orders up to 10000 are either prime powers or have the property that groups of that order have a normal Sylow p-subgroup.

Counting Elements of Prime Order

If p is a prime and $p \mid o(G)$ but $p^2 \nmid o(G)$, then each of the n_p distinct Sylow p-subgroups of G has order p and so the subgroups are pairwise essentially disjoint. Hence, G contains exactly $n_p \cdot (p - 1)$ distinct elements of order p. Sometimes this simple counting of elements (for different primes p) is enough to show than one of the Sylow subgroups is normal.

Example 8.17 Let $o(G) = 30 = 2 \cdot 3 \cdot 5$. Then based on the fact that $n_p = 1 + kp \mid o(G)$, we can conclude only that $n_3 \in \{1, 10\}$ and $n_5 \in \{1, 6\}$. However, if $n_3 = 10$ and $n_5 = 6$, then G contains at least $n_3 \cdot (3 - 1) = 20$ elements of order 3 and 24 elements of order 5, totalling 44 elements. Hence, one of Y_3 or Y_5 must be normal in G.□

Index Equal to Smallest Prime Divisor

If $o(G) = pq^k$ where $p < q$ are primes, then $Y_q \lhd G$, because $(G : Y_q) = p$ is the smallest prime dividing $o(G)$. Moreover, it is clear that

$$G = Y_q \rtimes Y_p$$

Example 8.18 If $o(G) = 3 \cdot 5^2 = 75$, then $Y_5 \lhd G$ and

$$G = Y_3 \rtimes Y_5$$

Also, $1 + 3k \mid 25$ holds only for $k = 0$ or $k = 8$ and so $n_3 = 1$ or $n_3 = 25$. Note that if $n_3 = 1$, then $G = Y_3 \bowtie Y_5$ is abelian.□

When $o(G) = pq$, we can give a fairly complete analysis as follows.

Theorem 8.19 *Let $o(G) = pq$, with $p < q$ primes. Then*

$$G = C_q(b) \rtimes C_p(a)$$

where

$$b^a = b^k$$

for some $1 \le k < q$ and $k^p \equiv 1 \bmod q$. Moreover, G is cyclic if and only if $p \nmid q - 1$.

Proof. We have seen that

$$G = Y_q \rtimes Y_p = C_q(b) \rtimes C_p(a)$$

Thus, $aba^{-1} = b^k$ for some $1 \le k < q$ and repeated conjugation by a gives

$$b = a^p b a^{-p} = b^{k^p}$$

which implies that $k^p \equiv 1 \bmod q$. Moreover, $n_p \mid q$ and so $n_p = 1$ or $n_p = q$. But $n_p = 1$ if and only if $Y_p \trianglelefteq G$, that is, if and only if G is cyclic and $n_p = q$ if and only if $1 + kp = q$, that is, if and only if $p \nmid q - 1$.□

Example 8.20 Let us return to the case $o(G) = 30 = 2 \cdot 3 \cdot 5$. We saw in Example 8.17 that one of Y_3 or Y_5 must be normal in G. It follows that $Y_3 Y_5 \trianglelefteq G$ has order 15 and so is cyclic. Hence, $Y_3, Y_5 \sqsubseteq Y_3 Y_5 \trianglelefteq G$ and so both Y_3 and Y_5 are normal in G.□

Using the Kernel of an Action

The kernel of an action $\lambda\colon G \to S_X$ is normal in G and this can be a useful technique for finding normal subgroups, although they need not be Sylow subgroups.

For example, if G acts on $\mathrm{Syl}_p(G)$ by conjugation, then the representation map $\lambda\colon G \to S_{kp+1}$ has kernel

$$K = \bigcap_{Y \in \mathrm{Syl}_p(G)} N_G(Y)$$

which is a normal subgroup of G. The problem is that it may be either trivial or equal to G.

Let $o(G) = p^m u$ and $o(K) = p^s v$, where $m \geq 1, u > 1$ and $p \nmid u$. Also, let $n_p = kp + 1$. It is clear that $K = G$ is equivalent to $n_p = 1$ and implies that $s = m$. Conversely, if $s = m$, then K contains a Sylow p-subgroup S of G. But the only Sylow p-subgroup of G in $N_G(Y)$ is Y itself and so $n_p = 1$ and $K = G$. Thus,

$$K = G \quad \Leftrightarrow \quad Y_p \lhd G \quad \Leftrightarrow \quad s = m$$

As to the nontriviality of K, the induced embedding of G/K into S_{kp+1} implies that

$$p^{m-s} \frac{u}{v} \,\Big|\, (kp+1)!$$

and so $p^{m-s} \mid (kp)!$. Hence, if $k < p$, then $m - k \leq s$. It follows that if $k < m$, then $s > 0$ and K is nontrivial. Thus,

$$k < \min\{m, p\} \quad \Rightarrow \quad K \neq \{1\}$$

We note finally that K has a somewhat simpler form if $n_p = u$, since then each Y_p is self-normalizing and

$$K = \bigcap_{Y \in \mathrm{Syl}_p(G)} Y$$

Theorem 8.21 *Let $o(G) = p^m u$ where p is prime, $m \geq 1$, $u > 1$ and $p \nmid u$. Let $n_p = 1 + kp$.*
1) If $k = 0$, then $Y_p \lhd G$.
2) If $0 < k < \min\{m, p\}$, then

$$K = \bigcap_{Y \in \mathrm{Syl}_p(G)} N_G(Y)$$

is a nontrivial proper normal subgroup of G of order $p^s v$, where $m - k \leq s \leq m - 1$ and $v \mid u$. In addition, if $k = (u-1)/p$, then

$$K = \bigcap_{Y \in \mathrm{Syl}_p(G)} Y$$

has order p^s. □

Example 8.22 If $o(G) = 108 = 3^3 \cdot 4$, then $1 + 3k \mid 4$ and so $k = 0$ or $k = 1$. Thus, this case is not amenable to the n_p-argument. However, if $k = 1$ then

Theorem 8.21 implies that

$$K = \bigcap_{Y \in \mathrm{Syl}_3(G)} Y$$

is a nontrivial proper normal subgroup of G of order 9. Thus, G is not simple.

If $o(G) = 189 = 3^3 \cdot 7$, then $1 + 3k \mid 7$ and so $k = 0$ or $k = 2$. If $k = 2$, then Theorem 8.21 implies that

$$K = \bigcap_{Y \in \mathrm{Syl}_3(G)} Y$$

is a nontrivial proper normal subgroup of G of order 3 or 9.

If $o(G) = 300 = 2^2 \cdot 3 \cdot 5^2$, then $n_5 = 1 + 5k \mid 12$ and so $k = 0$ or $k = 1$ (and $n_5 = 6$). But if $k = 1$, then Theorem 8.21 implies that G is not simple.□

Even when K is trivial and the previous theorem does not apply, we learn that $G \hookrightarrow S_{kp+1}$, which can sometimes be useful.

Example 8.23 If $o(G) = p(p+1)$, where p is prime. Then $n_p = 1$ or $n_p = p + 1$. While the previous theorem does not apply, if $n_p = p + 1$, then

$$K = \bigcap_{Y \in \mathrm{Syl}_p(G)} Y = \{1\}$$

Hence, $G \hookrightarrow S_{p+1}$. As an example, if $o(G) = 12 = 3 \cdot 4$, then either $Y_3 \triangleleft G$ or $G \hookrightarrow S_4$. But $o(G) = o(A_4)$ and so in the latter case, $G \approx A_4$. Thus, if $G \not\approx A_4$ then $Y_3 \triangleleft G$. We will use this fact later to determine all groups of order 12.□

The Normal Interior

If $H < G$, we have seen that

$$(G : H^\circ) \mid (G : H)!$$

and so if $o(G) \nmid (G : H)!$, then H° is a nontrivial proper normal subgroup of G. Hence, if

$$o(G) = p_1^{e_1} \cdots p_m^{e_m}$$

where $p_1 < \cdots < p_m$ are primes and $m \geq 2$, then for any k,

$$e_k < p_k, (G : H) < e_k p_k \quad \Rightarrow \quad p_k^{e_k} \nmid (G : H)! \quad \Rightarrow \quad o(G) \nmid (G : H)!$$

and so $H^\circ \triangleleft G$ is nontrivial.

Theorem 8.24 *Let $o(G) = p_1^{e_1} \cdots p_m^{e_m}$ where $p_1 < \cdots < p_m$ are primes and $m \geq 2$. Suppose that $e_k < p_k$ for some $1 \leq k \leq m$.*

1) *If $H < G$ has index $(G : H) < e_k p_k$, then H° is a nontrivial proper normal subgroup of G.*
2) *In particular, if $1 < n_{p_i} < e_k p_k$, then $N_G(Y_{p_i})^\circ$ is a nontrivial proper normal subgroup of G.*□

Example 8.25 Let $o(G) = 6201 = 3^2 \cdot 13 \cdot 53$. Then $n_3 = 1 + 3k \mid 13 \cdot 53$, which implies that $n_3 \in \{1, 13\}$. If $n_3 = 1$, then $Y_3 \lhd G$. If $n_3 = 13 < 53$, then Theorem 8.24 implies that $N_G(Y_3)^\circ$ is a nontrivial proper normal subgroup of G. Thus G is not simple.□

Using the Normalizer of a Sylow Subgroup

Let $p \neq q$ be primes dividing $o(G)$ and let $Y_q \in \mathrm{Syl}_q(G)$. Under the assumption that $n_q > 1$ and so $N_G(Y_q) < G$, suppose that $p \nmid n_q$, that is, $p \mid o(N_G(Y_q))$ and that $P \in \mathrm{Syl}_p(N_G(Y_q))$. There are various things we can say about $o(N_G(P))$.

First, if $P \lhd N_G(Y_q)$, then $N_G(Y_q) \leq N_G(P)$ and so

$$o(N_G(Y_q)) \mid o(N_G(P))$$

On the other hand, even if P is not normal in $N_G(Y_q)$, the fact that $Y_q \trianglelefteq N_G(Y_q)$ implies that PY_q is a subgroup of G. Hence, if PY_q is abelian, then $Y_q \leq N_G(P)$ and so

$$o(Y_q) \mid o(N_G(P))$$

In either case, if P is *not* a Sylow p-subgroup of G but $P \lhd P^* \in \mathrm{Syl}_p(G)$, then $P^* \leq N_G(P)$, whence

$$o(P^*) \mid o(N_G(P))$$

These conditions tend to make $N_G(P)$ large.

Example 8.26 If $o(G) = 3675 = 3 \cdot 5^2 \cdot 7^2$, then it is easy to see that $n_7 \in \{1, 15\}$. If $n_7 = 15$, then $o(N_G(Y_7)) = 5 \cdot 7^2$. Let P be a Sylow 5-subgroup of $N_G(Y_7)$. The number of such subgroups is $1 + 5k \mid 7^2$ and so $P \lhd N_G(Y_7)$. Hence,

$$5 \cdot 7^2 \mid o(N_G(P))$$

Also, P has index 5 in $P^* \in \mathrm{Syl}_5(G)$ and so $P \lhd P^*$, whence

$$5^2 \mid o(N_G(P))$$

and so

$$5^2 \cdot 7^2 \mid o(N_G(P))$$

Hence, either $N_G(P) = G$, in which case $P \lhd G$ or else $N_G(P)$ has index 3 in G and so is normal in G.□

Suppose that $o(G) = pqu$, where $p \neq q$ are primes that do not divide u. If $p \nmid n_q$, that is, if $p \mid o(N(Y_q))$, then $Y_p \leq N(Y_q)$ and so $Y_pY_q \leq G$ has order pq. Hence, if $p \nmid (q-1)$, then Y_pY_q is abelian (cyclic) and so $Y_q \leq N(Y_p)$. Thus, $q \mid o(N(Y_p))$ and so $n_p \mid o(G)/pq$.

Theorem 8.27 *If $o(G) = pqu$, where $p < q$ are primes that do not divide u, then*

$$p \nmid (q-1) \quad \textit{and} \quad p \nmid n_q \quad \Rightarrow \quad n_p \,\Big|\, \frac{o(G)}{pq}$$ □

Example 8.28 If $o(G) = 1785 = 3 \cdot 5 \cdot 7 \cdot 17$, then a routine calculation gives

$$n_3 \in \{1, 7, 85, 595\} \quad \text{and} \quad n_{17} \in \{1, 35\}$$

But

$$3 < 17, \quad 3 \nmid (17-1), \quad 3 \nmid n_{17} \quad \Rightarrow \quad n_3 \,\Big|\, \frac{o(G)}{3 \cdot 17} = 35$$

and so $n_3 = 1$ or $n_3 = 7$. Hence, Theorem 8.24 now implies that one of Y_3 or $N(Y_3)^\circ$ is a proper nontrivial normal subgroup of G.□

Groups of Small Order

We have already examined the groups of order 4, 6 and 8. Let us now look at all groups of order 15 or less. Of course, all groups of prime order are cyclic. We will again denote an arbitrary Sylow p-subgroup of G by Y_p.

Groups of Order 4

The groups of order 4 are (up to isomorphism):

1) C_4, the cyclic group
2) $V \approx C_2 \boxtimes C_2$, the Klein 4-group.

Groups of Order 6

The groups of order 6 are (up to isomorphism):

1) C_6, the cyclic group
2) $D_6 \approx S_3$, the nonabelian dihedral (and symmetric) group.

Groups of Order 8

The groups of order 8 are (up to isomorphism):

1) C_8, the cyclic group
2) $C_4 \boxtimes C_2$, abelian but not cyclic
3) $C_2 \boxtimes C_2 \boxtimes C_2$, abelian but not cyclic

4) D_8, the (nonabelian) dihedral group
5) Q, the (nonabelian) quaternion group.

Groups of Order 9

Theorem 7.9 implies that the groups of order 9 are (up to isomorphism):

1) C_9, the cyclic group
2) $C_3 \boxtimes C_3$, abelian but not cyclic.

Groups of Order 10

If $o(G) = 10 = 2 \cdot 5$, then Theorem 8.19 implies that

$$G = \langle a, b \rangle, o(a) = 2, o(b) = 5, aba^{-1} = b^k$$

where $k^2 \equiv 1 \bmod 5$, that is, $k = 1$ or 4. In the former case, a and b commute and G is cyclic. In the latter case, $G = D_{10}$. Thus, the groups of order 10 are (up to isomorphism):

1) $C_{10} \approx C_5 \boxtimes C_2$, the cyclic group
2) D_{10}, the nonabelian dihedral group.

Groups of Order 12

We have accounted for three nonabelian groups of order 12: the alternating group A_4, the dihedral group D_{12} and the semidirect product $T = C_3(a) \rtimes C_4(b)$, where

$$bab^{-1} = a^2$$

We wish to show that this completes the list of nonabelian groups of order 12.

Assume that $G \not\approx A_4$. Then Example 8.23 shows that $Y_3 = C_3(a) \trianglelefteq G$ and so $G = C_3(a) \rtimes Y_2$, where $Y_2 = C_4(b)$ or $Y_2 = C_2(x) \bowtie C_2(y)$.

If $G = C_3(a) \rtimes C_4(b)$, then

$$bab^{-1} = a \quad \text{or} \quad bab^{-1} = a^2$$

In the former case G is abelian and in the latter case $G = T$. If

$$G = C_3(a) \rtimes (C_2(x) \bowtie C_2(y))$$

then

$$a^x = a \text{ or } a^2 \quad \text{and} \quad a^y = a \text{ or } a^2$$

We consider three cases. If $a^x = a$ and $a^y = a$, then G is abelian. If $a^x = a$ and $a^y = a^2$, then $o(ax) = o(a)o(x) = 6$ and

$$(axy)^2 = axyaxy = ay(xax)y = ayay = a^3 = 1$$

Hence, $\langle axy, y\rangle$ is generated by two distinct involutions whose product has order 6 and so Theorem 2.36 implies that $G = D_{12}$. Finally, if $a^x = a^2 = a^y$, then $(ax)^2 = axax = a^3 = 1$ and $a^{xy} = (a^2)^y = a^4 = a$ and so $o(axy) = o(a)o(xy) = 6$. Thus, $\langle ax, y\rangle$ is dihedral and $G = D_{12}$.

Thus, the groups of order 12 are (up to isomorphism)

1) $C_{12} \approx C_4 \boxtimes C_3$, the cyclic group
2) $C_2 \boxtimes C_2 \boxtimes C_3$, abelian but not cyclic
3) A_4, the nonabelian alternating group
4) D_{12}, the nonabelian dihedral group
5) T, the nonabelian group described above.

Groups of Order 14

If $o(G) = 14 = 2 \cdot 7$, then Theorem 8.19 implies that

$$G = \langle a, b\rangle, o(a) = 2, o(b) = 7, aba^{-1} = b^k$$

where $k^2 \equiv 1 \bmod 7$, that is, $k = 1$ or 6. In the former case, G is cyclic. In the latter case, G is dihedral. Thus, the groups of order 14 are (up to isomorphism):

1) $C_{14} \approx C_7 \boxtimes C_2$, the cyclic group
2) D_{14}, the nonabelian dihedral group.

Groups of Order 15

Theorem 8.19 implies that all groups of order 15 are cyclic.

On the Existence of Complements: The Schur–Zassenhaus Theorem

In this section, we use group actions to prove the Schur–Zassenhaus Theorem, which gives a simple sufficient (but not necessary) condition under which a normal subgroup H of a group G has a complement K, thus giving a semidirect decomposition $G = H \rtimes K$.

Definition *Let G be a finite group. A* **Hall subgroup** *H of G is a subgroup with the property that its order $o(H)$ and index $(G : H)$ are relatively prime.* □

The Schur–Zassenhaus Theorem states that a *normal* Hall subgroup H has a complement. The tool that we will use in proving the Schur–Zassenhaus Theorem is the Frattini argument (Theorem 7.2). In particular, we consider the action of left translation by G on the set $\mathcal{R}$ of all right transversals of H and show that this action is regular, whence

$$G = H \rtimes \operatorname{stab}_G(R)$$

for any $R \in \mathcal{R}$. So let us take a closer look at transversals and their actions.

Transversals and Their Actions

Let G be a finite group and let H be a normal Hall subgroup of G, with right cosets

$$H\backslash G = \{H = H_1, \ldots, H_m\}$$

Let $\mathcal{R}$ be the set of all right transversals of H. If $R = \{r_1, \ldots, r_m\} \in \mathcal{R}$ where $r_i \in H_i$ for all i and if $a \in G$, then

$$Har_i = aHr_i$$

and so the cosets Har_i are distinct. Hence,

$$aR = \{ar_1, \ldots, ar_m\} \in \mathcal{R}$$

and it is clear that G acts on $\mathcal{R}$ by left translation.

Although H need not act transitively on $\mathcal{R}$, we can raise the action of G to an action on the congruence classes of an appropriate G-congruence on $\mathcal{R}$ so that H does act transitively. The G-congruence condition is

$$R \equiv S \quad \Rightarrow \quad aR \equiv aS$$

for all $a \in G$, that is,

$$\{r_1, \ldots, r_m\} \equiv \{s_1, \ldots, s_m\} \quad \Rightarrow \quad \{ar_1, \ldots, ar_m\} \equiv \{as_1, \ldots, as_m\}$$

This leads us to try the following. Assuming that R and S are indexed so that r_i and s_i are in the same right coset H_i, define a binary relation $\equiv$ by

$$R \equiv S \quad \text{if} \quad \prod_{i=1}^{m} r_i s_i^{-1} = 1$$

Letting

$$R|S = \prod_{i=1}^{m} r_i s_i^{-1}$$

the definition becomes

$$R \equiv S \quad \text{if} \quad R|S = 1$$

This relation is clearly reflexive. Also, since $r_i s_i^{-1} \in H$ for all i, if H is *abelian*, then for any $R, S, T \in \mathcal{R}$,

$$(R|S)^{-1} = S|R \quad \text{and} \quad (R|S)(S|T) = R|T$$

and so $\equiv$ is an equivalence relation on $\mathcal{R}$. Let $\mathcal{R}/\equiv$ denote the set of equivalence classes of $\mathcal{R}$ and let $[R]$ denote the equivalence class containing R.

Note that if $h \in H$, then $r_i s_i^{-1} \in H$ implies that

$$(hR)|S = \prod_{i=1}^{m} hr_i s_i^{-1} = h^m(R|S)$$

Moreover, since r_i and s_i are in the same right coset of H, so are ar_i and as_i and so for any $a \in G$,

$$aR|aS = \prod_{i=1}^{m}(ar_i)(as_i)^{-1} = a\left(\prod_{i=1}^{m} r_i s_i^{-1}\right)a^{-1} = (R|S)^a$$

Hence, $aR \equiv aS$ if and only if $R \equiv S$. Thus, $\equiv$ is a G-congruence on $\mathcal{R}$ and the induced action is

$$a[R] = [aR]$$

Now, H acts transitively if and only if for all $R, S \in \mathcal{R}$, there is an $h \in H$ for which $[hR] = [S]$, that is, for which

$$1 = (hR)|S = h^m(R|S)$$

or equivalently,

$$h^{-m} = R|S$$

But this equation always has a solution in H since $m = (G : H)$ and $o(H)$ are relatively prime.

The action of H on $\mathcal{R}/\equiv$ is also regular, since $[hR] = [R]$ if and only if

$$1 = (hR)|R = h^m(R|R) = h^m$$

which implies that $h = 1$. Hence, if H is a normal abelian subgroup of G, the Frattini argument implies that

$$G = H \rtimes \text{stab}_G([R])$$

for any $R \in \mathcal{R}$.

As to the conjugacy statement, any conjugate of a complement of H is also a complement of H. On the other hand, if

$$G = H \rtimes K$$

then Theorem 5.3 implies that $K \in \mathcal{R}$ and so

$$H \rtimes K = G = H \rtimes \text{stab}_G([K])$$

But $K \le \text{stab}_G([K])$ and so

$$K = \text{stab}_G([K])$$

Thus, the set of complements of H in G is precisely the set

$$\{\text{stab}_G([R]) \mid R \in \mathcal{R}\}$$

Also, since G acts transitively on $\mathcal{R}/\equiv$, all stabilizers are conjugate and so all complements of H are conjugate. We have proved the abelian version of the Schur–Zassenhaus Theorem.

Theorem 8.29 (Schur–Zassenhaus Theorem—abelian version) *If G is a finite group, then any normal abelian Hall subgroup H of G has a complement in G. Moreover, the complements of H in G form a conjugacy class of* $\text{sub}(G)$.□

The Schur–Zassenhaus Theorem

The condition of abelianness can be removed from the Schur–Zassenhaus Theorem. The proof of the general Schur–Zassenhaus theorem uses the Frattini argument as well, but also uses the Feit–Thompson theorem to establish the conjugacy portion of the theorem. Let us isolate the portion that uses the Feit–Thompson Theorem.

Theorem 8.30 *Let G be a nontrivial group of odd order.*
1) $G' < G$
2) *G has a normal subgroup K of prime index.*

Proof. Part 1) can be proved by induction on $o(G)$. If $o(G) = 3$, then G is abelian and $G' = \{1\} < G$. Assume the result holds for groups of odd order less than $o(G)$ and let $o(G)$ be odd. Then the Feit–Thompson theorem implies that G is either abelian or nonsimple. If G is abelian, then $G' = \{1\} < G$. If G is nonsimple, then there is a $\{1\} \lhd K \lhd G$ and so G/K has odd order less than $o(G)$. Hence, the inductive hypothesis implies that $(G/K)' < G/K$, which implies that $G' < G$. For part 2), since $G' < G$, it follows that G has a maximal normal subgroup K containing G'. Then G/K is simple and abelian and so has prime order.□

Theorem 8.31 (Schur–Zassenhaus theorem) *Any normal Hall subgroup H of finite group G has a complement in G. Moreover, the complements of H in G form a conjugacy class in G. In particular, any normal Sylow subgroup of G is complemented.*

Proof. We may assume that H is nontrivial and proper. The proof of the existence of a complement is by induction on $o(G)$. The result is true if $o(G) = 1$. Assume it is true for all orders less than $o(G)$. Let $H \trianglelefteq_{\text{Hall}} G$ denote the fact that H is a normal Hall subgroup of G.

If H has a proper supplement K, that is, if

$$G = HK$$

for some $K < G$, then $I = H \cap K \trianglelefteq_{\text{Hall}} K$ and so the inductive hypothesis implies that $K = I \rtimes J$, whence

$$G = HK = H(I \rtimes J) = H \rtimes J$$

and so H is complemented.

But since H is a Hall subgroup, if p is a prime dividing $o(H)$, then any $S \in \mathrm{Syl}_p(H)$ is also a Sylow p-subgroup of G and the Frattini argument implies that

$$G = HN$$

where $N = N_G(S)$. Thus, if $N < G$, then H is complemented. On the other hand, if $N = G$, then S is a normal Sylow p-subgroup of G and so $S \sqsubset G$, whence $Z := Z(S) \sqsubset G$. Since $H/Z \trianglelefteq_{\text{Hall}} G/Z$, the inductive hypothesis implies that

$$\frac{G}{Z} = \frac{H}{Z} \rtimes \frac{K}{Z}$$

for some $K \le G$ and so K is a supplement of H. But if $K < G$, then H is complemented and if $K = G$, then $H = Z$ is abelian and so H is complemented in this case as well.

We now turn to the statement about conjugacy. Since any conjugate of a complement of H is also a complement of H, we are left with showing that any two complements K_1 and K_2 of H are conjugate. The proof is by induction on $o(G)$. Of course, the result is true if $o(G) = 1$. Assume the result is true for groups of order less than $o(G)$.

If H contains a nontrivial proper subgroup N that is normal in G, then $K_i N/N$ is a complement of H/N in G/N for each i and so by the inductive hypothesis,

$$\frac{K_2 N}{N} = \left(\frac{K_1 N}{N}\right)^{aN} = \frac{K_1^a N}{N}$$

for some $a \in G$. Hence,

$$N \rtimes K_2 = N \rtimes K_1^a$$

and so the inductive hypothesis applied to the group $N \rtimes K_2 < G$ implies that K_2 and K_1^a are conjugates, whence so are K_2 and K_1.

If H does not contain a nontrivial proper subgroup that is normal in G, then we can easily dispatch the case $o(H)$ odd with the help of Theorem 8.30, which implies that $H' \sqsubset H \lhd G$ and so $H' \lhd G$, whence $H' = \{1\}$, that is, H is abelian. Then the abelian version of the Schur–Zassenhaus Theorem completes the proof. So we may assume that $o(H)$ is even, which implies that $o(G/H)$ is odd.

Then Theorem 8.30 implies that G has a normal subgroup A for which $H \le A \lhd G$ and $(G:A) = p$ is prime. Hence,

$$A = H \rtimes (K_1 \cap A) \quad \text{and} \quad A = H \rtimes (K_2 \cap A)$$

and the inductive hypothesis implies that

$$K_2 \cap A = (K_1 \cap A)^a$$

for some $a \in A$. But

$$(K_i : K_i \cap A) = (K_i A : A) = (G : A) = p$$

and so $K_i \cap A$ is a supplement of any $Y_i \in \mathrm{Syl}_p(K_i)$, that is,

$$K_i = (K_i \cap A)Y_i$$

Conjugating this for $i = 1$ by a gives

$$K_1^a = (K_2 \cap A)Y_1^a$$

and so it is clear that we need to relate Y_1^a to Y_2. But $K_1 \cap A \trianglelefteq K_1$ implies that

$$Y_1 \le K_1 \le N_G(K_1 \cap A)$$

and so

$$Y_1^a \le N_G(K_1 \cap A)^a = N_G(K_2 \cap A)$$

Hence, Y_2 and Y_1^a are Sylow p-subgroups of $N_G(K_2 \cap A)$, whence $Y_1^{ba} = Y_2$ for some $b \in N_G(K_2 \cap A)$. Conjugating by b gives

$$K_1^{ba} = (K_2 \cap A)Y_1^{ba} = (K_2 \cap A)Y_2 = K_2$$

as desired.□

The Schur–Zassenhaus Theorem leads to the following important corollary.

Corollary 8.32 *Let $o(G) = nm$ where $(n, m) = 1$. If G has a normal (Hall) subgroup N of order n, then any subgroup H of G that has order m' dividing m is contained in some complement of N.*
Proof. The Schur–Zassenhaus Theorem implies that there is a $K \le G$ for which $G = N \rtimes K$. Then $|NH \cap K| = m'$ and

$$N \rtimes H = N \rtimes (NH \cap K)$$

Hence, the Schur–Zassenhaus Theorem implies that there exists $a \in G$ for which

$$H = (NH \cap K)^a \le K^a$$

But K^a is also a complement of N in G.□

*Sylow Subgroups of S_n

In this section, we determine the Sylow subgroups of the symmetric group S_n in terms of wreath products. First, we need to compute the order of a Sylow p-subgroup of S_n.

Theorem 8.33 *Let p be a prime dividing $n!$ and let*

$$n = a_0 + a_1 p + \cdots + a_m p^m$$

be the base-p representation of n, that is, $0 \leq a_k < p$. The largest exponent e of p for which $p^e \mid n!$ is

$$L(m) = \sum_{k=1}^{m} a_k M(k)$$

where

$$M(k) = p^k + p^{k-1} + \cdots + 1 = \frac{p^k - 1}{p - 1}$$

and so the order of a Sylow p-subgroup of S_n is

$$p^{L(m)} = \prod_{k=1}^{m} p^{a_k M(k)}$$

In particular, the order of a Sylow p-subgroup of S_{p^k} is

$$p^{M(k)}$$

Proof. The number of factors in $n! = 1 \cdot 2 \cdots n$ that are multiples of p is $\lfloor n/p \rfloor$ where $\lfloor x \rfloor$ is the floor of x. Among these $\lfloor n/p \rfloor$ factors, there are $\lfloor n/p^2 \rfloor$ factors that are multiples of p. Thus,

$$L(m) = \lfloor n/p \rfloor + \lfloor n/p^2 \rfloor + \cdots + \lfloor n/p^m \rfloor$$

Using the base-p expansion of n, we can write this as

$$\begin{aligned} L(m) &= (a_1 + a_2 p + \cdots + a_m p^{m-1}) + (a_2 + a_3 p + \cdots + a_m p^{m-2}) + \cdots + a_m \\ &= a_1 + a_2(p+1) + \cdots + a_m(p^{m-1} + \cdots + 1) \\ &= a_1 M(1) + a_2 M(2) + \cdots + a_m M(m) \end{aligned}$$

□

Let us first determine the Sylow p-subgroups of the symmetric groups S_{p^n}. Since the order of such a Sylow p-subgroup is $p^{M(k)}$, all we need to do is find a subgroup of S_{p^k} of size

$$p^{M(k)} = p^{p^k + p^{k-1} + \cdots + 1}$$

Since

$$M(k+1) = pM(k) + 1$$

it follows that

$$p^{M(k+1)} = \left(p^{M(k)}\right)^p \cdot p$$

and this puts us in mind of the wreath product, since if D is a finite group and $|Q| = |\Omega| = p$, then

$$|D \wr_\Omega Q| = |D|^p \cdot p$$

Now, the cyclic group C_p acts faithfully on itself by left translation and so

$$W_1 := C_p \hookrightarrow S_{C_p} \approx S_p$$

and since the Sylow p-subgroups of S_p have order p, they are isomorphic to C_p. Since C_p acts faithfully on itself by left translation, the regular wreath product

$$W_2 = C_p \wr_r C_p$$

acts faithfully on $C_p \times C_p = C_p^2$, that is,

$$W_2 \hookrightarrow S_{C_p^2} \approx S_{p^2}$$

and since

$$o(W_2) = p^p \cdot p = p^{p+1} = p^{M(2)}$$

it follows that the Sylow p-subgroups of S_{p^2} are isomorphic to W_2.

Now, W_2 acts faithfully on C_p^2 and C_p acts faithfully on itself and so

$$W_3 = (C_p \wr_r C_p) \wr_r C_p$$

acts faithfully on C_p^3, that is,

$$W_3 \hookrightarrow S_{C_p^3} \approx S_{p^3}$$

and since

$$o(W_3) = (p^{p+1})^p \cdot p = p^{p^2+p+1} = p^{M(3)}$$

it follows that the Sylow p-subgroups of S_{p^3} are isomorphic to W_3.

In general, if W_{n-1} acts faithfully on C_p^{n-1}, then the n-fold regular wreath product

$$W_n = W_{n-1} \wr_r C_p = ((C_p \wr_r C_p) \wr_r C_p) \cdots \wr_r C_p$$

acts faithfully on C_p^n and so $W_n \hookrightarrow S_{p^n}$. Moreover, since

$$o(W_n) = |W_{n-1}|^p \cdot p = p^{M(n-1)p+1} = p^{M(n)}$$

it follows that the Sylow p-subgroups of S_{p^n} are isomorphic to W_n.

To determine the nature of the p-Sylow subgroups of S_n, note that if

$$\mathcal{P} = \{B_1, \ldots, B_m\}$$

is a partition of $I_n = \{1, \ldots, n\}$ and if σ_i is a permutation of B_i, then the map

$$\sigma = \sigma_1 \times \cdots \times \sigma_m$$

defined by $\sigma(k) = \sigma_j(k)$ if $k \in B_j$, is a permutation of I_n. It follows that if $|B_k| = n_k$, then

$$T = S_{n_1} \boxtimes \cdots \boxtimes S_{n_m}$$

is isomorphic to a subgroup of S_n.

We would like to find a partition $\mathcal{P}$ of $I_n = \{1, \ldots, n\}$ whose block sizes are powers of p. We can do this from the base-p representation of n:

$$n = a_0 + a_1 p + \cdots + a_m p^m$$

by letting

$$\mathcal{P} = \{B_{k,j} \mid j = 1, \ldots, a_k\}$$

be any partition of I_n consisting of a_k blocks of size p^k. Each symmetric group $S_{B_{k,j}}$ is isomorphic to a subgroup of S_n where we simply let $\sigma \in S_{B_{k,j}}$ be the identity on $I_n \setminus B_{k,j}$. If $Y_{k,j}$ is a Sylow p-subgroup of $S_{B_{k,j}}$, then the direct product

$$Y = \boxtimes_{k,j} Y_{k,j}$$

is isomorphic to a subgroup of S_n of order

$$\prod_{k=1}^{m} p^{a_k M(k)}$$

and since Y has the correct order, it is isomorphic to a Sylow p-subgroup of S_n.

Theorem 8.34 *Let p be a prime dividing $n!$ and let*

$$n = a_0 + a_1 p + \cdots + a_m p^m$$

be the base-p representation of n, that is, $0 \leq a_k < p$.

1) The Sylow p-subgroups of S_{p^n} are isomorphic to the n-fold regular wreath product

$$W_n = ((C_p \wr_r C_p) \wr_r C_p) \cdots \wr_r C_p$$

2) The Sylow p-subgroups of S_n are isomorphic to a direct product

$$Y = (Y_1)^{a_0} \boxtimes (Y_p)^{a_1} \boxtimes \cdots \boxtimes (Y_{p^m})^{a_m}$$

where Y_{p^k} is a Sylow p-subgroup of S_{p^k}. □

Exercises

1. Prove that if H is a normal p-subgroup of a finite group G, then H is contained in every Sylow p-subgroup of G.
2. Let the order of G be a product pqr of three distinct primes, with $p < q < r$. Show that if one of n_q or n_r is equal to 1, then G has a normal subgroup of order qr.
3. Find all Sylow subgroups of S_3. What are n_2 and n_3?
4. Find all Sylow subgroups of A_4. What are n_2 and n_3?
5. Show that A_5 has no subgroup of index 4.
6. Show that no group of order 56 is simple.
7. Let $o(G) = p^m(4p+1)$ for $m \geq 1$. Show that $n_p = 1$ or $n_p = 4p+1$ or else $p = 2$ and $n_p = 3$.
8. Show that there are no simple groups of order p^2q, where p and q are primes.
9. Let $n = p^k m$ with $(m,p) = 1$. Show that if $p^k \nmid (m-1)!$, then there is no simple group of order n. This holds if k is "sufficiently large" relative to $o(G)/p^k$.
10. Show that there is no simple group of order 858.
11. Show that there is no simple group of order 324.
12. Show that there is no simple group of order 3393.
13. Show that there is no simple group of order 4095.
14. Show that any group of order 561 is abelian.
15. Let $o(G) = 60$. Prove that if $n_5 > 1$, then G is simple. *Hint*: Use the fact that groups of order 15, 20 and 30 have a normal subgroup of order 5.
16. Let $k \geq 3$ be an odd integer. Show that every Sylow p-subgroup of the dihedral group D_{2k} is cyclic.
17. How many Sylow 2-subgroups does A_5 contain?
18. Let $N \trianglelefteq G$ and suppose that $S \trianglelefteq N$ is a normal Sylow p-subgroup of N. Prove that S is normal in G.
19. Let G be a finite group. Prove that G is the group generated by all of its Sylow subgroups.
20. Let G be a finite group and let $N \trianglelefteq G$. Let S be a Sylow p-subgroup of G not contained in N. Show that SN/N is a Sylow p-subgroup of G/N.
21. Let H be a p-subgroup of a finite group G. Let $a \notin H$ be a p-element. Is $\langle H, a \rangle$ necessarily a p-subgroup of G? Does it help to assume that $H \trianglelefteq G$?

22. A subgroup H of a group G is **abnormal** if
$$a \in \langle H, H^a \rangle$$
for all $a \in G$. Prove that if $K \trianglelefteq G$ and $S \in \mathrm{Syl}_p(K)$, then $N_G(S)$ is abnormal in G. In particular, the normalizer of a Sylow p-subgroup of G is abnormal.
23. Let G be a finite group and let $H, K \leq G$.
    a) Suppose that $H \trianglelefteq G$ and $K \trianglelefteq G$. Show that if S is a Sylow p-subgroup of G, then $S = (S \cap H)(S \cap K)$.
    b) Show that if we drop the condition that both subgroups be normal, then the conclusion of the previous part may fail.
    c) Assume that $H \trianglelefteq G$. Show that for each prime $p \mid o(G)$, there is some Sylow p-subgroup S for which $S \cap H$ is a Sylow p-subgroup of H, $S \cap K$ is a Sylow p-subgroup of K and $S \cap H \cap K$ is a Sylow p-subgroup of $H \cap K$.
    d) Assume that $H \trianglelefteq G$. Show that for each prime $p \mid o(G)$, there is some Sylow p-subgroup S for which $S = (S \cap H)(S \cap K)$.
24. (S. Abhyankar) Let G be a finite group and let p be a prime dividing $o(G)$. Let $p(G)$ be the subgroup of G generated by the union of the Sylow p-subgroups of G. Show that $p(G) \trianglelefteq G$. A finite group G is a **quasi *p*-group** if $p(G) = G$. Prove that the following are equivalent:
    a) G is a quasi p-group.
    b) G is generated by all of its p-elements.
    c) G has no nontrivial quotient group whose order is relatively prime to p.

Chapter 9
The Classification Problem for Groups

The Classification Problem for Groups

One of the most important outstanding problems of group theory is the problem of classifying all groups up to isomorphism. This is the **classification problem** for groups. More precisely, isomorphism of groups is an equivalence relation. Therefore, a set of canonical forms or a complete invariant constitutes a theoretical solution to the classification problem for groups. Of course, it may not be a *practical* solution unless some form of "algorithm" is available for determining the canonical form or invariant of any group.

The classification problem for groups is unsolved and seems to be exceedingly difficult. It is even beyond present day ability to classify all finite groups. All finite *abelian* groups have been classified. Indeed, we shall see in a later chapter that a finite group is abelian if and only if it is the direct sum of cyclic groups of prime power orders. Moreover, the multiset of prime powers (which need not be distinct) is a complete invariant for isomorphism. All finite simple groups *seem* to have been classified. We will elaborate on this in more detail below.

The Classification Problem for Finite Simple Groups

The classification problem for finite simple groups is generally believed by experts in the field to have been solved. The classification theorem is the following:

Up to isomorphism, a finite simple group is one of the following:
1) A cyclic group C_p of prime order.
2) An alternating group A_n for $n \geq 5$.
3) A classical linear group.
4) An exceptional or twisted group of Lie type.
5) A sporadic simple group (these include Mathieu groups, Janko groups, Conway groups, Fischer groups, Monster groups and more).

The effort to solve this problem spanned the years from roughly 1950 to 1980 and involves something on the order of 15,000 pages of mathematics produced by a variety of researchers, some of which is as yet unpublished. As a result, a second effort, led by three group theorists: Daniel Gorenstein, Richard Lyons, and Ron Solomon has been underway to collect this massive effort into a single source, which now spans five volumes and will, when finished, treat most (but not all) of the overall project.

The shear massiveness of this work has prompted some to believe that it is too soon to say categorically that the classification theorem as it is currently formulated is indeed a theorem. This viewpoint is further supported by the fact that, in the ensuing years since 1980 several gaps, some of which were quite serious, have appeared. Fortunately, all of the known gaps have since been filled.

As an example, Michael Aschbacher [2] writes in his 2004 article *The Status of the Classification of Finite Simple Groups* as follows:

> I have described the Classification as a theorem, and at this time I believe that to be true. Twenty years ago I would also have described the Classification as a theorem. On the other hand, ten years ago, while I often referred to the Classification as a theorem, I knew formally that that was not the case, since experts had by then become aware that a significant part of the proof had not been completely worked out and written down. More precisely, the so-called "quasithin groups" were not dealt with adequately in the original proof. Steve Smith and I worked for seven years, eventually classifying the quasithin groups and closing this gap in the proof of the Classification Theorem. We completed the write-up of our theorem last year; it will be published (probably in 2004) by the AMS.

Let us take a *very* broad look at the approach taken to solve the classification problem for finite simple groups. The abelian case is easily settled, so we will concentrate our remarks on finite nonabelian simple groups.

We have already mentioned the famous result of Feit–Thompson (whose proof runs about 255 pages itself) that says that every nonabelian finite simple group has even order. This implies that any nonabelian finite simple group G contains an involution b. Moreover, the centralizer $C(b)$ is a nontrivial *proper* subgroup of G, since the center of G is trivial.

Thus, every nonabelian finite simple group has a nontrivial proper *involution centralizer* $C(b)$. At least this gives us a starting point for an investigation: Perhaps one can relate the structure of the whole group G to that of $C(b)$.

Indeed, we will show that if G is a nonabelian finite simple group with involution centralizer $C(b)$, then

$$o(G) \,\Big|\, \binom{|C(b)|+1}{2}!$$

The significance of this result is that $|C(b)|$, and therefore the right-hand side above, does not depend on the structure of G nor on the nature of $C(b)$ beyond its size. Thus, for a given even number q, there are only a finite number of possible orders of groups that have an involution centralizer of size q. But for each finite order, there are only finitely many isomorphism classes of groups of that order, because there are only finitely many multiplication tables of that size. It follows that there are only finitely many isomorphism classes of nonabelian finite simple groups that have an involution centralizer of size q.

Thus, the size of an involution centralizer does at least restrict the number of possible isomorphism classes of its parent groups. This raises the question of whether more details about the structure of an involution centralizer (and related substructures) might do more than just restrict the number of isomorphism classes for its parent.

Indeed, in 1954, Richard Brauer proposed that for a finite nonabelian simple group G with involution centralizer $C(b)$, the possible isomorphism classes of G are determined by the isomorphism class of $C(b)$. Moreover, during the period of 1950–1965, Brauer and others developed methods for determining the isomorphism classes of all finite simple groups that have an involution centralizer isomorphic to a given group H. However, involution centralizers alone prove not to be sufficient to solve the classification problem for nonabelian finite simple groups.

Note that an involution centralizer $C(b)$ is also the normalizer $N_G(\langle b\rangle)$ of the 2-subgroup $\langle b\rangle$. More generally, the normalizer $N_G(H)$ of a p-subgroup H of a group G is called a ***p*-local subgroup** of G. Thus, an involution centralizer is a special type of 2-local subgroup of G. The search for nonisomorphic nonabelian finite simple groups involves looking at the entire p-local structure of a group and can be roughly described as follows:

1) If the current list of nonisomorphic nonabelian finite simple groups is not complete, let G be a minimal counterexample. Thus, any proper subgroup of G is on the list.
2) Show that the p-local structure of G resembles that of a simple group S that is already on the list.
3) Use this resemblance to show that G is isomorphic to S. If not, then perhaps G must be added to the list.

So let us proceed to the promised theorem.

Involutions

Let G be an even-order group whose set $\mathcal{I}$ of involutions has size $n \geq 1$ and let

$$\mathcal{I}' = \mathcal{I} \cup \{1\}$$

If s is an involution, then for any involution t, we have

$$t = s(st) = sx$$

where $x = st$ has the property that

$$x^s = x^{-1}$$

This property of x is very important.

Definition *Let G be a group.*
1) *An element $x \in G$ is* **real** *by $a \in G$ if*

$$x^a = x^{-1}$$

   Also, x is **real** *if it is real by some element a. Let $\mathcal{R}$ be the set of real elements of G.*
2) *An element $x \in G$ is* **strongly real** *by s if*

$$x^s = x^{-1}$$

   where s is an involution. Also, x is **strongly real** *if it is strongly real by some involution s. Let $\mathcal{S}$ denote the set of strongly real elements of G.*□

Thus, every pair of involutions is related by a strongly real element. It is not hard to see that $\mathcal{I}$, $\mathcal{S}$ and $\mathcal{R}$ are each closed under conjugation and so each set is a union of conjugacy classes. Associated with the equation

$$x^a = x^{-1}$$

are some important sets.

Definition *Let G be a group.*
1) *For $x \in G$, let*

$$C'(x) = \{a \in G \mid x^a = x^{-1}\}$$

   be the set of all elements by which x is real. The **extended centralizer** *of $x \in G$ is*

$$C^*(x) = C(x) \cup C'(x)$$

   where $C(x)$ is the centralizer of x.

2 *a)* *For $x \in \mathcal{S}$, let*

$$A(x) = \{s \in \mathcal{I} \mid x^s = x^{-1}\}$$

be the set of involutions by which x is strongly real.

b) *For $s \in \mathcal{I}$, let*

$$\mathcal{S}(s) = \{x \in G \mid x^s = x^{-1}\}$$

be the set of strongly real elements by the involution s.□

Let us take a look at these sets. Note first that $x \in G$ is real if and only if $C'(x)$ is nonempty and $x \in G$ is strongly real if and only if $A(x)$ is nonempty.

$C^*(x)$

If $x \in \mathcal{R}$ and $a, b \in C'(x)$, then

$$x^a = x^{-1} = x^b$$

and so $a^{-1}b \in C(x)$. Thus, $C'(x)$ is a coset of $C(x)$,

$$C'(x) = aC(x)$$

and so

$$|C'(x)| = |C(x)|$$

Thus, for $x \in \mathcal{R}$ and $a \in C'(x)$,

$$C^*(x) = C(x) \cup aC(x) = \begin{cases} C(x) & \text{if } x \in \mathcal{I}' \\ C(x) \sqcup aC(x) & \text{if } x \in \mathcal{R} \setminus \mathcal{I}' \end{cases}$$

where if $x \in \mathcal{R} \setminus \mathcal{I}'$, then

$$(C^*(x) : C(x)) = 2$$

$A(x)$

If $x \in \mathcal{S}$, then $A(x) \subseteq C'(x)$ and so

$$|A(x)| \le |C(x)|$$

But we can do a bit better in some cases. Since $A(1) = \mathcal{I}$, we have

$$|A(1)| = n$$

and if $x \in \mathcal{I}$, then $A(x) \subseteq C(x) \setminus \{1\}$ and so

$$|A(x)| \le |C(x)| - 1$$

Also, for any $g \in G$, it is easy to see that

$$A(x^g) = A(x)^g$$

and so

$$|A(x^g)| = |A(x)|$$

that is, $|A(x)|$ is constant on conjugacy classes of G.

***S*(*s*)**

If $s \in \mathcal{I}$, then we have seen that

$$\mathcal{I}' \subseteq s\mathcal{S}(s)$$

Conversely, if $x \in \mathcal{S}(s)$ for $s \in \mathcal{I}$, then

$$(sx)^2 = sxsx = x^{-1}x = 1$$

and so $t = sx \in \mathcal{I}'$. Hence,

$$\mathcal{I}' = s\mathcal{S}(s)$$

and so

$$|\mathcal{S}(s)| = |s\mathcal{S}(s)| = n + 1$$

The Fundamental Relation

Now we can count the size of the set

$$U = \{(s, x) \in \mathcal{I} \times \mathcal{S} \mid x^s = x^{-1}\}$$

in two ways. From the point of view of an $s \in \mathcal{I}$,

$$|U| = \sum_{s \in \mathcal{I}} |\mathcal{S}(s)| = n(n+1)$$

From the point of view of an $x \in \mathcal{S}$,

$$|U| = \sum_{x \in \mathcal{S}} |A(x)|$$

Thus,

$$n^2 + n = \sum_{x \in \mathcal{S}} |A(x)| \tag{9.1}$$

To split up the sum on the right, we choose a system of distinct representatives (SDR)

$$M = \{x_0, x_1, \ldots, x_t\}$$

for the conjugacy classes of G, where

1) $\{x_0 = 1\}$
2) $\{x_1, \ldots, x_u\}$ is an SDR for the conjugacy classes in $\mathcal{I}$

3) $\{x_{u+1}, \ldots, x_v\}$ is an SDR for the conjugacy classes in $\mathcal{S} \setminus \mathcal{I}'$

Then since $|A(x)|$ is constant on conjugacy classes and since $|x^G| = (G : C(x))$, equation (9.1) can be written

$$n^2 + n = \sum_{i=0}^{v} |A(x_i)|(G : C(x_i))$$

Splitting this sum further gives

$$\begin{aligned}
n^2 + n &= n + \sum_{i=1}^{u} |A(x_i)|(G : C(x_i)) + \sum_{i=u+1}^{v} |A(x_i)|(G : C(x_i)) \\
&\le n + \sum_{i=1}^{u} (|C(x_i)| - 1)(G : C(x_i)) + \sum_{i=u+1}^{v} |C(x_i)|(G : C(x_i)) \\
&\le n + u|G| - \sum_{i=1}^{u} (G : C(x_i)) + (v-u)|G| \\
&= n + v|G| - \sum_{i=1}^{u} |x_i^G| \\
&= v|G|
\end{aligned}$$

and so

$$n^2 + n \le v|G| \tag{9.2}$$

To get further estimates, note that

$$n = |\mathcal{I}| = \sum_{i=1}^{u} (G : C(x_i)) \tag{9.3}$$

Now, if we assume that the center $Z(G)$ of G has odd order, then it contains no involutions and so $C(x_i) < G$ for $x_i \in \mathcal{I}$. Hence, if m is the smallest index among all *proper* subgroups of G, we have

$$n = |\mathcal{I}| \ge mu \tag{9.4}$$

We can make a similar estimate for

$$|\mathcal{S}| = 1 + n + \sum_{i=u+1}^{v} (G : C(x_i))$$

and since $x_i \in \mathcal{S} \setminus \mathcal{I}'$, the terms in the final sum satisfy

$$(G : C(x_i)) = (G : C^*(x_i))(C^*(x_i) : C(x_i)) = 2(G : C^*(x_i))$$

Therefore, if we assume that $C^*(x_i)$ is also proper in G, then

$$(G : C(x_i)) \geq 2m$$

and so

$$|S| \geq 1 + n + 2m(v - u) \tag{9.5}$$

The condition that $C^*(x_i)$ is proper in G is a bit awkward, but is satisfied if we assume that G has no subgroups of index 2.

The inequalities (9.4) and (9.5) together imply that

$$v \leq \frac{|S| - 1 + n}{2m}$$

and so (9.2) implies that

$$n^2 + n \leq \frac{|S| - 1 + n}{2m}|G|$$

Some elementary algebra, using the fact that $|S| \leq |G|$, gives

$$m \leq \frac{1}{2}\left(\frac{|G|}{n}\right)\left(\frac{|G|}{n} + 1\right) = \binom{|G|/n + 1}{2}$$

We have proved a key theorem.

Theorem 9.6 (R. Brauer and K. A. Fowler [4], 1955) *Let G be a group of even order with exactly $n \geq 1$ involutions. Assume that $Z(G)$ has odd order. Then either G has a subgroup of index 2 (which must be normal) or G has a* proper *subgroup H with*

$$(G : H) \leq \binom{|G|/n + 1}{2}$$

□

Equation (9.3) implies that for any involution b, which we can assume is x_1, we have

$$\frac{n}{|G|} = \sum_{i=1}^{u} \frac{1}{|C(x_i)|} \geq \frac{1}{|C(b)|}$$

that is,

$$\frac{|G|}{n} \leq |C(b)|$$

Hence, if G fails to have a (normal) subgroup of index 2, then it has a subgroup H for which

$$(G:H)\leq\binom{|C(b)|+1}{2}$$

Since H is a proper subgroup of G, it follows that the normal interior H° is a proper *normal* subgroup with $(G:H^\circ)\mid(G:H)!$. In particular, if G is simple, then $o(G)\mid(G:H)!$ and so

$$o(G)\,\bigg|\,\binom{|C(b)|+1}{2}!$$

Thus, we arrive at our final goal.

Theorem 9.7 *Let G be a finite nonabelian simple group and let $b\in G$ be an involution. Then the centralizer $C(b)$ is a* proper *subgroup of G and*

$$o(G)\,\bigg|\,\binom{|C(b)|+1}{2}! \qquad \square$$

This is the result that we promised at the beginning of this section and is as far as we propose to take our discussion of the classification problem for nonabelian finite simple groups.

Exercises

1. Let G be a group.
   a) Under what conditions does the set $S=\mathcal{I}\cup\{1\}$ of elements of G of exponent 2 form a subgroup of G?
   b) Under what conditions does the set S form a normal subgroup of G?
   c) If S is a subgroup of G, what can you say about the strongly real elements of the group?
2. Prove that $\mathcal{S}$ is closed under conjugation.
3. Prove that if a finite group G has a nontrivial real element, then G has even order.
4. Find the real elements in the symmetric group S_n. Find the strongly real elements.
5. In the alternating group A_n, show that any permutation that is a product of disjoint cycles of length congruent to 1 modulo 4 is real.
6. Find the real elements of the dihedral group D_{2n}. Find the strongly real elements.
7. Find the real elements of the quaternion group Q. Find the strongly real elements.

Chapter 10
Finiteness Conditions

There are many forms of finiteness that a group can possess, the most obvious of which is being a finite set. However, as we have observed, chain conditions are also a form of finiteness condition. Another type of finiteness condition on a group G is the condition that G has a finite direct sum decomposition

$$G = D_1 \bowtie \cdots \bowtie D_n$$

that cannot be further refined by decomposing any of the factors D_i into a direct sum; that is, for which each D_i is *indecomposable.* In this chapter, we explore these finiteness conditions. First, however, we generalize the notion of a group.

Groups with Operators

As we have seen, a group G has several important families of subgroups, in particular, the families of all subgroups, all normal subgroups, all characteristic subgroups and all fully-invariant subgroups. Each of these families can be characterized as being the family of all subgroups that are *invariant* under a certain subset of $\text{End}(G)$. In particular, a subgroup H of G is

1) normal if and only if it is invariant under $\text{Inn}(G)$,
2) characteristic if and only if it is invariant under $\text{Aut}(G)$,
3) fully invariant if and only if it is invariant under $\text{End}(G)$.

We can also say that the subgroups of G are invariant under the empty subset of $\text{End}(G)$.

This point of view leads us to define the concept of *groups with operators*, which will include all of these special cases. Intuitively speaking, a group with operators is a group G with a distinguished family $\mathcal{E}$ of endomorphisms of G. Rather than associate a subgroup of $\text{End}(G)$ with G directly, we use a function $f\colon \Omega \to \text{End}(G)$ from a set Ω into $\text{End}(G)$. Here is the formal definition.

Definition *Let Ω be a set. An* **Ω-group** *is a pair*

$$(G, f: \Omega \to \operatorname{End}(G))$$

where G is a group and $f: \Omega \to \operatorname{End}(G)$ is a function. It is customary to denote the endomorphism $f(\omega)$ simply by ω and thus write $f(\omega)a$ as ωa (some authors write a^{ω}). An Ω-group is called a **group with operators** *and Ω is called the* **operator domain**. *Let G be an Ω-group.*

1) *An* **Ω-subgroup** *H of G is an Ω-invariant subgroup H of G. We use the notations $H \leq_{\Omega} G$ and $H \trianglelefteq_{\Omega} G$ to denote an Ω-subgroup and a normal Ω-subgroup of G, respectively. We also use the notations*

$$\Omega\text{-sub}(G) \quad \textit{and} \quad \Omega\text{-nor}(G)$$

to denote the set of all Ω-subgroups of G and the set of all normal Ω-subgroups of G, respectively.

2) *If G and H are Ω-groups, an* **Ω-homomorphism** *from G to H is a homomorphism $\sigma: G \to H$ that is compatible with the group operators, that is,*

$$\sigma(\omega a) = \omega(\sigma a)$$

for all $a \in G$. A bijective Ω-homomorphism is an **Ω-isomorphism**, *and similarly for the other types of homomorphisms. We write $\sigma: G \approx_{\Omega} H$ to denote an Ω-isomorphism from G to H. The existence of an Ω-isomorphism from G to H is denoted by $G \approx_{\Omega} H$.*

3) *If $H \in \Omega\text{-nor}(G)$, then the* **$\Omega$-quotient group** *(or* **Ω-factor group**) *is the quotient group G/H with operators $\omega \in \Omega$ defined by*

$$\omega(aH) = (\omega a)H$$

for all $a \in G$, that is, defined so that the canonical projection π_H is an Ω-homomorphism. □

Let $(G, f: \Omega \to \operatorname{End}(G))$ be an Ω-group. Then a subgroup $H \leq G$ is an Ω-subgroup of G if and only if the restricted operators

$$\Omega|_H = \{f(\omega)|_H \mid \omega \in \Omega\}$$

are operators on H, that is, if and only if H is an $\Omega|_H$-group. We will always think of an Ω-subgroup H of G as an $\Omega|_H$-group, although we will use notation such as $K \leq_{\Omega} H$ in place of $K \leq_{\Omega|_H} H$. Thus, Ω-subgroupness is transitive, that is,

$$H \leq_{\Omega} G, \quad K \leq_{\Omega} H \quad \Rightarrow \quad K \leq_{\Omega} G$$

Note that an Ω-subgroup H of G may be an operator group in other related ways. To illustrate, if $\Omega = \operatorname{Inn}(G)$, then H is an $\Omega|_H$-group as well as an operator group under its own family $\operatorname{Inn}(H)$ of inner automorphisms. But in general, $\mathrm{Inn}(H)$ is a *proper* subset of $\mathrm{Inn}(G)|_H$, since conjugation by $a \in G \setminus H$ need not be an inner automorphism of H.

If Ω is the empty set, then an Ω-group is nothing more than an ordinary group and it is customary to drop the prefix "Ω-". Also, in the most important cases, Ω is a subset of $\mathrm{End}(G)$ and $f: \Omega \to \mathrm{End}(G)$ is the inclusion map, in which case f is suppressed. This applies to the cases $\Omega = \mathrm{Inn}(G)$, $\Omega = \mathrm{Aut}(G)$ and $\Omega = \mathrm{End}(G)$.

Example 10.1 Let V be a vector space over a field F. Each $\alpha \in F$ defines an endomorphism of the abelian group V by scalar multiplication. Thus, a vector space over F is a group with operators F. An F-subgroup is a subspace and an F-homomorphism is a linear transformation.□

The Lattice of Ω-Subgroups of an Ω-Group

Let G be an Ω-group. Then the intersection and the join of any family $\mathcal{F} = \{H_i \mid i \in I\}$ of Ω-subgroups of G is also an Ω-subgroup of G. Hence, the meet and join in $\Omega\text{-sub}(G)$ is the same as the meet and join in $\mathrm{sub}(G)$. In other words, $\Omega\text{-sub}(G)$ is a complete sublattice of $\mathrm{sub}(G)$.

We leave it to the reader to show that the Ω-subgroup $\langle X \rangle_\Omega$ generated by a nonempty subset X of G is

$$\langle X \rangle_\Omega = \langle \omega x \mid x \in X, \omega \in \Omega \rangle$$

An Ω-subgroup H of an Ω-group G is **finitely Ω-generated** if $H = \langle X \rangle_\Omega$ for some finite set X. This generalizes the normal closure of a subset X of G.

The Ω-Isomorphism and Ω-Correspondence Theorems

The concept of universality given in Theorem 4.5 and the consequent isomorphism theorems have direct generalizations to groups with operators. Let G be a Ω-group and let $K \le_\Omega G$. Let $\mathcal{F}_\Omega(G; K)$ be the family of all pairs $(H, \sigma: G \to H)$, where $\sigma: G \to H$ is an Ω-homomorphism and $K \subseteq \ker(\sigma)$. Then the pair

$$(G/K, \pi_K: G \to G/K)$$

is universal in $\mathcal{F}_\Omega(G; K)$, in the sense that for any pair $(H, \sigma: G \to H)$ in $\mathcal{F}_\Omega(G; K)$, there is a unique mediating Ω-homomorphism $\tau: S \to H$ for which

$$\tau \circ \pi_K = \sigma$$

To see this, note that Theorem 4.5 guarantees the existence of a mediating homomorphism τ. But if π_K and σ are Ω-homomorphisms, then for any $\omega \in \Omega$ and $a \in G$,

$$\tau(\omega(aK)) = \tau((\omega a)K) = \sigma(\omega a) = \omega(\sigma a) = \omega\tau(aK)$$

and so τ is also an Ω-homomorphism. Also, Ω-universality enjoys the same uniqueness up to isomorphism as ordinary universality ($\Omega = \emptyset$). The Ω-isomorphism theorems now follow in the same manner as before.

Theorem 10.2 (The Ω-isomorphism theorems) *Let G be an Ω-group.*

1) **(First Ω-isomorphism theorem)** *Every Ω-homomorphism $\sigma: G \to H$ induces an Ω-embedding $\overline{\sigma}: G/\ker(\sigma) \hookrightarrow H$ defined by*

$$\overline{\sigma}(g\ker(\sigma)) = \sigma(g)$$

and so

$$\frac{G}{\ker(\sigma)} \approx_\Omega \operatorname{im}(\sigma)$$

2) **(Second Ω-isomorphism theorem)** *If $H, K \in \Omega\text{-sub}(G)$ with $K \trianglelefteq G$, then $H \cap K \in \Omega\text{-nor}(H)$ and*

$$\frac{HK}{K} \approx_\Omega \frac{H}{H \cap K}$$

3) **(Third Ω-isomorphism theorem)** *If $H \le K \le G$ with $H, K \in \Omega\text{-nor}(G)$, then $K/H \in \Omega\text{-nor}(G/H)$ and*

$$\frac{G}{H} \Big/ \frac{K}{H} \approx_\Omega \frac{G}{K}$$

4) **(The Ω-correspondence theorem)** *Let $N \in \Omega\text{-nor}(G)$. If $\Omega\text{-sub}(N; G)$ denotes the lattice of all Ω-subgroups of G that contain N, then the map $\overline{\pi}: \text{sub}(N; G) \to \text{sub}(G/N)$ defined by*

$$\overline{\pi}(H) = H/N$$

preserves Ω-invariance in both directions and so maps $\Omega\text{-sub}(N; G)$ bijectively onto $\Omega\text{-sub}(G/N)$.□

Ω-Series and Ω-Subnormality

We can now generalize the notion of series and subnormality to groups with operators. This will provide considerable time savings in our later work.

Definition *Let G be an Ω-group.*

1) *An* **Ω-series** *in G is a series*

$$\mathcal{G}: G_0 \trianglelefteq G_1 \trianglelefteq \cdots \trianglelefteq G_n$$

in which every term is an Ω-subgroup of G.

2) *A* **refinement** *of an Ω-series $\mathcal{G}$ is an Ω-series $\mathcal{H}$ obtained from $\mathcal{G}$ by including zero or more additional Ω-subgroups between the endpoints. A* **proper refinement** *is a refinement that includes at least one new subgroup.*□

We review the usual suspects in the context of Ω-series:
a) If $\Omega = \emptyset$, an Ω-series is just a series.
b) If $\Omega = \operatorname{Inn}(G)$, an Ω-series is a normal series.
c) If $\Omega = \operatorname{Aut}(G)$, an Ω-series is a characteristic series.
d) If $\Omega = \operatorname{End}(G)$, an Ω-series is a fully invariant series.

Of course, if $\{G_i\}$ is a nonproper Ω-series for G, we can *dedup* the series by removing any duplicate subgroups to obtain a proper series.

Definition *Let G be an Ω-group. Two equal-length Ω-series*

$$\mathcal{G}: G_0 \trianglelefteq G_1 \trianglelefteq \cdots \trianglelefteq G_{n-1} \trianglelefteq G_n$$

and

$$\mathcal{H}: H_0 \trianglelefteq H_1 \trianglelefteq \cdots \trianglelefteq H_{n-1} \trianglelefteq H_n$$

in G with common endpoints $G_0 = H_0$ and $G_n = H_n$ are **Ω-isomorphic** (*also called* **Ω-equivalent**) *if there is a bijection f of the index set $\{0, \ldots, n-1\}$ for which*

$$G_{i+1}/G_i \approx_\Omega H_{f(i)+1}/H_{f(i)}$$ □

As usual, when $\Omega = \emptyset$, we use the term *isomorphic.* Thus, for example, the series

$$\{1\} \triangleleft C_p \triangleleft (C_p \bowtie C_q)$$

and

$$\{1\} \triangleleft C_q \triangleleft (C_p \bowtie C_q)$$

are isomorphic.

Ω-Subnormality

Ω-subnormality of a subgroup $H \leq_\Omega G$ requires not just that H be subnormal and Ω-invariant, but that the *entire series* that witnesses subnormality be an Ω-series.

Definition *Let G be an Ω-group. An Ω-subgroup H of G is* **Ω-subnormal** *in G, denoted by $H \trianglelefteq\trianglelefteq_\Omega G$, if there is an Ω-series from H to G. We use the notations*

$$\operatorname{subn}_\Omega(G) \quad \textit{and} \quad \operatorname{subn}_\Omega(N; G)$$

to denote the family of all Ω-subnormal subgroups of G and the family of all Ω-subnormal subgroups of G that contain N, respectively. □

We can now generalize Theorem 4.24.

Theorem 10.3 *Let G be an Ω-group and let $H, K \in \Omega\text{-sub}(G)$.*

1) **(Transitivity)**

$$H \trianglelefteq\trianglelefteq_\Omega K \quad and \quad K \trianglelefteq\trianglelefteq_\Omega G \quad \Rightarrow \quad H \trianglelefteq\trianglelefteq_\Omega G$$

2) **(Intersection)** *If $L \in \Omega\text{-sub}(G)$, then*

$$H \trianglelefteq\trianglelefteq_\Omega K \quad \Rightarrow \quad H \cap L \trianglelefteq\trianglelefteq_\Omega K \cap L$$

In particular,

$$H \le K \le G, \quad H \trianglelefteq\trianglelefteq_\Omega G \quad \Rightarrow \quad H \trianglelefteq\trianglelefteq_\Omega K$$

and

$$H \trianglelefteq\trianglelefteq_\Omega G, \quad K \trianglelefteq\trianglelefteq_\Omega G \quad \Rightarrow \quad H \cap K \trianglelefteq\trianglelefteq_\Omega G$$

3) **(Normal lifting)** *If $N \in \Omega\text{-nor}(G)$, then*

$$H \trianglelefteq\trianglelefteq_\Omega K \quad \Rightarrow \quad HN \trianglelefteq\trianglelefteq_\Omega KN$$

4) **(Quotient/unquotient)** *If $N \in \Omega\text{-nor}(K)$ and $N \le H \le K$, then*

$$H \trianglelefteq\trianglelefteq_\Omega K \quad \Leftrightarrow \quad H/N \trianglelefteq\trianglelefteq_\Omega K/N$$ □

Composition Series

If $G_0, G_n \le_\Omega G$, then refinement is a partial order on the set of all *proper* Ω-series from G_0 to G_n (assuming that this set is nonempty). Maximal proper Ω-series are particularly important.

Definition *Let G be an Ω-group and let $G_0, G_n \le_\Omega G$. An* **Ω-composition series** *from G_0 to G_n is a proper Ω-series*

$$G_0 \triangleleft G_1 \triangleleft \cdots \triangleleft G_n$$

that is maximal in the family of all proper Ω-series from G_0 to G_n, under refinement. If there is an Ω-composition series from G_0 to G_n, we will write

$$\exists\text{CompSer}_\Omega(G_0; G_n) \quad or \quad \exists\text{CompSer}_\Omega(G_n) \text{ when } G_0 = \{1\}$$

The factor groups of an Ω-composition series are called **Ω-composition factors**.

1) A maximal series ($\Omega = \emptyset$) is simply called a **composition series** *and the factor groups are called* **composition factors**.

2) A maximal normal series ($\Omega = \text{Inn}(G)$) in G is called a **chief series** *or* **principal series** *and the factor groups are called* **chief factors** *or* **principal factors**. □

When the endpoints of a series are clear from the description of the series, we will drop the "from-to" terminology. To characterize maximal series, we use the following concept.

Definition *A nontrivial Ω-group G is* **Ω-simple** *if G has no nontrivial proper normal Ω-subgroups.*$\square$

The Ω-correspondence theorem implies the following.

Theorem 10.4 *A proper Ω-series is an Ω-composition series if and only if its factor groups are Ω-simple.*$\square$

Thus, a series

$$\mathcal{G}: G_0 \lhd G_1 \lhd \cdots \lhd G_n$$

in G is a composition series if and only if its factor groups G_{k+1}/G_k are simple and $\mathcal{G}$ is a chief series if and only if each factor group G_{k+1}/G_k is a *minimal normal subgroup* of G/G_k.

It is clear from Theorem 10.4 that any Ω-series that is Ω-isomorphic to an Ω-composition series is also an Ω-composition series. Also, if we remove an endpoint from an Ω-composition series, the result is also an Ω-composition series (with different endpoints, of course). Finally, if

$$G_0 \lhd \cdots \lhd G_k \quad \text{and} \quad G_k \lhd \cdots \lhd G_n$$

are Ω-composition series in G, then so is their concatenation

$$G_0 \lhd \cdots \lhd G_n$$

The Extension Problem

An **extension** of a pair (N, Q) of groups is a group G that has a normal subgroup N' isomorphic to N and for which G/N' is isomorphic to Q. The **extension problem** for the pair (N, Q) is the problem of determining (up to isomorphism) all possible extensions of (N, Q). Note that any external semidirect product $G = N \rtimes_\theta Q$ is an extension of (N, Q). However, semidirect products alone do not solve the extension problem. For example, $\mathbb{Z}$ is an extension of $(2\mathbb{Z}, \mathbb{Z}_2)$ but $\mathbb{Z}$ is not a semidirect of any of its nontrivial proper subgroups.

The importance of the extension problem can be clearly seen in the light of composition series. Suppose that we can solve the extension problem and that we can determine (up to isomorphism) all simple groups. The simple groups are precisely the groups that have a composition series of length 1:

$$\{1\} = G_0 \lhd G_1$$

Next, for each G_1, we solve the extension problem for all pairs of the form (G_1, H_1), where H_1 ranges over the simple groups. The solutions G_2 are precisely the groups that have composition series of length 2:

$$G_0 \lhd G_1 \lhd G_2$$

where $G_2/G_1 \approx H_1$. Continuing to extend by all possible simple groups produces all possible groups that have composition series, and this includes all finite groups. Thus, in particular, if we can solve the extension problem and if we can determine all finite simple groups, we can determine all finite groups. Unfortunately, a practical solution to the extension problem does not exist at this time.

The Zassenhaus Lemma and the Schreier Refinement Theorem

Let G be an Ω-group. Our next goal is to show that any two Ω-series in G with the same endpoints have refinements that are Ω-isomorphic. This result is called the *Schreier refinement theorem* and has two extremely important consequences, as we will see.

First, let us recall that the projection

$$(A \trianglelefteq B) \to (H \trianglelefteq K)$$

of $A \trianglelefteq B$ into $H \trianglelefteq K$ is the extension

$$H(A \cap K) \trianglelefteq H(B \cap K)$$

Moreover, when A, B, H and K are Ω-subgroups, then so are $H(A \cap K)$ and $H(B \cap K)$ and the Zassenhaus lemma (Theorem 4.12) generalizes directly to Ω-subgroups.

Theorem 10.5 (Zassenhaus lemma [37], 1934) *Let G be an Ω-group and let*

$$A \trianglelefteq B \quad \textit{and} \quad H \trianglelefteq K$$

where $A, B, H, K \in \Omega\text{-sub}(G)$. Then the reverse projections

$$(A \trianglelefteq B) \to (H \trianglelefteq K) \quad \textit{and} \quad (H \trianglelefteq K) \to (A \trianglelefteq B)$$

have Ω-isomorphic factor groups, that is,

$$\frac{H(B \cap K)}{H(A \cap K)} \approx_\Omega \frac{A(K \cap B)}{A(H \cap B)}$$ □

Now we can prove the Schreier refinement theorem. Let $H \le K$ be Ω-subgroups of an Ω-group G and consider a pair of Ω-series from H to K:

$$\mathcal{G}: H = G_0 \trianglelefteq G_1 \trianglelefteq \cdots \trianglelefteq G_n = K$$

and

$$\mathcal{H}: H = H_0 \trianglelefteq H_1 \trianglelefteq \cdots \trianglelefteq H_m = K$$

Projecting each of the m steps of $\mathcal{H}$ into each of the n steps of $\mathcal{G}$ creates a new Ω-series with mn steps, some of which may be trivial. In view of the preceeding remarks, the new Ω-series is a refinement of $\mathcal{G}$. Similarly, projecting each of the n steps of G into each of the m steps of $\mathcal{H}$ creates a new Ω-series with mn steps. Moreover, since the two sets of projections consist of inverse pairs, the Zassenhaus lemma implies that the resulting series are Ω-isomorphic.

Theorem 10.6 (Schreier refinement theorem, Schreier [29], 1928) *Let G be an Ω-group. Then any two Ω-series in G with the same endpoints have Ω-isomorphic refinements.* □

Consequences of the Schreier Refinement Theorem

The Schreier refinement theorem has two very important consequences. First, suppose that there is an Ω-composition series $\mathcal{C}$ in G from H to K. Then any proper Ω-series

$$\mathcal{H}: H_0 \lhd H_1 \lhd \cdots \lhd H_n$$

with Ω-subnormal endpoints between H and K can be refined to an Ω-composition series. To see this, note that since H_0 and H_n are Ω-subnormal, the series $\mathcal{H}$ can be expanded to an Ω-series $\mathcal{K}$ from H to K and the Schreier refinement theorem implies that $\mathcal{C}$ and $\mathcal{K}$ have Ω-isomorphic refinements to proper series $\mathcal{C}'$ and $\mathcal{K}'$, respectively. But $\mathcal{C}' = \mathcal{C}$ is an Ω-composition series and therefore so is $\mathcal{K}'$, which contains a refinement of $\mathcal{H}$.

The second consequence of the Schreier refinement theorem is that any two Ω-composition series with the same endpoints are Ω-isomorphic.

Theorem 10.7 *Let G be an Ω-group and let $H \le K$ be Ω-subgroups of G.*

1) *Suppose that there is an Ω-composition series from H to K of length n. If $H_0 \le H_n$ are Ω-subnormal subgroups of G between H and K, then any Ω-series from H_0 to H_n can be refined to an Ω-composition series and so has length at most n.*
2) **(The Jordan–Hölder Theorem)** *Every two Ω-composition series from H to K are Ω-isomorphic. In particular, they have the same length.* □

The Jordan–Hölder Theorem allows us to make the following definition.

Definition *Let G be an Ω-group and let H and K be Ω-subgroups of G. If $\exists\mathrm{CompSer}_\Omega(H;K)$, then the* **$\Omega$-composition distance** *$d(H,K)$ from H to K is the length of any Ω-composition series from H to K. The distance $d(K) = d(\{1\},K)$ is called the* **Ω-composition length** *of K. For chief series, the terms* **chief distance** *and* **chief length** *are also employed.* □

Of course, the composition distance is *positive definite* (when it is defined), that is, if $\exists\mathrm{CompSer}_\Omega(H;K)$, then $d(H,K) \ge 0$ and

$$d(H,K) = 0 \quad \Leftrightarrow \quad H = K$$

The Existence of Ω-Composition Series

Any finite group has an Ω-composition series. For *abelian* groups, composition series and chief series coincide and an abelian group has a composition series if and only if it is finite, since each factor group must have prime order. Thus, not all groups have composition or chief series.

The existence of an Ω-composition series between Ω-subnormal subgroups H and K of an Ω-group G can be characterized in terms of chain conditions on Ω-subnormal subgroups. If $\exists\text{CompSer}_\Omega(H;K)$, then any Ω-series from H to K has length at most $d(H,K)$. Hence, Theorem 1.5 gives the following.

Theorem 10.8 *Let G be an Ω-group and let $H,K \in \text{subn}_\Omega(G)$. Then*

$$\exists\text{CompSer}_\Omega(H;K) \quad \Leftrightarrow \quad \text{subn}_\Omega(H;K) \textit{ has BCC}$$ □

Proof of the following is left to the reader.

Theorem 10.9 *Let G be an Ω-group.*

1) **(Subgroup)** *If $K \in \text{subn}_\Omega(G)$ and $H \le K$, then*

$$\exists\text{CompSer}_\Omega(H;G) \quad \Rightarrow \quad \exists\text{CompSer}_\Omega(H;K) \textit{ and } \exists\text{CompSer}_\Omega(K;G)$$

and

$$d(H,G) = d(H,K) + d(K,G)$$

2) **(Quotients)** *If $N \in \Omega\text{-nor}(G)$, then*

$$\exists\text{CompSer}_\Omega(N;G) \quad \Leftrightarrow \quad \exists\text{CompSer}_\Omega(G/N)$$

and

$$d(N,G) = d(G/N)$$

3) **(Extensions)** *If $N \in \Omega\text{-nor}(G)$, then*

$$\exists\text{CompSer}_\Omega(N), \quad \exists\text{CompSer}_\Omega(G/N) \quad \Rightarrow \quad \exists\text{CompSer}_\Omega(G)$$

4) **(Direct products)** *If G and H are Ω-groups, then*

$$\exists\text{CompSer}_\Omega(G) \textit{ and } \exists\text{CompSer}_\Omega(H) \quad \Leftrightarrow \quad \exists\text{CompSer}_\Omega(G \boxtimes H)$$

and

$$d(G \boxplus H) = d(G) + d(H)$$ □

The Remak Decomposition

Let us recall Theorem 5.12.

Theorem 10.10 (Remak) *If a group G has either (and therefore both) chain condition on direct summands, then G has a Remak decomposition*

$$G = R_1 \bowtie \cdots \bowtie R_n$$

that is, each R_k is indecomposible.□

The question of uniqueness of a Remak decomposition is rather more complicated than the question of existence. Recall that if

$$G = H_1 \bowtie \cdots \bowtie H_n$$

then the kth projection map $\rho_{H_k}: G \to H_k$ is defined by

$$\rho_{H_k}(a) = [a]_{H_k}$$

where $[a]_{H_k}$ is the kth coordinate of a. Moreover, ρ_{H_i} is idempotent and normal as an endomorphism of G. Note also that the sum $\rho_{H_{i_1}} + \cdots + \rho_{H_{i_k}}$ is projection onto $H_{i_1} \bowtie \cdots \bowtie H_{i_{kj}}$ and so is an endomorphism of G.

The Krull–Remak–Schmidt Theorem

Suppose now that

$$G = H_1 \bowtie \cdots \bowtie H_n \tag{10.11}$$

with projection maps $\pi_{H_1}, \ldots, \pi_{H_n}$ and

$$G = K_1 \bowtie \cdots \bowtie K_m \tag{10.12}$$

with projection maps $\kappa_{K_1}, \ldots, \kappa_{K_m}$, where the factors H_k and K_k are indecomposable and $n \leq m$.

In searching for possible isomorphisms between the H-factors and the K-factors, we recall Theorem 5.25 as it applies to the restricted projection maps

$$\pi_{H_i}|_{K_j}: K_j \to H_i \quad \text{and} \quad \kappa_{K_j}|_{H_i}: H_i \to K_j$$

Namely, if $(\pi_{H_i}|_{K_j})(\kappa_{K_j}|_{H_i}) \in \operatorname{Aut}(H_i)$ and $\operatorname{im}(\kappa_{K_j}|_{H_i}) \trianglelefteq K_j$, then $\pi_{H_i}|_{K_j}$ and $\kappa_{K_j}|_{H_i}$ are isomorphisms.

Note however that $\operatorname{im}(\kappa_{K_j}|_{H_i}) \trianglelefteq K_j$, since if $h \in H_i$ and $k \in K_j$, then

$$k[\kappa_{K_j}|_{H_i}(h)]k^{-1} = \kappa_{K_j}(k)\kappa_{K_j}(h)\kappa_{K_j}(k)^{-1} = \kappa_{K_j}(khk^{-1}) \in \operatorname{im}(\kappa_{K_j}|_{H_i})$$

Thus, we have the following.

Theorem 10.13 *Let*

$$G = H_1 \bowtie \cdots \bowtie H_n = K_1 \bowtie \cdots \bowtie K_m$$

where the factors H_k and K_k are indecomposable. If the composition

$$\alpha_{i,j} = (\pi_{H_i}|_{K_j}) \circ (\kappa_{K_j}|_{H_i}): H_i \to H_i$$

of the restricted projection maps is an automorphism of H_i, then the maps

$$\pi_{H_i}|_{K_j}: K_j \approx H_i \quad \textit{and} \quad \kappa_{K_j}|_{H_i}: H_i \approx K_j$$

are isomorphisms.□

In attempting to show that a composition is an automorphism, we are reminded of Fitting's lemma. We have assumed that H_i is indecomposable. Also, since the restriction and composition of normal maps is normal, $\alpha_{i,j}$ is normal and so has normal higher images. Thus, if we assume that G has BCC on normal subgroups, then Fitting's lemma implies that $\alpha_{i,j}$ is either nilpotent or an automorphism for all i, j.

To see that for each i, not all of the maps $\alpha_{i,j}$ can be nilpotent, note that for $j \neq k$,

$$\alpha_{i,j} + \alpha_{i,k} = (\pi_{H_i}\kappa_{K_j} + \pi_{H_i}\kappa_{K_k})|_{H_i} = \pi_{H_i}(\kappa_{K_j} + \kappa_{K_k})|_{H_i}$$

which is an endomorphism of H_i and so the images $\operatorname{im}(\alpha_{i,j})$ and $\operatorname{im}(\alpha_{i,k})$ commute elementwise. Also,

$$\pi_{H_i}\kappa_{K_1} + \cdots + \pi_{H_i}\kappa_{K_m} = \pi_{H_i}(\kappa_{K_1} + \cdots + \kappa_{K_m}) = \pi_{H_i}$$

and so

$$\alpha_{i,1} + \cdots + \alpha_{i,m} = \pi_{H_i}|_{H_i} = \iota_{H_i}$$

which is not nilpotent. Hence, Theorem 4.3 implies that $\alpha_{i,j}$ is not nilpotent for some j. It follows from Fitting's lemma that for each i, there is a j for which $\alpha_{i,j} \in \operatorname{Aut}(H_i)$ and so

$$\pi_{H_i}|_{K_j}: K_j \approx H_i \quad \text{and} \quad \kappa_{K_j}|_{H_i}: H_i \approx K_j$$

Now, we can make a significant improvement to this by noticing that if $K_j = H_k$ for some $k \neq i$, then

$$\alpha_{i,j} = (\pi_{H_i}|_{K_j})(\kappa_{K_j}|_{H_i}) = (\pi_{H_i}|_{H_k})(\kappa_{H_k}|_{H_i}) = 0$$

and so if we delete from the sum $\sum_j \alpha_{i,j}$ all terms indexed by a j for which $K_j \in \{H_1, \ldots, H_n\} \setminus \{H_i\}$, the sum remains unchanged and so is not nilpotent. Hence,

$$\pi_{H_i}|_{K_j}: K_j \approx H_i \quad \text{and} \quad \kappa_{K_j}|_{H_i}: H_i \approx K_j$$

for some j for which $K_j \notin \{H_1, \ldots, H_n\} \setminus \{H_i\}$.

Now suppose that after possible reindexing of the K's, there is a $k \geq 1$ for which

$$G = K_1 \bowtie \cdots \bowtie K_{k-1} \bowtie H_k \bowtie \cdots \bowtie H_n \tag{10.14}$$

with projections $\mu_1, \ldots, \mu_n$, where

$$\lambda_i := \kappa_{K_i}|_{H_i}: H_i \approx K_i$$

for all $1 \le i \le k-1$. We may also assume that $H_k \ne K_j$ for all $j \ge k$ or else we can replace H_k by K_j (reindexed to K_k). This certainly holds for $k = 1$. Then there is a $j \ge k$ and we may assume that $j = k$, for which

$$\pi_{H_k}|_{K_k}: K_k \approx H_k \quad \text{and} \quad \lambda_k := \kappa_{K_k}|_{H_k}: H_k \approx K_k$$

Moreover, since π_{H_k} maps K_k isomorphically onto H_k and maps the complement

$$H_{(k)} = H_1 \bowtie \cdots \bowtie H_{k-1} \bowtie H_{k+1} \bowtie \cdots \bowtie H_n$$

to $\{1\}$, it follows that

$$H_{(k)} \cap K_k = \{1\}$$

Thus, if we replace H_k by K_k in (10.14), the result is a direct sum

$$G_1 = K_1 \bowtie \cdots \bowtie K_k \bowtie H_{k+1} \bowtie \cdots \bowtie H_n$$

and our goal is to show that $G_1 = G$. To this end, if

$$\mu_{(k)} = \mu_1 + \cdots + \mu_{k-1} + \mu_{k+1} + \cdots + \mu_n$$

then $\mu_k + \mu_{(k)} = \iota_G$ and the map

$$\theta = \kappa_{K_k}\mu_k + \mu_{(k)}$$

is a normal endomorphism of G with $\operatorname{im}(\theta) \subseteq K_k H_{(k)}$. To show that θ is surjective and so $G = K_k H_{(k)}$, it is sufficient to show that θ is injective.

Now, any $a \in G$ has the form $a = xh$ for $x \in H_{(k)}$ and $h \in H_k$. But for any $x \in H_{(k)}$,

$$\theta(x) = \kappa_{K_k}\mu_k(x)\mu_{(k)}(x) = x$$

and any $h \in H_k$,

$$\theta(h) = \kappa_{K_k}\mu_k(h)\mu_{(k)}(h) = \kappa_{K_k}(h)$$

and so

$$\theta(xh) = x\kappa_{K_k}(h)$$

and since $H_{(k)} \cap K_k = \{1\}$, it follows that $\theta(xh)$ implies $x = 1$ and $\kappa_{K_k}(h) = 1$. But $\kappa_{K_k}|_{H_k}: H_k \approx K_k$ and so $h = 1$. Thus, θ is injective.

It follows that

$$G = K_1 \bowtie \cdots \bowtie K_k \bowtie H_{k+1} \bowtie \cdots \bowtie H_n$$

for all $1 \le k \le n$ and $\lambda_k = \kappa_{K_k}|_{H_k}: H_k \approx K_k$ and so this holds for all $k = 1, \ldots, n$. In particular, $n = m$.

Note also that since $\kappa_{K_i}(H_i) = K_i$, the map

$$\lambda = \lambda_1 \pi_{H_1} + \cdots + \lambda_n \pi_{H_n} = \kappa_{K_1} \pi_{H_1} + \cdots + \kappa_{K_n} \pi_{H_n}$$

is a surjective normal endomorphism of G and so $\lambda \in \operatorname{Aut}(G)$. Moreover,

$$\lambda(H_k) = K_k$$

We can now summarize.

Theorem 10.15 (The Krull–Remak–Schmidt Theorem) *Let G be a group that has BCC on normal subgroups. Suppose that*

$$G = H_1 \bowtie \cdots \bowtie H_n = K_1 \bowtie \cdots \bowtie K_m$$

where all factors H_i and K_j are indecomposable. Then $n = m$ and there is a reindexing of the K_i's and a normal automorphism λ of G for which

$$\lambda: H_i \approx K_i$$

for all $i = 1, \ldots, n$ and for each $1 \le k \le n$,

$$G = K_1 \bowtie \cdots \bowtie K_k \bowtie H_{k+1} \bowtie \cdots \bowtie H_n \qquad \square$$

True Uniqueness

The Krull–Remak–Schmidt Theorem gives uniqueness of the terms of a Remak decomposition up to isomorphism. Let us now consider the question of when a group G has an **essentially unique** Remak decomposition, that is, a Remak decomposition that is unique up to the order of the factors. First suppose that

$$G = H_1 \bowtie \cdots \bowtie H_n = K_1 \bowtie \cdots \bowtie K_n$$

are Remak decompositions of G (where $n > 1$). Then the Krull–Remak–Schmidt Theorem implies that there is a normal automorphism $\lambda: G \to G$ for which $\lambda H_k = K_k$ (after reindexing). Hence, if H_k is invariant under every normal automorphism of G, then $K_k = \lambda H_k \le H_k$ and so $K_k = H_k$ for all k and G has an essentially unique Remak decomposition.

By way of converse, suppose that G has an essentially unique Remak decomposition

$$G = H_1 \bowtie \cdots \bowtie H_n$$

with projections $\{\pi_1, \ldots, \pi_n\}$, but that there is a normal *endomorphism* λ of G

for which $\lambda H_1 \not\leq H_1$. Then there is a $k > 1$ for which $\pi_k \lambda \neq 0$ on H_1, that is, there is an $x \in H_1$ for which $1 \neq \pi_k \lambda(x) \in H_k$.

The set

$$K_1 = \{h \cdot \pi_k \lambda(h) \mid h \in H_1\}$$

is easily seen to be a normal subgroup of G and $K_1 \neq H_1$, since

$$x \cdot \pi_k \lambda(x) \in K_1 \setminus H_1$$

But it is easy to see that

$$G = K_1 \bowtie H_2 \bowtie \cdots \bowtie H_n$$

Moreover, if K_1 is decomposable, then there would be a Remak decomposition of G consisting of more than n terms, which is false. Hence, this is a Remak decomposition of G that is distinct from the previous decomposition. We have proved the following.

Theorem 10.16 *Let G have BCC on normal subgroups and let*

$$G = H_1 \bowtie \cdots \bowtie H_n$$

be a Remak decomposition of G. The following are equivalent:
1) This Remak decomposition of G is essentially unique .
2) H_k is invariant under all normal endomorphisms of G.
3) H_k is invariant under all normal automorphisms of G.□

If $\alpha\colon H_i \to Z(G)$ is a nonzero homomorphism, then we can build a normal endomorphism $\lambda\colon G \to G$ by specifying that

$$\lambda|_{H_k} = \begin{cases} \alpha & \text{if } k = i \\ 0 & \text{if } k \neq i \end{cases}$$

The map λ is normal since for any $a \in G$, $h_i \in H_i$, $h_k \in H_k$ where $k \neq i$,

$$\lambda(h_i^a) = \lambda h_i = (\lambda h_i)^a \quad \text{and} \quad \lambda(h_k^a) = 1 = (\lambda h_k)^a$$

Thus, if H_i is not λ-invariant, then Theorem 10.16 implies that the Remak decomposition of G is not unique.

Conversely, suppose that the Remak decomposition of G is not unique and so there is a normal endomorphism λ of G for which $\pi_j \lambda|_{H_i} \neq 0$ for some $j \neq i$. Then for $h_i \in H_i$ and $h_j \in H_j$,

$$[\lambda(h_i)]^{h_j} = \lambda(h_i^{h_j}) = \lambda(h_i)$$

which shows that $\lambda|_{H_i}\colon H_i \to Z(G)$. Hence, $\pi_j \lambda|_{H_i}\colon H_i \to Z(H_j)$ is a nonzero homomorphism.

Theorem 10.17 *Let G have BCC on normal subgroups and let*

$$G = H_1 \bowtie \cdots \bowtie H_n$$

be a Remak decomposition of G.

1) *The following are equivalent:*
   a) *This Remak decomposition of G is essentially unique.*
   b) *Every homomorphism $\alpha: H_i \to Z(G)$ satisfies $\alpha: H_i \to Z(H_i)$.*
   c) *There are no nonzero homomorphisms $\lambda: H_i \to Z(H_j)$ for $i \neq j$.*
2) *If G is either perfect or centerless, then G has an essentially unique Remak decomposition.*

Proof. For part 2), if $G = G'$, then $H_i = H_i'$ for all i and so if $\lambda: H_i \to Z(H_j)$ for $j \neq i$, then for $a, b \in H_i$,

$$\lambda([a,b]) = [\lambda a, \lambda b] = 1$$

and so $\lambda|_{H_i} = 0$. If G is centerless, the result is immediate.□

Exercises

1. Give an example of an infinite group with a composition series.
2. Find isomorphic refinements of the two series

   $$\{0\} \lhd p\mathbb{Z} \lhd \mathbb{Z}$$

   and

   $$\{0\} \lhd q\mathbb{Z} \lhd \mathbb{Z}$$

   where p and q are distinct primes.
3. Prove that if G and H are groups and if

   $$\mathcal{G}: \{1\} = G_0 \lhd G_1 \lhd \cdots \lhd G_n = G$$

   and

   $$\mathcal{H}: \{1\} = H_0 \lhd H_1 \lhd \cdots \lhd H_m = H$$

   are composition series for G and H, respectively, then the series

   $$(\{1\} \boxplus \{1\}) \lhd (G_1 \boxplus \{1\}) \lhd \cdots \lhd (G_n \boxplus \{1\}) \lhd (G_n \boxplus H_1) \lhd \cdots \lhd (G_n \boxplus H_m)$$

   is a composition series for $G \boxplus H$.
4. How many composition series does a cyclic group of order p^n have?
5. Prove that the multiset of composition factors of a group is an invariant under isomorphism, but not a complete invariant.
6. Prove the uniqueness part of the fundamental theorem of arithmetic using the Jordan–Hölder Theorem.
7. Let $G = C_p^n$ be the direct product of n cyclic groups of prime order p. How many compositions series does G have? *Hint*: G is a vector space.

8. Prove that a subgroup of a group with a composition series need not have a composition series as follows. Let $I = \{1, 2, \dots\}$ and let $G = S_{(I)}$ be the restricted symmetric group on I. (Recall from an earlier exercise that this is the set of all permutations of I that have finite support.) For each $n \geq 1$, let

$$G_n = \{\sigma \in G \mid \sigma x = x \text{ for } x > n\}$$

and let

$$H_n = \{\sigma \in G_n \mid \sigma|_{\{1,\dots,n\}} \text{ is even}\}$$

   a) Show that $G = \bigcup_{n>1} G_n$.
   b) Show that $H = \bigcup_{n>1} H_n$ has index 2 in G and so $H \trianglelefteq G$.
   c) Show that H is simple.
   d) Show that H contains an infinite abelian subgroup A.
   e) Show that A has no composition series but that H does.
9. Let $\mathcal{P}$ be an isomorphic-invariant property of finite groups. Let G be a group that has $\mathcal{P}$ and for which if $N \trianglelefteq G$, then G/N has $\mathcal{P}$. Prove that the following are equivalent:
   a) G has a normal series

$$\{1\} = G_0 \trianglelefteq G_1 \trianglelefteq \cdots \trianglelefteq G_n = G$$

   whose factor groups have $\mathcal{P}$.
   b) G has the property that any nontrivial quotient group of G has a nontrivial normal subgroup that has $\mathcal{P}$.

   Such a group G is said to be **hyper-$\mathcal{P}$**.
10. Prove the following facts:
   a) For any $a, b \in G$,

$$[ab]_{H_i} = [a]_{H_i}[b]_{H_i}$$

   and so the projection map $\rho_k\colon G \to H_k$ is a homomorphism (and an endomorphism of G).
   b) For any $a \in G$,

$$a = [a]_{H_1} \cdots [a]_{H_s}$$

   c) The projection map commutes with any inner automorphism γ_g of G and therefore preserves normality.
11. Let G have BCC on normal subgroups and suppose that H is a group for which $G \boxtimes G \approx H \boxtimes H$. Show that $G \approx H$.

Chapter 11
Solvable and Nilpotent Groups

Classes of Groups

By a **class** $\mathcal{K}$ of groups, we mean a subclass of the class of all groups with the following two properties:

1) $\mathcal{K}$ contains a trivial group
2) $\mathcal{K}$ is closed under isomorphism, that is,

$$G \in \mathcal{K} \quad \text{and} \quad H \approx G \quad \Rightarrow \quad H \in \mathcal{K}$$

For example, the abelian groups form a class of groups. A group of class $\mathcal{K}$ is called a **$\mathcal{K}$-group** and $\mathcal{K}$-group H that is a subgroup of a group G is called a **$\mathcal{K}$-subgroup** of G. A class $\mathcal{K}$ is a **trivial class** if it contains only one-element groups.

Closure Properties

We will be interested in the following closure properties for a class $\mathcal{K}$ of groups:

1) **(Subgroup)**

$$G \in \mathcal{K}, \quad H \le G \quad \Rightarrow \quad H \in \mathcal{K}$$

2) **(Intersection and Cointersection)** For $H, K \le G$,

$$\begin{aligned} H, K \in \mathcal{K} \quad &\Rightarrow \quad H \cap K \in \mathcal{K} \\ \frac{G}{H}, \frac{G}{K} \in \mathcal{K} \quad &\Rightarrow \quad \frac{G}{H \cap K} \in \mathcal{K} \end{aligned}$$

3) **(Quotient and Extension)** For $N \trianglelefteq G$,

$$\begin{aligned} G \in \mathcal{K} \quad &\Rightarrow \quad G/N \in \mathcal{K} \\ N, G/N \in \mathcal{K} \quad &\Rightarrow \quad G \in \mathcal{K} \end{aligned}$$

4) **(Seminormal Join, Normal Join and Cojoin)** For $H, K \le G$,

$$\begin{aligned} H, K \in \mathcal{K}, \text{one normal in } G \quad &\Rightarrow \quad HK \in \mathcal{K} \\ H, K \in \mathcal{K}, \text{both normal in } G \quad &\Rightarrow \quad HK \in \mathcal{K} \\ \frac{G}{H}, \frac{G}{K} \in \mathcal{K} \quad &\Rightarrow \quad \frac{G}{HK} \in \mathcal{K} \end{aligned}$$

5) **(Direct product)**

$$H, K \in \mathcal{K} \quad \Rightarrow \quad H \boxtimes K \in \mathcal{K}$$

These properties are not independent.

Theorem 11.1 *The following implications hold for a class $\mathcal{K}$ of groups:*
1) *subgroup $\Rightarrow$ intersection*
2) *quotient $\Rightarrow$ cojoin*
3) *seminormal join $\Rightarrow$ normal join $\Rightarrow$ direct product*
4) *subgroup and direct product $\Rightarrow$ cointersection*

Thus, a class that is closed under

subgroup, quotient, seminormal join, extension

is closed under all nine properties above.

Proof. Part 1) is clear. For part 2), we have

$$\frac{G}{HK} \approx \frac{G}{H} \bigg/ \frac{HK}{H} \in \mathcal{K}$$

For part 3), the direct product $H \boxplus K$ is the seminormal join of $H \boxplus \{1\}$ and $\{1\} \boxplus K$, each of which is in $\mathcal{K}$ if $H, K \in \mathcal{K}$. For part 4), if $G/H, G/K \in \mathcal{K}$, then

$$\frac{G}{H \cap K} \hookrightarrow \frac{G}{H} \boxtimes \frac{G}{K} \in \mathcal{K}$$

via the map $\sigma: g(H \cap K) \mapsto (gH, gK)$.□

The following definition will prove very convenient.

Definition *Let $\mathcal{K}$ be a class of groups.*
1) *A* **$\mathcal{K}$-series** *is a series whose factor groups belong to the class $\mathcal{K}$.*
2) *A* **$\mathcal{K}_s$-group** *is a group that has a $\mathcal{K}$-series and a* **$\mathcal{K}_n$-group** *is a group that has a normal $\mathcal{K}$-series.*
3) *The* **$\mathcal{K}_s$-class** *is the class of all $\mathcal{K}_s$-groups and the* **$\mathcal{K}_n$-class** *is the class of all $\mathcal{K}_n$-groups.*□

Our main interest is in the $\mathcal{K}_s$ and $\mathcal{K}_n$ classes in which $\mathcal{K}$ is either the class of cyclic groups or the class of abelian groups. However, we are also interested in a class of groups that is not a $\mathcal{K}_s$ or $\mathcal{K}_n$ class, namely, the nilpotent groups.

Definition

1) *a)* *A* **cyclic series** *is a series that has cyclic factor groups.*

b) *An* **abelian series** *is a series that has abelian factor groups.*

c) *A* **central series** *for a group G is a normal series*

$$\{1\} = G_0 \trianglelefteq G_1 \trianglelefteq \cdots \trianglelefteq G_n = G$$

for which

$$G_{k+1}/G_k \leq Z(G/G_k)$$

for all k.

2) *As shown in the table below, we have the following definitions.*

	Series	**Normal Series**
Abelian	Solvable	= Solvable
Cyclic	Polycyclic	Supersolvable
Central		Nilpotent

a) *A group is* **solvable** *if it has an abelian series.*

b) *A group is* **polycyclic** *if it has a cyclic series.*

c) *A group is* **supersolvable** *if it has a normal cyclic series.*

d) *A group G is* **nilpotent** *if it has a central series.* □

Note that in previous chapters (see Theorem 7.10), we have found it convenient to use the term *central series* even when the series does not start at the trivial group, which requires that we distinguish carefully between series *in G* and series *for G*.

As the title of this chapter suggests, our primary interest is in nilpotent and solvable groups. It is clear that nilpotent groups are solvable. Also, all abelian groups are nilpotent and Theorem 7.10 shows that all finite p-groups are nilpotent:

$$\left.\begin{array}{c}\text{finite } p\text{-group}\\ \text{or}\\ \text{abelian}\end{array}\right\} \Rightarrow \text{nilpotent} \Rightarrow \text{solvable}$$

Note, however, that S_3 is solvable but not nilpotent. Also, the symmetric groups S_n are solvable if and only if $n < 5$. In fact, if $n \geq 5$, then S_n has only one nontrivial proper normal subgroup A_n, which is not abelian.

Operations on Series

In order to study the closure properties of series-based classes of groups and of the nilpotent class, we must consider various operations on series. Indeed, the operations of intersection, normal lifting, quotient and unquotient as defined in

Theorem 4.10 can be applied to each step in a series. Specifically, we have the following operations on series. Let

$$\mathcal{G} : G_0 \trianglelefteq \cdots \trianglelefteq G_n$$

be a series in G.

1) The **intersection** of $\mathcal{G}$ with $H \leq G$ is

$$\mathcal{G} \cap H : G_0 \cap H \trianglelefteq \cdots \trianglelefteq G_n \cap H = H$$

2) The **normal lifting** of $\mathcal{G}$ by $N \trianglelefteq G$ is

$$\mathcal{G}N : G_0N \trianglelefteq \cdots \trianglelefteq G_nN$$

3) The **quotient** of $\mathcal{G}$ by $N \trianglelefteq G$ is

$$\frac{\mathcal{G}N}{N} : \frac{G_0N}{N} \trianglelefteq \cdots \trianglelefteq \frac{G_nN}{N}$$

4) The **unquotient** of the series

$$\mathcal{G} : \frac{G_0}{N} \trianglelefteq \cdots \trianglelefteq \frac{G_n}{N}$$

where $N \leq G_0$ is

$$\mathcal{G} \uparrow N : G_0 \trianglelefteq \cdots \trianglelefteq G_n$$

5) For $H_i \leq G$, the **concatenation** of the series

$$\mathcal{H} : H_0 \trianglelefteq \cdots \trianglelefteq H_m$$

and

$$\mathcal{K} : H_m \trianglelefteq \cdots \trianglelefteq H_{m+n}$$

is the series

$$\mathcal{H} * \mathcal{K} : H_0 \trianglelefteq \cdots \trianglelefteq H_m \trianglelefteq \cdots \trianglelefteq H_{m+n}$$

6) For the series

$$\mathcal{H}\colon H_0 \trianglelefteq \cdots \trianglelefteq H_n$$

and

$$\mathcal{G}\colon G_0 \trianglelefteq \cdots \trianglelefteq G_m$$

in G, the **interleaved series** $\mathcal{H} \sim \mathcal{G}$ is formed as follows. First, if $s = \max\{n, m\}$, then we extend whichever series is shorter by repeating the upper endpoint (H_n or G_m) an appropriate number of times to make the resulting series of equal length s. Then

$$\mathcal{H} \sim \mathcal{G}\colon (H_0 \boxplus G_0) \trianglelefteq (H_1 \boxplus G_0) \trianglelefteq (H_1 \boxplus G_1) \trianglelefteq (H_2 \boxplus G_1) \trianglelefteq (H_2 \boxplus G_2) \trianglelefteq \cdots \trianglelefteq (H_s \boxplus G_s)$$

Note that the intersection, normal lifting, quotient, unquotient and interleave of normal series is normal. However, the concatenation of two normal series need not be normal.

These operations are used as follows.

Theorem 11.2 *Let G be a group and let $H \leq G$ and $N \trianglelefteq G$. Let*
a) $\mathcal{G}$ be a series for G
b) $\mathcal{H}$ be a series for H
c) $\mathcal{N}$ be a series for N
d) $\mathcal{Q}$ be a series for G/N.
Then
*1) **(Subgroup)** $\mathcal{G} \cap H$ is a series for H*
*2) **(Seminormal join)** $\mathcal{N} * \mathcal{H}N$ is a series for HN*
*3) **(Quotient)** $\mathcal{G}N/N$ is a series for G/N*
*4) **(Extension)** $\mathcal{N} * (\mathcal{Q} \uparrow N)$ is a series for G*
*5) **(Direct product)** If $\mathcal{G}_i$ is a series for G_i for $i = 1, 2$, then $\mathcal{G}_1 \sim \mathcal{G}_2$ is a series for $G_1 \boxplus G_2$.*□

Closure Properties of Groups Defined by Series

Theorem 4.10 and the previous theorems provide the following facts about closure of $\mathcal{K}_s$-classes and $\mathcal{K}_n$-classes.

Theorem 11.3 *Let $\mathcal{K}$ be a class of groups.*
*1) **(Subgroup)** If $\mathcal{K}$ is closed under subgroup, then the $\mathcal{K}_s$ and $\mathcal{K}_n$ classes are closed under subgroup.*
*2) **(Seminormal join)** If $\mathcal{K}$ is closed under quotient, then the $\mathcal{K}_s$-class is closed under seminormal join.*
*3) **(Quotient)** If $\mathcal{K}$ is closed under quotient, then the $\mathcal{K}_s$ and $\mathcal{K}_n$ classes are closed under quotient.*
*4) **(Extension)** The $\mathcal{K}_s$-class is closed under extension.*
*5) **(Direct product)** The $\mathcal{K}_s$ and $\mathcal{K}_n$ classes are closed under direct product.*

In particular, if $\mathcal{K}$ is closed under subgroup and quotient, then the $\mathcal{K}_s$-class is closed under the following eight operations:
6) subgroup
7) quotient
8) intersection
9) cointersection
10) cojoin
11) direct product

12) seminormal join
13) extension
and the $\mathcal{K}_n$-class is closed under all of these operations except seminormal join and extension.

Proof. For 1), if $\mathcal{K}$ is closed under subgroup, then (normal) $\mathcal{K}$-series are closed under intersection and so the $\mathcal{K}_s$ and $\mathcal{K}_n$ classes are closed under subgroup. For 2) and 3), if $\mathcal{K}$ is closed under quotient, then $\mathcal{K}$-series are closed under normal lifting. Hence, $\mathcal{N} * \mathcal{H}N$ and $\mathcal{G}N/N$ are $\mathcal{K}$-series. It follows that the $\mathcal{K}_s$-class is closed under seminormal join and the $\mathcal{K}_s$ and $\mathcal{K}_n$ classes are closed under quotient. For 5), if $\mathcal{G}_i$ is a (normal) $\mathcal{K}$-series for G_i, then $\mathcal{G}_1 \sim \mathcal{G}_2$ is a (normal) $\mathcal{K}$-series for $G_1 \boxplus G_2$.□

Thus, since the classes of cyclic groups and abelian groups are closed under subgroup and quotient, we have the following.

Theorem 11.4 *The polycyclic, solvable and supersolvable classes are closed under the following eight operations (except where noted):*
1) subgroup
2) quotient
3) intersection
4) cointersection
5) cojoin
6) direct product
7) seminormal join (except for supersolvable)
8) extension (except for supersolvable).□

Let us now turn to the closure properties of nilpotent groups.

Theorem 11.5
1) *Central series are closed under intersection, normal lifting, quotient and unquotient.*
2) *The class of nilpotent groups has the following seven closure properties:*
   a) *subgroup*
   b) *quotient*
   c) *normal join (this is Fitting's theorem, to be proved a bit later)*
   d) *direct product*
   e) *intersection*
   f) *cointersection*
   g) *cojoin*

   but not extension.

Proof. Part 1) follows from Theorem 4.10 and Theorem 4.11. The statement about extensions in part 3) can be verified by looking at S_3.□

Nilpotent Groups

We now undertake a closer look at nilpotent groups. We will prove that a finite group G is nilpotent if and only if it is the direct product of p-groups and so Theorem 8.11 shows that finite nilpotent groups have many other interesting characterizations.

Higher Centers

An extension $H \trianglelefteq K$ in G is central in G if and only if

$$\frac{K}{H} \leq Z\left(\frac{G}{H}\right)$$

and so the largest subgroup K of G for which $H \trianglelefteq K$ is central in G is the subgroup K for which

$$\frac{K}{H} = Z\left(\frac{G}{H}\right)$$

With this in mind, we define a function $\zeta = \zeta_G$ on $\operatorname{nor}(G)$ by

$$Z\left(\frac{G}{N}\right) = \frac{\zeta_G(N)}{N}$$

for $N \trianglelefteq G$. Note also that

$$\frac{\zeta_G(N)}{N} = Z\left(\frac{G}{N}\right) \sqsubseteq \frac{G}{N}$$

and so $\zeta_G(N) \sqsubseteq G$.

Writing $\zeta^k(\{1\})$ as $\zeta^k(1)$, the proper series

$$\zeta^0(1) \sqsubset \zeta^1(1) \sqsubset \zeta^2(1) \sqsubset \cdots$$

is called the **upper central series** for G and each $\zeta^k(1)$ is called a **higher center** of G. The first higher center is the center $Z(G)$ of G.

To see that the upper central series ascends more rapidly than any other central series of the form

$$\{1\} = G_0 \trianglelefteq G_1 \trianglelefteq G_2 \trianglelefteq \cdots$$

we have $\zeta(G_k) \geq G_{k+1}$ for all k and it is clear that

$$G_k \leq \zeta^k(1)$$

holds for $k = 0$. If this holds for a particular value of k, then the monotonicity of ζ implies that

$$G_{k+1} \le \zeta(G_k) \le \zeta(\zeta^k(1)) = \zeta^{k+1}(1)$$

Hence, $G_k \le \zeta^k(1)$ for all k.

Theorem 11.6 *Let G be a nilpotent group. The upper central series*

$$\zeta^0(1) \sqsubset \zeta^1(1) \sqsubset \zeta^2(1) \sqsubset \cdots$$

for G is characteristic and ascends more rapidly than any other central series for G, that is, if

$$\{1\} = G_0 \trianglelefteq G_1 \trianglelefteq G_2 \trianglelefteq \cdots$$

is central in G, then

$$G_k \le \zeta^k(1)$$

for all $k \ge 0$.
1) *G is nilpotent if and only if the upper central series reaches G.*
2) *If G is nilpotent, then all central series for G have length greater than or equal to the length of the upper central series for G.*□

The higher centers have some important applications.

Theorem 11.7 *Let G be nilpotent, with higher centers $\zeta^k(1)$.*
1) *If $H \le G$, then*

$$H\zeta^k(1) \trianglelefteq H\zeta^{k+1}(1)$$

2) *If $N \trianglelefteq G$, then*

$$N \cap \zeta^k(1) = \{1\} \quad \Rightarrow \quad N \cap \zeta^{k+1}(1) \le Z(G)$$

As a consequence,
3) *G has the property that every subgroup is subnormal*
4) *G has the center-intersection property*
5) *Every chief series for G is central.*

Proof. For part 1), since $[G, \zeta^{k+1}(1)] \le \zeta^k(1)$, Theorem 3.41 implies that

$$[H\zeta^k(1), H\zeta^{k+1}(1)] = [H\zeta^k(1), H][H\zeta^k(1), \zeta^{k+1}(1)]^H$$

But each factor on the right is contained in $H\zeta^k(1)$ and so $H\zeta^k(1) \trianglelefteq H\zeta^{k+1}(1)$. Part 3) follows from part 1), since we may lift the upper central series for G by H to get a series from H to G, whence H is subnormal in G.

For part 2), since $[\zeta^{k+1}(1), G] \le \zeta^k(1)$ and $[N, G] \le N$, it follows that

$$[N \cap \zeta^{k+1}(1), G] \le N \cap \zeta^k(1) = \{1\}$$

and so $N \cap \zeta^{k+1}(1) \le Z(G)$. For part 4), there is a largest k for which

$N \cap \zeta^k(1) = \{1\}$ and so $N \cap \zeta^{k+1}(1)$ is a nontrivial subgroup of $Z(G)$. For part 5), the factor group G_{k+1}/G_k of a chief series $\mathcal{G}$ for G is a minimal normal subgroup of the nilpotent group G/G_k and so the center-intersection property implies that G_{k+1}/G_k is central. Thus, $\mathcal{G}$ is central.□

We can now augment Theorem 8.11 by adding the nilpotent condition.

Theorem 11.8 *The following are equivalent for a finite group G:*
1) G is nilpotent.
2) Every Sylow subgroup of G is normal.
3) G is the direct product of its Sylow p-subgroups

$$G = \bowtie_{p \in \mathcal{P}} Y_p$$

4) If $H \leq G$, then

$$H = \bowtie_{p \in \mathcal{P}} (H \cap Y_p)$$

5) **(Strong converse of Lagrange's theorem)** *If $n \mid o(G)$, then G has a normal subgroup of order n.*
6) G is the direct product of p-subgroups.
7) Every subgroup of G is subnormal.
8) G has the normalizer condition.
9) Every maximal subgroup of G is normal.
10) $G/\Phi(G)$ is abelian.

Proof. Theorem 8.11 states that 2)–10) are equivalent. Moreover, Theorem 7.10 implies that a finite p-group is nilpotent and therefore so is any direct product of finite p-groups. Hence, 6) implies 1). Theorem 11.7 shows that 1) implies 7).□

Lower Centers

In terms of commutators, an extension $H \trianglelefteq K$ in G is central in G if and only if

$$[K, G] \leq H$$

and so $H = [K, G]$ is the smallest subgroup K of G for which $H \trianglelefteq K$ is central in G; in fact, $[K, G] \sqsubseteq K$. Thus, if we define the "commutator with G" function $\Gamma = \Gamma_G$ on $\text{sub}(G)$ by

$$\Gamma_G(K) = [K, G]$$

then $\Gamma_G(K)$ is the smallest subgroup of G for which $\Gamma_G(K) \leq K$ is central in G. The (possibly infinite) descending central series

$$\cdots \sqsubset \Gamma^2(G) \sqsubset \Gamma^1(G) \sqsubset \Gamma^0(G) = G$$

is called the **lower central series** for G and each $\Gamma^k(G)$ is called a **lower center** of G. The first lower center is the commutator subgroup $\Gamma(G) = G'$.

An induction argument shows that the lower central series descends more rapidly than any other central series, in the sense that if

$$\cdots \trianglelefteq G_2 \trianglelefteq G_1 \trianglelefteq G_0 = G$$

is central in G, then

$$\Gamma^k(G) \leq G_k$$

for all $k \geq 0$. For if $\Gamma^k(G) \leq G_k$, then the monotonicity of Γ implies that

$$\Gamma^{k+1}(G) = \Gamma(\Gamma^k G) \leq \Gamma(G_k) \leq G_{k+1}$$

Theorem 11.9 *Let G be a group. The lower central series*

$$\cdots \sqsubset \Gamma^2(G) \sqsubset \Gamma^1(G) \sqsubset \Gamma^0(G) = G$$

for G is characteristic and descends more rapidly than any central series for G, that is, if

$$\cdots \trianglelefteq G_2 \trianglelefteq G_1 \trianglelefteq G_0 = G$$

is central in G, then

$$\Gamma^k(G) \leq G_k$$

for all $k \geq 0$.

1) *G is nilpotent if and only if the lower central series reaches $\{1\}$.*
2) *If G is nilpotent, then all central series for G have length greater than or equal to the length of the lower central series for G.*□

The commutator function Γ_G has some simple but useful properties that are consequences of Theorems 3.40 and 3.41.

Theorem 11.10 *Let G be a group and let $H, K, L \trianglelefteq G$.*

1) $\Gamma_H(K) \trianglelefteq G$
2) $\Gamma_H(K) = \Gamma_K(H)$
3) $\Gamma_H(KL) = \Gamma_H(K)\Gamma_H(L)$ *and* $\Gamma_{HK}(L) = \Gamma_H(L)\Gamma_K(L)$
4) Γ_H *is deflationary, that is,*

$$\Gamma_H(K) \leq H$$

In fact,

$$\Gamma_H(K) \leq H \cap K$$

Hence, for all $k \geq 1$,

$$\Gamma_H^k(H) \leq \Gamma_G^k(G)$$

5) If $N \trianglelefteq G$ and $N \le H \cap K$, then for any $k \ge 1$,

$$\Gamma^k_{K/N}\left(\frac{H}{N}\right) = \frac{\Gamma^k_K(H)N}{N}$$

6) For $k \ge 1$,

$$\Gamma^k_{HK}(HK) = \prod_{A_1,\ldots,A_k,B\in\{H,K\}} \Gamma_{A_1}\cdots\Gamma_{A_k}(B)$$

Proof. Part 5) holds for $k = 1$ since

$$\Gamma_{K/N}\left(\frac{H}{N}\right) = \left[\frac{H}{N},\frac{K}{N}\right] = \frac{[H,K]N}{N} = \frac{\Gamma_K(H)N}{N}$$

and if it holds for a particular value of k, then

$$\begin{aligned}\Gamma^{k+1}_{K/N}\left(\frac{H}{N}\right) &= \left[\Gamma^k_{K/N}\left(\frac{H}{N}\right),\frac{K}{N}\right]\\ &= \left[\frac{\Gamma^k_K(H)N}{N},\frac{K}{N}\right]\\ &= \frac{[\Gamma^k_K(H),K]N}{N}\\ &= \frac{\Gamma^{k+1}_K(H)N}{N}\end{aligned}$$

and so this holds for all $k \ge 1$.

For part 6), we have

$$\Gamma_{HK}(HK) = \Gamma_H(H)\Gamma_K(H)\Gamma_H(K)\Gamma_K(K) = \prod_{A,B\in\{H,K\}} \Gamma_A(B)$$

and an easy induction proves the general formula.□

Nilpotency Class

Theorem 11.9 and Theorem 11.6 imply that for a nilpotent group, the upper and lower central series have the same length.

Definition *Let G be a nilpotent group. The common length of the upper and lower central series is called the* **nilpotency class** *of G, which we denote by* $\operatorname{nilclass}(G)$.□

Moreover, if G is nilpotent and

$$\mathcal{G} : \{1\} = G_0 \lhd \cdots \lhd G_m = G$$

is a central series for G of length m, then

$$\Gamma_G^{m-k}(G) \le G_k \le \zeta_G^k(1)$$

for all $k = 0, \ldots, m$, where $\Gamma^i(G) = \{1\}$ and $\zeta^i(1) = G$ if $i \ge \text{nilclass}(G)$.

The nilpotent groups of class 0 are the trivial groups and the nilpotent groups of class 1 have $Z(G) = G$ or equivalently, $G' = \{1\}$ and so are the nontrivial abelian groups. A group G has nilpotency class 2 if and only if either of the following conditions holds:

1) $\{1\} < G' < G$ and $[G', G] = \{1\}$, or equivalently,
$$\{1\} < G' \le Z(G) < G$$
2) $\{1\} < Z(G) < G$ and
$$Z\left(\frac{G}{Z(G)}\right) = \frac{G}{Z(G)}$$
or equivalently, G is not abelian but $G/Z(G)$ is abelian.

Theorem 3.42 implies that $G' \le Z(G)$ if and only if

$$[[a, b], c] = [a, [b, c]]$$

for all $a, b, c \in G$. Hence, for nonabelian groups, this condition is equivalent to being nilpotent of class 2.

Theorem 11.11 *Let G be nilpotent.*
1) If $H \le G$, then
$$\text{nilclass}(H) \le \text{nilclass}(G)$$
2) If $N \trianglelefteq G$, then
$$\text{nilclass}(G/N) \le \text{nilclass}(G)$$
3) **(Fitting's theorem)** *The join of two normal nilpotent subgroups of a group G is nilpotent. In fact, if $H, K \trianglelefteq G$, then*
$$\text{nilclass}(HK) \le \text{nilclass}(H) + \text{nilclass}(K)$$

Proof. The first two parts follow from Theorem 11.10. For part 3), Theorem 11.10 implies that

$$\Gamma_{HK}^k(HK) = \prod_{A_1, \ldots, A_k, B \in \{H, K\}} \Gamma_{A_1} \cdots \Gamma_{A_k}(B)$$

for all $k \ge 1$. Now, suppose that $\text{nilclass}(H) = c$ and $\text{nilclass}(K) = d$ and let $k = c + d$. Then each factor on the right above has the form

$$L = \Gamma_{A_1} \cdots \Gamma_{A_{c+d}}(B)$$

Among the subgroups $A_1, \ldots, A_{c+d}, B$, suppose there are h H's and k K's, where $h + k = c + d + 1$. Then either $h \geq c + 1$ or $k \geq d + 1$ and we may assume without loss of generality that $h \geq c + 1$. Since Γ_K is deflationary, removing all Γ_K's from the expression for L results in a possibly larger subgroup

$$M = \begin{cases} \Gamma_H^h(K) & \text{if } B = K \\ \Gamma_H^{h-1}(H) & \text{if } B = H \end{cases}$$

However, in the former case,

$$\Gamma_H^h(K) = \Gamma_H^{h-1}\Gamma_H(K) = \Gamma_H^{h-1}\Gamma_K(H) \leq \Gamma_H^{h-1}(H)$$

and so

$$L \leq M \leq \Gamma_H^{h-1}(H) \leq \Gamma_H^c(H) = \{1\}$$

Hence,

$$\Gamma_{HK}^{c+d}(HK) = \{1\}$$

which implies that HK is nilpotent of class at most $c + d$.□

An Example

We now describe a family of groups showing that for any $c \geq 0$, there are nilpotent groups of nilpotency class c. Let $\mathcal{M}_n(R)$ be the family of all $n \times n$ matrices over a commutative ring R with identity and let $U = UT(n, R)$ be the unitriangular matrices over R. (Recall that a matrix is **unitriangular** if it is upper triangular, with 1's on the main diagonal.) Denote the (i, j)th entry in M by $M_{i,j}$. For $k \geq 0$, the kth **superdiagonal** $M(k)$ of a matrix M are the elements of the form $M_{i,i+k}$.

For $0 \leq k \leq n - 1$, let N_k be the set of all $n \times n$ matrices over R with 0's on or below the kth superdiagonal, that is, for $A \in \mathcal{M}_n(R)$,

$$A \in N_k \quad \Leftrightarrow \quad A_{i,j} = 0 \text{ for all } j \leq i + k$$

It is routine to confirm that

$$N_k N_m \subseteq N_{k+m+1}$$

In particular,

$$N_k N_k \subseteq N_k$$

For $k \geq 0$, let

$$U_k = \{I\} + N_k = \{I + A \mid A \in N_k\}$$

To see that U_k is a subgroup of U, we have

$$U_kU_k = (\{I\} + N_k)(\{I\} + N_k) \subseteq \{I\} + N_k + N_k + N_kN_k \subseteq U_k$$

and if $A \in N_k$, then since $A^u = 0$ for some $u \geq 0$, it follows that

$$(I + A)^{-1} = I - A + A^2 - \cdots \pm A^{u-1} \in U_k$$

As to commutators, if $A \in N_k$ and $B \in N_m$, then

$$\begin{aligned}[I + A, I + B] &= (I + A)(I + B)((I + B)(I + A))^{-1} \\ &= (I + A + B + AB)(I + B + A + BA)^{-1} \\ &= (I + X)(I + Y)^{-1} \\ &= I + (X - Y)(I + Y)^{-1}\end{aligned}$$

where

$$X = A + B + AB \quad \text{and} \quad Y = B + A + BA$$

Moreover,

$$X - Y = AB - BA \in N_{k+m+1}$$

implies that $[I + A, I + B] \in U_{k+m+1}$ and so

$$[U_k, U_m] \leq U_{k+m+1}$$

Taking $m = 0$ gives

$$[U_k, U] \leq U_{k+1} \leq U_k$$

which shows both that U_k is normal in U and that $U_{k+1} \lhd U_k$ is central in U. Thus, the series

$$\{I\} = U_{n-1} \lhd \cdots \lhd U_1 \lhd U_0 = U$$

is central and so U is nilpotent of class at most $n - 1$.

On the other hand, let $E_{i,j}$ denote the matrix with all 0s except for a 1 in the $(i, j)$the position. Then for $i \neq j + 1$, an easy calculation shows that

$$[I + E_{i,j}, I + E_{j,j+1}] = I + E_{i,j+1}$$

In particular, if $n \geq 3$, then

$$[I + E_{1,2}, I + E_{2,3}] = I + E_{1,3}$$

and so $\Gamma(G) = [G, G] \neq \{I\}$. Also, if $n \geq 4$, then

$$[I + E_{1,3}, I + E_{3,4}] = I + E_{1,4}$$

and so $\Gamma^2(G) = [\Gamma(G), G] \neq \{I\}$. More generally,

$$[I + E_{1,n-1}, I + E_{n-1,n}] = I + E_{1,n} \neq I$$

and so $\Gamma^{n-2}(G) \neq \{I\}$, which shows that U is nilpotent of class $n-1$.

Solvability

We now turn to a discussion of solvable groups.

Perspective on Solvability

Solvable groups have played an extremely important role in the study of the location of the roots of polynomials over a field F. Let us pause to describe this role in general terms. For more details, we refer to reader to Roman, *Field Theory* [27].

If F is a subfield of a field E, we say that $F \leq E$ is a **field extension**. Associated to each field extension $F \leq E$ is a group $G_F(E)$, called the **Galois group** of the extension and defined as the group of all (ring) automorphisms σ of E that fix the elements of F, that is, for which $\sigma a = a$ for all $a \in F$. It turns out that the properties of the "simpler" Galois group can often shed considerable light on properties of the "more complex" field extension.

To illustrate, one of the principal motivations for the development of abstract algebra since, oh say 3000 B.C., has been the desire (expressed in one form or another) to find the roots of a polynomial $p(x)$ with coefficients from a given **base field** F.

In fact, we now know that for any nonconstant polynomial $p(x)$ of degree d, there is an extension $\overline{F}$ of F, called an **algebraic closure** of F, that contains a full set of d roots for $p(x)$. Moreover, lying between the fields F and $\overline{F}$ is the smallest field E containing these roots of $p(x)$, called a **splitting field** for $p(x)$. The desire to express the roots of $p(x)$ by arithmetic formula (similar to the quadratic formula) or to show that this could not be done is what motivated Galois to first define some version of what we now know as a group.

The idea of expressing the roots of a polynomial $p(x)$ *by formula* means that, starting with the elements of the base field F, we can "capture" all of the roots of $p(x)$ through a finite number of special types of extensions of F. In particular, for each extension, we are allowed to include an nth root of an existing element and only whatever else is required in order to make a field.

Specifically, for the first extension, we may choose any $u_0 \in F$ and any root $r_0 = \sqrt[n_0]{u_0}$ of the polynomial $x^{n_0} - u_0$ where $n_0 \geq 1$. Then the first extension is

$$F \leq F(\sqrt[n_0]{u_0})$$

where $F(\sqrt[n_0]{u_0})$ is the smallest subfield of $\overline{F}$ containing F and r_0. Repeated extensions produce a tower of fields of the form

$$F \le F(\sqrt[n_0]{u_0}) \le F(\sqrt[n_0]{u_0}, \sqrt[n_1]{u_1}) \le \cdots \le F(\sqrt[n_0]{u_0}, \ldots, \sqrt[n_k]{u_k})$$

where each u_i is an element of the immediately preceeding field. This type of tower of fields is called a **radical series**. If all of the roots of a polynomial $p(x)$ lie within a radical series over F, then we say that the polynomial equation $p(x) = 0$ is **solvable by radicals**. (For simplicity, we assume that F has characteristic 0.)

It is not hard to show that a polynomial equation $p(x) = 0$ is solvable by radicals if and only if the roots of $p(x)$ can be captured within a finite tower of fields

$$F \le F(\alpha_1) \le F(\alpha_1, \alpha_2) \le \cdots \le F(\alpha_1, \ldots, \alpha_m)$$

where each field has *prime* degree over the previous field, that is, the dimension of each field as a vector space over the previous field is a prime number.

Now let E be the splittting field for $p(x)$ in $\overline{F}$. If the radical series above does capture the roots of $p(x)$, that is, if $E \le F(\alpha_1, \ldots, \alpha_m)$, then taking Galois groups (which reverses inclusion) gives the descending sequence

$$\mathcal{G}: G_F(E) \ge G_{F(\alpha_1)}(E) \ge \cdots \ge G_{F(\alpha_1,\ldots,\alpha_m)}(E)$$

But

$$G_{F(\alpha_1,\ldots,\alpha_m)}(E) < G_E(E) = \{\iota\}$$

and so the sequence $\mathcal{G}$ reaches the trivial group. Galois showed that (in modern terminology) $\mathcal{G}$ is an abelian series and so $G_F(E)$ is solvable. Thus, Galois proved that if $p(x) = 0$ is solvable by radicals, then its Galois group $G_F(E)$ is solvable. He also proved the converse.

Now, an element $\sigma \in G_F(E)$ of the Galois group fixes the coefficients of $p(x)$, since they lie in F. Hence, σ must send a root r of $p(x)$ in E to another root of $p(x)$ in E, that is, the elements of $G_F(E)$ are *permutations* of the set of roots of $p(x)$ in E. Moreover, since E is generated by these roots, each $\sigma \in G_F(E)$ is uniquely determined by its behavior on these roots. Thus, one often thinks of $G_F(E)$ simply as a subgroup of the permutation group S_n, where $n = \deg(p(x))$.

In fact, there are many cases in which $G_F(E)$, thought of as a subgroup of S_n, is actually S_n itself. For example, it can be shown for any prime p, the Galois group of the splitting field of any irreducible polynomial $f(x) \in \mathbb{Q}[x]$ of degree p with exactly two nonreal roots is S_p.

However, S_n is not solvable for $n \geq 5$, since A_n is simple and so the only nontrivial series for S_n is $\{\iota\} \lhd A_n \lhd S_n$, which is not abelian. Thus, the roots of the polynomials described above cannot be captured within a radical series, that is, these polynomials are not solvable by radicals.

Note that this shows that there are *individual polynomials* that are not solvable by radicals. Thus, not only is there no *general formula*, similar to the quadratic, cubic and quartic formulas, for the solutions of arbitrary quintic equations, but there are even individual quintic equations whose solutions are not obtainable by formula!

Galois used his remarkable theory in his paper *Memoir on the Conditions for Solvability of Equations by Radicals* of 1831 (but not published until 1846!), to show that the general equation of degree 5 or larger is not solvable by radicals. (Proofs that the 5th degree equation is not solvable by radicals were offered earlier: An incomplete proof by Ruffini in 1799 and a complete proof by Abel in 1826.)

Thus, the notion of solvability arose through the desire to settle the question of whether we could solve all polynomial equations by simple formula. Of course, solvable groups are important for other reasons. In fact, we will see that the class of solvable groups has a sort of super-Sylow theorem, to wit, if G is solvable of order mn where $(m,n)=1$, then G has a (Hall) subgroup of order m and all subgroups of order m are conjugate.

The Derived Series

For any solvable group, there is an abelian series that descends more rapidly than any other abelian series. Moreover, this series is also characteristic in G. A series

$$\{1\} = G_0 \trianglelefteq G_1 \trianglelefteq \cdots \trianglelefteq G_m = G$$

is abelian if and only if

$$G'_{k+1} \leq G_k$$

for all $k = 0, \ldots, m-1$, where the prime denotes commutator subgroup.

Definition *Let G be a group. The subgroups defined by*

$$G^{(0)} = G, \quad G^{(1)} = G'$$

and, in general for $n \geq 1$,

$$G^{(n)} = (G^{(n-1)})'$$

are called the **higher commutators** *of G. The group $G^{(n)}$ is called the nth* **commutator subgroup** *of G. The series of higher commutators:*

$$\cdots \sqsubset G^{(3)} \sqsubset G^{(2)} \sqsubset G^{(1)} \sqsubset G$$

is called the **derived series** *for* G.□

The monotonicity of the commutator operation implies that the derived series descends from G more rapidly than any other abelian series. Moreover, since $G^{(k+1)} \sqsubseteq G^{(k)}$, the derived series is characteristic.

Theorem 11.12 *Let* G *be a group.*

1) *The derived series*

$$\cdots \sqsubset G^{(3)} \sqsubset G^{(2)} \sqsubset G^{(1)} \sqsubset G$$

is the abelian series of steepest descent, in the sense that if the series

$$\cdots G_3 \trianglelefteq G_2 \trianglelefteq G_1 \trianglelefteq G_0 = G$$

is abelian, then

$$G^{(k)} \leq G_k$$

for all k.

2) G *is solvable if and only if its derived series reaches the trivial group, that is, if and only if there is a* $k \geq 1$ *for which* $G^{(k)} = \{1\}$. *The smallest integer* n *for which* $G^{(n)} = \{1\}$ *is called the* **derived length** *of* G, *which we denote by* $\operatorname{derlen}(G)$.
3) *A group* G *is solvable if and only if it has a normal abelian series.*
4) *The length of any abelian series for* G *is greater than or equal to the derived length of* G.□

We will have use for the following fact about higher commutators of quotient groups.

Theorem 11.13 *Let* G *be a group and let* $N \trianglelefteq G$. *Then*

$$\left(\frac{G}{N}\right)^{(n)} = \frac{G^{(n)}N}{N}$$

for all $n \geq 0$.

Proof. For any $N \leq H \trianglelefteq G$, we have

$$\left(\frac{H}{N}\right)' = \left[\frac{H}{N}, \frac{H}{N}\right] = \frac{[H,H]N}{N} = \frac{H'N}{N}$$

In particular, for $H = G$, we have

$$\left(\frac{G}{N}\right)' = \left[\frac{G}{N}, \frac{G}{N}\right] = \frac{[G,G]N}{N} = \frac{G'N}{N}$$

and so the result holds for $n = 1$. Assuming that the result holds for an arbitrary k, we have with $H = G^{(n)}N$,

$$\left(\frac{G}{N}\right)^{(n+1)} = \left[\left(\frac{G}{N}\right)^{(n)}\right]' = \left(\frac{G^{(n)}N}{N}\right)' = \frac{(G^{(n)}N)'N}{N}$$

Finally, Theorem 3.41 implies that

$$(G^{(n)}N)'N = [G^{(n)}N, G^{(n)}N]N = [G^{(n)}, G^{(n)}]N = G^{(n+1)}N$$

and so the result follows.□

Theorem 11.14 *If G is solvable, $A, B, H \leq G$ and $N, K \trianglelefteq G$, then*
1) $\operatorname{derlen}(H) \leq \operatorname{derlen}(G)$
2) $\operatorname{derlen}(G/N) \leq \operatorname{derlen}(G)$
3) $\operatorname{derlen}(G) \leq \operatorname{derlen}(N) + \operatorname{derlen}(G/N)$
4) $\operatorname{derlen}(A \bowtie B) \leq \max\{\operatorname{derlen}(A), \operatorname{derlen}(B)\}$.

Proof. For part 1), since $H^{(k)} \leq G^{(k)}$ for all $k \geq 0$, if $n = \operatorname{derlen}(G)$ then $H^{(n)} \leq G^{(n)} = \{1\}$. Thus, $\operatorname{derlen}(H) \leq n$. For part 2), Theorem 11.13 implies that if $G^{(n)} = \{1\}$, then $(G/N)^{(n)} = \{N\}$. For part 3), if G/N has derived length d, then $G^{(d)}N/N = \{N\}$ and so $G^{(d)} \leq N$. If the derived length of N is e, then

$$G^{(d+e)} \leq N^{(e)} = \{1\}$$

and so the derived length of G is at most $d + e$. Part 4) follows from the fact that $(A \bowtie B)' = A' \bowtie B'$.□

Properties of Solvable Groups

If N is a minimal normal subgroup of a group G and if

$$\mathcal{G}: \{1\} = G_d \trianglelefteq \cdots \trianglelefteq G_0 = G$$

is any normal series in G, then $N \leq G_i$ or else $N \cap G_i = \{1\}$ for all i and so there is an index k for which

$$N \leq G_k \quad \text{and} \quad N \cap G_{k+1} = \{1\}$$

Therefore, if G is solvable and if $\mathcal{G}$ is the derived series for G, then

$$N \leq G^{(k)} \quad \text{and} \quad N \cap G^{(k+1)} = \{1\}$$

and so $N' \leq N \cap G^{(k+1)} = \{1\}$, whence N is abelian.

Theorem 11.15 *Let G be solvable. Any minimal normal subgroup N of G is abelian. Moreover, if N contains a nontrivial element of finite order, then N is elementary abelian.*

Proof. For the final statement, N has an element of prime order p and since

$$N_p := \{x \in N \mid x^p = 1\} \trianglelefteq N$$

it follows that $N = N_p$ is an elementary abelian group.□

If G is solvable and has a composition series, then the factor groups of the composition series are both simple and solvable and therefore cyclic of prime order.

Theorem 11.16 *The following are equivalent for a group G that has a composition series.*

1) *G is solvable.*
2) *Every composition series for G has prime order factor groups.*
3) *G has a cyclic series in which each factor group has prime order.*
4) *G has a cyclic series, that is, G is polycyclic.*
5) *Every chief series for G has factor groups that are elementary abelian.*

Proof. We have seen that 1) implies 2) and it is clear that 2) implies 3), that 3) implies 4) and that 4) implies 1). Thus, 1)–4) are equivalent.

It is clear that 5) implies 1). If G is solvable, the factor groups G_{k+1}/G_k of a chief series are minimal normal in the solvable group G/G_k and so are elementary abelian by Theorem 11.15.□

The following theorem contains some sufficient (but not necessary) conditions for solvability. The proof of the Feit–Thompson Theorem is quite involved, running almost 300 pages. For a proof of the Burnside result, we refer the reader to Robinson [26].

Theorem 11.17

1) **(Feit–Thompson Theorem)** *Any group of odd order is solvable; equivalently, every finite nonabelian simple group has even order.*
2) **(Burnside *pq* Theorem)** *Every group of order $p^m q^n$ where p and q are primes, is solvable.*

Proof. The equivalence in part 1) is left as an exercise.□

Hall's Theorem on Solvable Groups

Let G be a finite group. Recall that a **Hall subgroup** H of G is a subgroup with the property that its order $o(H)$ and index $(G:H)$ are relatively prime. The Schur–Zassenhaus Theorem tells us that every *normal* Hall subgroup has a complement and that all such complements are conjugate.

As to the existence of Hall subgroups, the Sylow theorems tell us that if $o(G) = p^k m$ with p prime and $(p^k, m) = 1$, then G has a Hall (Sylow) subgroup of order p^k and that all such subgroups are conjugate. In 1928, Philip

Hall showed that for a finite *solvable* group, this result applies not just to prime power factors.

Theorem 11.18 (Hall's theorem, 1928) *Let G be a finite solvable group with $o(G) = ab$, where $(a,b) = 1$. Then G has a Hall subgroup of order a and all subgroups of order a are conjugate.*
Proof. We may assume that $a, b > 1$. The proof is by induction on $o(G)$. If $o(G) = 1$, the result holds trivially. Assume that it holds for all groups of order less than $o(G)$. If G does not have a minimal normal subgroup, then G is simple and solvable and therefore cyclic of prime order and so $a, b > 1$ is false. Thus, G has a minimal normal subgroup N, which as we have seen, is an elementary abelian group of prime power order p^m. There are cases to consider, based on whether $p^m \mid a$ or $p^m \mid b$.

Case 1: $o(N) = p^m = b$

In this case, N is a normal Hall subgroup of G. Hence, the Schur–Zassenhaus Theorem implies that N has a complement and all such complements are conjugate. But the complements of N are precisely the subgroups of order a.

Case 2: $o(N) = p^m \mid b$ and $p^m < b$

If $p^m \mid b$ but $p^m \neq b$, then $o(G/N) = a(b/p^m) < ab$ and so the inductive hypothesis implies that G/N has a subgroup K/N of order a. Hence, $o(K) = ap^m < ab$ and the inductive hypothesis applied to K shows that K (and hence G) has a subgroup H of order a.

As to conjugation, if $o(H_1) = o(H_2) = a$, then $H_i \cap N = \{1\}$ and so $o(H_iN/N) = a$. Hence, the inductive hypothesis implies that

$$\left(\frac{H_2N}{N}\right)^{xN} = \frac{H_1N}{N}$$

for some $x \in G$ and so

$$H_2^x N = H_1 N$$

But $o(H_1N) = o(G)$ and H_1 and H_2^x are Hall subgroups of H_1N of order a. Hence, the induction hypothesis implies that H_1 and H_2^x are conjugate in H_1N, whence in G.

Case 3: $o(N) = p^m \mid a$

If $o(N) = p^m \mid a$, then $o(G/N) = (a/p^m)b < ab$ and the inductive hypothesis implies that G/N has a subgroup K/N of order a/p^m, whence $o(K) = a$.

As to conjugacy, if $o(H) = a$, then $N \leq H$, for if not, then

$$o(HN) \mid o(H)o(N) = ap^m$$

and so NH is a subgroup of G of order greater than a but relatively prime to b, which contradicts Lagrange's theorem. Therefore, if $o(H_1) = o(H_2) = a$, then H_1/N and H_2/N are Hall subgroups of G/N of order a/p^m and so H_1/N and H_2/N are conjugate in G/N, whence H_1 and H_2 are conjugate in G.□

A sort of converse of the previous theorem also holds. The proof uses the Burnside pq theorem (Theorem 11.16).

Definition *Let G be a finite group and let p be a prime for which $n = p^k m$, where $(m, p) = 1$ and $k \geq 1$. Then a* **Hall p'-subgroup** *of G is a subgroup H of order m.*□

Theorem 11.19 *If a finite group G has a Hall p'-subgroup for every prime p dividing $o(G)$, then G is solvable.*
Proof. Assume that the theorem is false and let G be a counterexample of smallest order. If G is not simple, then let N be a nontrivial proper normal subgroup of G. We leave it as an exercise to show that $N \cap H$ is a Hall p'-subgroup of N and HN/N is a Hall p'-subgroup of G/N. Hence, N and G/N are solvable and therefore so is G, a contradiction. Hence, G is simple.

Now suppose that $o(G) = p_1^{e_1} \cdots p_n^{e_n}$, where the p_i's are distinct primes and $e_i \geq 1$. The Burnside pq theorem implies that $k \geq 3$. If G_i is a Hall p_i'-subgroup of G, then $o(G_i) = o(G)/p_i^{e_i}$ and the Poincaré theorem and the fact that the indices $(G : G_i)$ are pairwise relatively prime imply that for any k of the groups G_i,

$$(G : G_{i_1} \cap \cdots \cap G_{i_k}) = \prod\nolimits_j p_{i_j}^{e_{i_j}}$$

and so

$$|G_{i_1} \cap \cdots \cap G_{i_k}| = \frac{o(G)}{\prod_j p_{i_j}^{e_{i_j}}}$$

Also, for any i,

$$\left|G_i \bullet \bigcap\nolimits_{j \neq i} G_j\right| = |G_i| \left|\bigcap\nolimits_{j \neq i} G_j\right| = \frac{o(G)}{p_i^{e_i}} \frac{o(G)}{\prod_{j \neq i} p_j^{e_j}} = o(G)$$

and so

$$G_i \bullet \bigcap\nolimits_{j \neq i} G_j = G$$

If $H = G_3 \cap \cdots \cap G_n$, then $o(H) = p_1^{e_1} p_2^{e_2}$, which is solvable by the Burnside pq theorem. If H is simple, then H is abelian and so H is cyclic of prime order, which is false. Hence, let N be a minimal normal subgroup of H. Then N is elementary abelian of exponent, say p_1 and so is contained in any Sylow p_1-subgroup of H. But $G_2 \cap H$ has order $p_1^{e_1}$ and so $N \trianglelefteq G_2 \cap H$. Now,

$N^{G_1 \cap H} \leq N$ and so

$$N^G \leq N^{G_2(G_1 \cap H)} \leq N^{G_2} \leq G_2$$

and so the normal closure N^G is a proper nontrivial normal subgroup of G, a contradiction.□

Exercises

1. Show that the classes of cyclic groups, abelian groups and nilpotent groups do not have the extension property.
2. Show that if $A \lhd B$ is abelian or cyclic, then any refinement

$$A \trianglelefteq H \trianglelefteq B$$

is also abelian or cyclic.

Nilpotent Groups

3. Can a nontrivial centerless group be nilpotent?
4. a) Prove that any finite nilpotent group is supersolvable.
   b) Find an example to show that not every finite supersolvable group is nilpotent.
5. Let G be a finite group. Prove that G is nilpotent if and only if every nontrivial quotient group of G has a nontrivial center.
6. Let G be nilpotent but not abelian. Let $A \leq G$ be maximal with respect to being normal in G and abelian. Prove that $A = C_G(A)$. *Hint*: Show that $C/A \trianglelefteq G/A$ and that $C_G(A)/A \cap Z(G/A) \neq \{1\}$.
7. For any even positive integer n, prove that every group of order n is nilpotent if and only if n is a power of 2.
8. Prove that a nilpotent group is supersolvable if and only if it satisfies the ascending chain condition on subgroups.
9. If H is nilpotent of class c and K is nilpotent of class d, prove that $H \boxtimes K$ is nilpotent of class $\max(c, d)$.

Solvable Groups

10. Prove that S_4 is solvable but not supersolvable.
11. Prove that S_n is solvable for $n \leq 4$.
12. Assuming that A_5 is the smallest nonabelian simple group (which it is), prove that every group of order less than 60 is solvable.
13. Let N be a nontrivial proper normal subgroup of a group G. Let p be a prime and $p \mid o(N)$. Let H be a Hall p'-subgroup of G. Prove that $N \cap H$ and NH/N are Hall p'-subgroups of N and G/N, respectively.
14. Prove that the following are equivalent for a finite group G:
    a) G is solvable.
    b) Every nontrivial normal subgroup of G has a nontrivial abelian quotient group.
    c) Every nontrivial quotient group of G has a nontrivial abelian normal subgroup.

15. Let G be a finite group of order $n = p_1^{e_1} \cdots p_m^{e_m}$. Prove that G is solvable if and only if the composition length of G is $c = \sum e_i$.
16. Prove that a solvable group with a composition series must be finite.
17. Let G be a nontrivial finite solvable group.
    a) Prove that $\mathcal{O}_p(G)$ is nontrivial for some prime p, that is, G has a nontrivial normal p-subgroup.
    b) Prove that $\mathcal{O}^q(G)$ is nontrivial for some prime q, that is, G has a normal subgroup H for which G/H is a nontrivial q-group.
18. Let G be a finite group. Prove directly that an abelian series can always be refined into a cyclic series with prime order factor groups.
19. Prove that the following are equivalent:
    a) Any finite group of odd order is solvable.
    b) Any finite nonabelian simple group has even order.
20. a) Prove that a finite group G is solvable if and only if $S' \neq S$ for all subgroups $S \neq \{1\}$ of G.
    b) Prove that if G contains a nonabelian simple subgroup S, then G is not solvable.
    c) Show that $S' \neq S$ for all subgroups of the dihedral group D_{2n}, showing that D_{2n} is solvable. *Hint*: Find an abelian subgroup N of index 2. How do subgroups interact with N?
21. A subgroup H of a group G is **abnormal** if

$$a \in \langle H, H^a \rangle$$

for all $a \in G$. Prove that the normalizer of a Hall subgroup of a solvable group is abnormal.

Polycyclic Groups

22. a) Let G be a polycyclic group with a cyclic series of length n. Prove that G is n-generated.
    b) Let A be an n-generated abelian group for $n \geq 1$. Prove that A is polycyclic.
23. Prove that the following are equivalent:
    a) G is polycyclic.
    b) Every subgroup of G is finitely generated and solvable.
    c) Every normal subgroup of G is finitely generated and solvable.
24. Prove that a group G is polycyclic if and only if it is solvable and satisfies the maximal condition on subgroups, that is, if and only if every nonempty collection of subgroups of G as a maximal member.
25. a) Let $A < B$ have an infinite cyclic factor group. Let

$$A = H_0 < H_1 < \cdots < H_m = B$$

be a proper refinement of $A < B$. Describe the factor groups of this refinement.

b) Let G be polycyclic. Prove that the number of steps whose factor group is infinite is the same for all cyclic series for G. *Hint*: Any two cyclic series have isomorphic refinements.

Supersolvable Groups

26. Prove that any supersolvable group is countable.
27. Prove that if G/N is supersolvable and N is cyclic, then G is supersolvable.
28. Prove that a group is supersolvable if and only if it has a series in which each factor group is cyclic of prime order or cyclic of infinite order. *Hint*: Recall that $A \sqsubseteq B$ and $B \trianglelefteq G$ implies $A \trianglelefteq G$.
29. Let G be supersolvable. Prove that if H is a maximal subgroup of G, then $(G : H)$ is prime. *Hint*: First assume that $H \lhd G$ and look at G/H. Then assume that H is not normal in G and factor by the normal interior H°. Conclude that it is sufficient to consider $H^\circ = \{1\}$. Consider the subgroup A in the first step $\{1\} < A$ in a normal cyclic series and how it interacts with H.
30. Prove that if G is supersolvable, then G' is nilpotent. *Hint*: Let

$$\{1\} = G_0 < \cdots < G_n = G$$

be a normal cyclic series for G and consider the series

$$\{1\} = G_0 \cap G' < \cdots < G_n \cap G' = G'$$

which is normal and cyclic as well. Let $B = G_{k+1}$ and $A = G_k$. To show that the series is central, it is sufficient to show that

$$\frac{X}{A} := \frac{A(B \cap G')}{A} \leq Z\left(\frac{AG'}{A}\right)$$

But X/A is cyclic and normal in G/A. What can be said about $(G/A)'$?

Radicals and Residues

Definition *Let $\mathcal{K}$ be a class of groups. Let G be a group.*

1) If the partially ordered set

$$\mathcal{K}(G) = \{H \trianglelefteq G \mid H \in \mathcal{K}\}$$

has a top element, it is called the **$\mathcal{K}$-radical** *for G and is denoted by $\mathcal{O}_{\mathcal{K}}(G)$.*

2) If the partially ordered set

$$G/\mathcal{K} = \{H \trianglelefteq G \mid G/H \in \mathcal{K}\}$$

has a bottom element, it is called the **$\mathcal{K}$-residue** *for G and is denoted by $\mathcal{O}^{\mathcal{K}}(G)$.* □

31. Show that the $\mathcal{K}$-radical $\mathcal{O}_{\mathcal{K}}(G)$ and the $\mathcal{K}$-residue $\mathcal{O}^{\mathcal{K}}(G)$ are characteristic subgroups of G (if they exist).
32. Let $\mathcal{K}$ be a class of groups closed under subgroup, quotient and join if at least one factor is normal. Let G be a group and let $H \trianglelefteq G$. Assume that all mentioned radicals and residues exist. Prove the following:
    a) If $H \le \mathcal{O}_{\mathcal{K}}(G)$, then
    $$\frac{\mathcal{O}_{\mathcal{K}}(G)}{H} \le \mathcal{O}_{\mathcal{K}}\left(\frac{G}{H}\right)$$
    and the inclusion may be proper.
    b) If $H \le \mathcal{O}^{\mathcal{K}}(G)$, then
    $$\mathcal{O}^{\mathcal{K}}\left(\frac{G}{H}\right) = \frac{\mathcal{O}^{\mathcal{K}}(G)}{H}$$
33. Let $\mathcal{K}$ be a class with the extension property: $N \in \mathcal{K}, G/N \in K$ implies $G \in \mathcal{K}$. Prove that the following hold for any group G:
    a) The $\mathcal{K}$-radical of $G/\mathcal{O}_{\mathcal{K}}(G)$ is trivial, that is,
    $$\mathcal{O}_{\mathcal{K}}\left(\frac{G}{\mathcal{O}_{\mathcal{K}}(G)}\right) = \{\mathcal{O}_{\mathcal{K}}(G)\}$$
    b) The $\mathcal{K}$-residue of $\mathcal{O}^{\mathcal{K}}(G)$ is $\mathcal{O}^{\mathcal{K}}(G)$, that is,
    $$\mathcal{O}^{\mathcal{K}}(\mathcal{O}^{\mathcal{K}}(G)) = \mathcal{O}^{\mathcal{K}}(G)$$
34. Let $\mathcal{K}$ be the class of finite groups.
    a) Show that there are groups with no $\mathcal{K}$-radical.
    b) Show that there are groups with no $\mathcal{K}$-residue.
35. Let G be a nontrivial finite group. Prove that the following are equivalent:
    a) G is solvable.
    b) For every proper normal subgroup $K \triangleleft G$, the factor group G/K has a nontrivial p-radical $\mathcal{O}_p(G/K)$ for some prime p, that is, G/K has a nontrivial normal p-subgroup.
    c) For every nontrivial characteristic subgroup K of G, the q-residue $\mathcal{O}^q(K)$ of K is proper in K for some prime q, that is, there is a proper normal subgroup A of K such that K/A is a nontrivial a q-group.

Additional Problems

36. Let G be a group. Let $\Gamma = \Gamma_G$.
    a) Prove that $G^{(k)} \le \Gamma_G^k(G)$.
    b) Prove that $\mathrm{derlen}(G) \le \mathrm{nilclass}(G)$.
37. Let G be a group. Let $\Gamma = \Gamma_G$. Prove the following:
    a) $[\Gamma^k(G), \Gamma^j(G)] \le \Gamma^{k+j+1}(G)$ *Hint*: Use induction. Use the three subgroups lemma on $[\Gamma^{k+1}(G), \Gamma^{j+1}(G)] = [[\Gamma^k(G), G], \Gamma^{j+1}(G)]$.
    b) $\Gamma^k_{\Gamma^j(G)}(G) \le \Gamma^{kj+1}(G)$ for $k, j \ge 1$.

c) $[\Gamma^k(G), \zeta^j(G)] \leq \zeta^{j-k-1}(G)$ for $j \geq k+1$.
d) $G^{(k)} \leq \Gamma^{\ell_k}(G)$, where $\ell_k = 2^k - 1$. Hence, $\operatorname{derlen}(G)$ is less than or equal to the smallest integer k for which $\ell_k \geq c := \operatorname{nilclass}(G)$ and so

$$\operatorname{derlen}(G) \leq \lceil \log_2(c+1) \rceil$$

Chapter 12
Free Groups and Presentations

Throughout this chapter, X denotes a nonempty set of formal symbols and X^{-1} denotes the set of formal symbols $\{x^{-1} \mid x \in X\}$. Further, we assume that X and X^{-1} are disjoint and write $X' = X \sqcup X^{-1}$.

Free Groups

The idea of a free group F_X on a nonempty set X is that F_X should be the "most general" possible group containing X, that is, the elements of X should generate F_X but have no relationships within F_X. In this case, X is referred to as a set of *free generators* or a *basis* for the free group F_X.

To draw an analogy, if V is a vector space, then a subset $\mathcal{B}$ of V is a basis for V if and only if for any vector space W and any assignment of vectors in W to the vectors in $\mathcal{B}$, there is a unique linear transformation from V to W that extends this assignment. This property and the analogous property that defines free groups are best described using univerality.

Definition *Let X be a nonempty set. A pair $(F, \kappa: X \to F)$ where F is a group has the* **universal mapping property** *for X (or is* **universal** *for X) if, as pictured in Figure 12.1, for any function $f: X \to G$ from X to a group G, there is a unique group homomorphism $\tau_f: F \to G$ for which*

$$\tau_f \circ \kappa = f$$

The map τ_f is called the **mediating morphism** *for f. In this case, we say that f can be* **factored through** *κ or that f can be* **lifted** *to F. The group F is called a* **free group** *on X and X is called a set of* **free generators** *or a* **basis** *for F. The map κ is called the* **universal map** *for the pair (F, κ). We use the notation F_X to denote a free group on X.*□

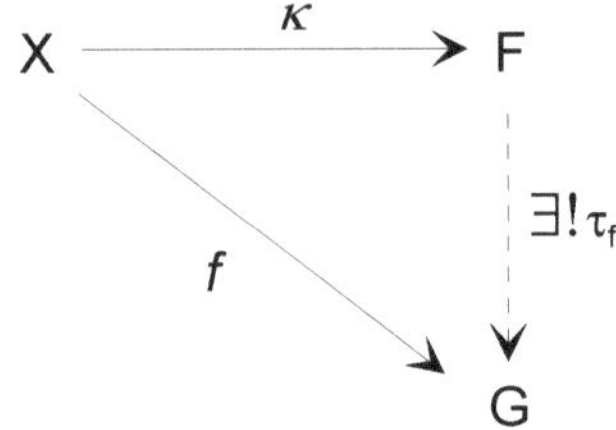

Figure 12.1

It is clear that the universal map κ is injective. Moreover, κX generates F_X, for if $\langle \kappa X \rangle < F_X$, then Theorem 4.17 implies that there are distinct group homomorphisms $\sigma, \tau: F_X \to G$ into some group G that agree on $\langle \kappa X \rangle$. Therefore, if $f = \sigma|_X = \tau|_X$, then the uniqueness condition of mediating morphisms is violated. Hence, κX generates F. For these reasons, it is common to suppress the map κ and think of X as a subset of F_X.

The following theorem says that the universal property characterizes groups up to isomorphism.

Theorem 12.1 *Let X be a nonempty set. If (F, κ) and (G, λ) are universal for X, then there is an isomorphism $\sigma: F \approx G$ connecting the universal maps, that is, for which*

$$\sigma \circ \kappa = \lambda$$

Proof. There are unique mediating morphisms $\tau_\lambda: F \to G$ and $\tau_\kappa: G \to F$ for which

$$\tau_\lambda \circ \kappa = \lambda \quad \text{and} \quad \tau_\kappa \circ \lambda = \kappa$$

and so

$$\tau_\lambda \circ \tau_\kappa \circ \lambda = \lambda \quad \text{and} \quad \tau_\kappa \circ \tau_\lambda \circ \kappa = \kappa$$

But the identity maps ι_G and ι_F are the *unique* mediating morphisms for which

$$\iota_G \circ \lambda = \lambda \quad \text{and} \quad \iota_F \circ \kappa = \kappa$$

and so

$$\tau_\lambda \circ \tau_\kappa = \iota_G \quad \text{and} \quad \tau_\kappa \circ \tau_\lambda = \iota_F$$

which shows that the maps τ_λ and τ_κ are inverse isomorphisms.□

Definition *Let X be a nonempty set. A word $w(x_1, \ldots, x_n)$ over X' (or the equation $w(x_1, \ldots, x_n) = 1$) is a* **law** *of groups if*

$$w(a_1, \ldots, a_n) = 1$$

for all groups G and all $a_i \in G$.□

The following theorem says that what's true in a free group is true in all groups.

Theorem 12.2 *Let X be a nonempty set and let $w(x_1,\ldots,x_n)$ be a word over X'. If (K_X,κ) is universal on X, then the following are equivalent:*
1) $w(\kappa x_1,\ldots,\kappa x_n)=1$ *in* F_X
2) $w(x_1,\ldots,x_n)$ *is a law of groups.*□

Proof. Let G be a group and let $a_i \in G$ for $i=1,\ldots,n$. The function sending x_i to a_i can be lifted to a unique homomorphism $\sigma\colon F_X \to G$ for which $\sigma\kappa x_i = a_i$ and so

$$w(a_1,\ldots,a_n)=\sigma w(\kappa x_1,\ldots,\kappa x_n)=\sigma 1 = 1$$
□

Cauchy's theorem says that any group is isomorphic to a *subgroup* of a symmetric group. There is an analog for *quotients* of free groups, but first we require a definition.

Definition *Let $\mathcal{F}=\{G_i \mid i\in I\}$ be a family of groups. A subgroup K of the direct product $\boxtimes\mathcal{F}$ is called a* **subdirect product** *of the family $\mathcal{F}$ if the restricted projection maps $\rho_i|_K\colon K \to G_i$ are surjective for all $i \in I$.*□

If G is a group, then the identity map $\iota\colon G \to G$ can be lifted to an epimorphism $\sigma\colon F_G \twoheadrightarrow G$ and so G is isomorphic to a quotient of the free group F_G. More generally, if X is a set for which $\operatorname{card}(X) \geq \operatorname{card}(G)$, then any surjection $f\colon X \twoheadrightarrow G$ can be lifted to an epimorphism $\sigma\colon F_X \twoheadrightarrow G$ and the induced map $\overline{\sigma}\colon F_X/K \approx G$ shows that G is isomorphic to a quotient group of the free group F_X. Moreover, we have freedom to choose the values of $\tau(xK)$ for $x \in X$ arbitrarily, but K depends on that choice.

More generally, if $\mathcal{F}=\{G_i \mid i\in I\}$ is a nonempty family of groups and if X is a set for which $\operatorname{card}(X)\geq \operatorname{card}(G_i)$ for all $i \in I$, then there are isomorphisms $\tau_i\colon F_X/K_i \approx G_i$ for all $i\in I$, where $\tau_i(xK_i)$ can be specified arbitrarily, but K_i depends on that choice. Now, the “Chinese” map

$$\sigma\colon F_X \to \boxtimes_{i\in I} F_X/K_i$$

defined by $\sigma(w)(i)=wK_i$ for $w\in F_X$ has kernel $I=\bigcap K_i$. Hence, if $\overline{\sigma}$ is the induced embedding, then the composition

$$\boxtimes\tau_i \circ \overline{\sigma}\colon F_X/I \hookrightarrow \boxtimes_{i\in I} G_i$$

shows that F_X/I is isomorphic to a subdirect product of the family $\mathcal{F}$. Moreover, we can specify that the element xI be sent to any element of $\boxtimes G_i$, for all $x\in X$ (again at the expense of K_i).

Theorem 12.3 *Let G be a group, let $\mathcal{F}=\{G_i \mid i\in I\}$ be a nonempty family of groups and let X be a set.*

1) *If* card$(X) \geq$ card(G), *then there is an isomorphism* $\tau: F_X/N \approx G$ *where we can choose the elements* τx *for* $x \in X$ *arbitrarily, but* N *depends on that choice.*
2) *Suppose that* card$(X) \geq$ card(G_i) *and that we have specified isomorphisms* $\tau_i: G_i \approx F_X/K_i$ *for all* $i \in I$. *Let* $I = \bigcap K_i$. *Then* F_X/I *is isomorphic to a subdirect product of the family* $\mathcal{F}$ *and the isomorphisms* τ_i *can be used to specify the elements* xI *for* $x \in X$ *arbitrarily in* $\boxtimes G_i$, *but* I *depends on that choice.* □

Construction of the Free Group

The notion of a free group can be defined constructively, without appeal to the universal mapping property. When a constructive approach is taken, one usually hastens to verify the universal mapping property, since this is arguably the most useful property of free groups. On the other hand, since we have chosen to define free groups via universality, we should hasten to give a construction for free groups.

Let $W = (X')^*$ be the set of all words over the alphabet X'. As a shorthand, we allow the use of exponents, writing

$$x^n = \begin{cases} \underbrace{x\cdots x}_{n \text{ factors}} & \text{if } n > 0 \\ \underbrace{x^{-1}\cdots x^{-1}}_{-n \text{ factors}} & \text{if } n < 0 \\ \epsilon & \text{if } n = 0 \end{cases}$$

where ϵ is the empty word. It is important to keep in mind that this is only a shorthand *notation*. Thus, for example, x^4x^2 and x^6 are both shorthand for $xxxxxx$ and so $x^4x^2 = x^6$. However, x^4x^{-2} is shorthand for $xxxxx^{-1}x^{-1}$ but x^2 is shorthand for xx and so $x^4x^{-2} \neq x^2$.

Since the operation of juxtaposition on W is associative and since the empty word ϵ is the identity, the set W is a monoid under juxatpostion. In an effort to form a group, we also want to require that $xx^{-1} = \epsilon = x^{-1}x$ for all $x \in X$. More specifically, consider the following rules that can be applied to members of W:

1) **Removal rules**: For $\omega, \mu \in W$ and $x \in X$,

$$\begin{aligned} \omega x x^{-1} \mu &\to \omega\mu \\ \omega x^{-1} x \mu &\to \omega\mu \\ x x^{-1} &\to \epsilon \\ x^{-1} x &\to \epsilon \end{aligned}$$

where one of ω or μ may be missing.

2) **Insertion rules**: For $\omega, \mu \in W$ and $x \in X$,

$$\begin{aligned}\omega\mu &\longrightarrow \omega x x^{-1}\mu \\ \omega\mu &\longrightarrow \omega x^{-1} x\mu \\ \epsilon &\longrightarrow x x^{-1} \\ \epsilon &\longrightarrow x^{-1} x\end{aligned}$$

where one of ω or μ may be missing.

Let us refer to a finite sequence $s_1, \ldots, s_k$ of applications of these rules as a **reduction** of u to v of **length** k (even though v may have greater length than u). The **trivial reduction** is an application of no rules and so u is obtained from itself by the trivial reduction. Since the removal and insertion rules come in inverse pairs, reduction defines an equivalence relation $\equiv$ on W. Let $W/\equiv$ denote the set of equivalence classes of W, with $[w]$ denoting the equivalence class containing w.

Since equivalent words must represent the same group element, it is really the equivalence classes that are the candidates for the elements of the free group F_X on X. Moreover, since

$$u \equiv r, \quad v \equiv s \quad \Rightarrow \quad uv \equiv rs$$

the equivalence relation $\equiv$ is a monoid *congruence relation* on W and so we may raise the operation of juxtaposition from W to $W/\equiv$, that is, the operation

$$[u][v] = [uv]$$

is well-defined on $W/\equiv$ and makes $W/\equiv$ into a group, with identity $[\epsilon]$ and for which

$$[x_1^{e_1} \cdots x_k^{e_k}]^{-1} = [x_k^{-e_k} \cdots x_1^{-e_1}]$$

However, since it is easier to work with elements of W rather than equivalence classes, we prefer to use a system of distinct representatives for $W/\equiv$. A desirable choice would be the set consisting of the *unique* word of shortest length from each equivalence class, and so we must prove that such words exist.

Let us say that a word w is **reduced** if it is not congruent to a word of shorter length. It is clear that a removal rule can be applied to a word $w \in W$ if and only if w contains a subword of the form xx^{-1} or $x^{-1}x$ for $x \in X$. Further, a reduced word w has no such subword and so no removal rules can be applied to w. We want to prove that the converse also holds, that is, a word w is reduced if and only if no removal rules can be applied to w. Then we can show that the set of reduced words form a system of distinct representatives for $W/\equiv$.

Theorem 12.4 *Let $r \in W$ be a word that does not contain a subword of the form xx^{-1} or $x^{-1}x$ for $x \in X$. If $w \equiv r$, then there is a reduction from w to r that involves only removal rules.*

1) *A word is reduced if and only if it does not contain a subword of the form xx^{-1} or $x^{-1}x$ for $x \in X$.*
2) *A word is reduced if and only if it can be written in the form $x_1^{e_1}\cdots x_n^{e_n}$ with $x_i \neq x_{i+1}$ and $e_i \neq 0$ for all i.*
3) *The set of reduced words is a system of distinct representatives for $W/\equiv$.*

Proof. Among all reductions from w to r, select a reduction with the fewest number of steps and suppose that there is at least one insertion step. Denote the steps by $s_1, s_2, \ldots, s_m$ and suppose that step s_k results in the word u_k. Let s_k be the last insertion step, say

$$\begin{aligned} u_{k-1} &= \alpha\beta \\ u_k &= \alpha(\overline{xx}^{-1})\beta \end{aligned}$$

where we have marked x with an overbar to distinguish it from any other occurrences of the symbol x. (A similar argument will work for the insertion of $x^{-1}x$.)

Since there are no further insertion steps, the pair $\overline{xx}^{-1}$ is never separated during the remaining steps, but must be altered at some point by a subsequent removal rule. Suppose that $\overline{xx}^{-1}$ is unaltered until step s_{k+j} and so the intermediate steps are

$$\begin{aligned} u_{k-1} &= \alpha\beta \\ u_k &= \alpha(\overline{xx}^{-1})\beta \\ u_{k+1} &= \alpha_1(\overline{xx}^{-1})\beta_1 \\ &\vdots \\ u_{k+j-1} &= \alpha_{j-1}(\overline{xx}^{-1})\beta_{j-1} \end{aligned}$$

There are three possibilities for step s_{k+j}. First, if $\overline{xx}^{-1}$ is removed, that is, if

$$u_{k+j} = \alpha_{j-1}\beta_{j-1}$$

then the insertion (and subsequent deletion) of $\overline{xx}^{-1}$ could have been omitted from the reduction process, in which case steps s_k and s_{k+j} do nothing can be removed from the reduction, resulting in a shorter reduction, which is a contradiction.

The other possibilities involve an interaction of $\overline{xx}^{-1}$ with either α_{j-1} or β_{j-1}. One possibility is that $\alpha_{j-1} = \alpha'_{j-1}x^{-1}$ and

$$\begin{aligned} u_{k+j-1} &= \alpha'_{j-1}x^{-1}(\overline{xx}^{-1})\beta_{j-1} \\ u_{k+j} &= \alpha'_{j-1}(\overline{x}^{-1})\beta_{j-1} \end{aligned}$$

But since $u_{k+j} = \alpha_j\beta_j$, step s_{k+j} can be replaced by the removal of $\overline{xx}^{-1}$, resulting in the same reduction as in the first case. Similarly, if $\beta_{j-1} = x\beta'_{j-1}$ and

$$\begin{aligned} u_{k+j-1} &= \alpha_{j-1}(\overline{x x}^{-1})x\beta'_{j-1} \\ u_{k+j} &= \alpha_{j-1}(\overline{x})\beta'_{j-1} \end{aligned}$$

then since $u_{k+j} = \alpha_{j-1}\beta_{j-1}$, again we can replace this reduction by the first reduction. Hence, a shortest reduction of w to r cannot have any insertion steps.

For part 1), suppose that no removal rules can be applied to $r \in W$ and that $w \equiv r$. Then there is a reduction from w to r that involves only removal steps and so $\text{len}(r) \leq \text{len}(w)$. Hence, r is reduced. The converse is clear.

For part 2), if $w = x_1^{e_1} \cdots x_n^{e_n}$ is reduced but $x_i = x_{i+1}$ where e_i and e_{i+1} have opposite signs, then we can apply a removal rule to w to produce a shorter equivalent word, which is not possible. Thus, if $x_i = x_{i+1}$, we may add the corresponding exponents. Conversely, if $w = x_1^{e_1} \cdots x_n^{e_n}$ is a word for which $x_i \neq x_{i+1}$ and $e_i \neq 0$, then no removal rule can be applied to w and so part 1) implies that w is reduced.

For part 3), any $w \in W$ can be reduced using only removal rules to a word r for which no removal rules apply and so r is reduced. Thus, every word is congruent to a reduced word. Moreover, if $u \neq v$ are congruent reduced words, then there must be a reduction consisting of zero or more removal steps that brings u to v, but no removal steps can be applied to u and so $u = v$.□

Thus, each word $w \in W$ is equivalent to a *unique* reduced word ω^r and we may use the bijection $[w] \leftrightarrow w^r$ to transfer the group structure from $W/\equiv$ to the set F_X of reduced words, specifically, if $u, v \in F_X$, then

$$uv = (uv)^r$$

It will be convenient to refer to the set R_X of reduced words on X' by the following name (which is not standard terminology).

Definition *Let X be a nonempty set. The* **concrete free group** *R_X on X is the set*

$$R_X = \{\epsilon\} \cup \{x_1^{e_1} \cdots x_n^{e_n} \mid x_i \neq x_{i+1} \in X, e_i \neq 0\}$$

of all reduced words over the alphabet X', under the operation of juxtaposition followed by reduction. The **rank** $\text{rk}(R_X)$ *of R_X is the cardinality of X.*□

Note that the use of the term "concrete" is not standard. Most authors would refer to R_X simply as the free group on X, which is justified by the following.

Theorem 12.5 *Let F_X be the concrete free group on a set X and let $j\colon X \to F_X$ be the inclusion map. Then the pair (F_X, j) is universal for X and so F_X is a free group on X.*
Proof. Let $f\colon X \to G$. If $\tau\colon F_X \to G$ is defined by $\tau\epsilon = 1$ and

$$\tau(x_1^{e_1}\cdots x_n^{e_n}) = f(x_1)^{e_1}\cdots f(x_n)^{e_n}$$

for $x_1^{e_1}\cdots x_n^{e_n} \in F_X$, then $\tau \circ j = f$ on X. Also, it is clear that reduction can take place before or after application of τ without affecting the final result. However, reduction has no effect in G and so if $u * v$ denotes the operation of F_X, then

$$\tau(u * v) = \tau((uv)^r) = [\tau(uv)]^r = \tau(uv) = \tau(u)\tau(v)$$

which shows that τ is a mediating morphism for f. As to uniqueness, if $\tau' \circ j = \tau \circ j$, then $\tau'$and τ agree on X, which generates F_X and so $\tau' = \tau$.□

Relatively Free Groups

Freedom can also come in the context of some restrictions. For example, the free *abelian* group on a set X is the most "universal" abelian group generated by X. More generally, if $\mathcal{K}$ is any class of groups, we can ask if there is a most universal $\mathcal{K}$-group generated by a nonempty set X. If so, such a group is referred to as a *free $\mathcal{K}$-group* or a *relatively free group*. For the class $\mathcal{K}$ of abelian groups, free $\mathcal{K}$-groups do exist, but this is not the case for all classes of groups, as we will see.

The definition of a free $\mathcal{K}$-group generalizes that of a free group.

Definition *Let $\mathcal{K}$ be a class of groups and let X be a nonempty set. A pair $(K_X, \kappa\colon X \to K_X)$ where K_X is a $\mathcal{K}$-group generated by κX, has the* **$\mathcal{K}$-universal property mapping** *for X (or is* **$\mathcal{K}$-universal** *for X) if for any function $f\colon X \to G$ from X to a $\mathcal{K}$-group G, there is a unique group homomorphism $\tau_f\colon K_X \to G$ for which*

$$\tau_f \circ j = f$$

The map τ_f is called the **mediating morphism** *for f. The group K_X is called a* **free $\mathcal{K}$-group** *on X with* **free generators** *or* **basis** *X and κ is called the $\mathcal{K}$-* **universal map** *for the pair (K_X, κ).*□

Note that if (K_X, κ) is $\mathcal{K}$-universal, then the $\mathcal{K}$-universal map κ is injective. For this reason, one often thinks of X as a subset of K_X. Proof of the following theorems is left to the reader.

Theorem 12.6 *Let X be a nonempty set. If (K, κ) and (G, λ) are $\mathcal{K}$-universal for X, then there is an isomorphism $\sigma\colon K \approx G$ for which*

$$\sigma \circ \kappa = \lambda$$

□

Definition *Let $\mathcal{K}$ be a class of groups and let X be a nonempty set. A word $w(x_1,\dots,x_n)$ over X' (or the equation $w(x_1,\dots,x_n)=1$) is a* **law** *of $\mathcal{K}$-groups if*

$$w(a_1,\dots,a_n)=1$$

for all $\mathcal{K}$-groups K and all $a_i\in K$. If $\{x_1,x_2,\dots\}\subseteq X$, then we denote the set of all laws of $\mathcal{K}$-groups in F_X by $\mathcal{L}(\mathcal{K})$ or $\mathcal{L}_X(\mathcal{K})$.□

Theorem 12.7 *Let $\mathcal{K}$ be a class of groups and let X be an infinite set. Then $\mathcal{L}_X(\mathcal{K})$ is a fully invariant subgroup of the free group F_X.*
Proof. It is clear that the product of two laws of $\mathcal{K}$-groups is a law of $\mathcal{K}$-groups, as is the inverse of a law of $\mathcal{K}$-groups. Also, if $\sigma\in\operatorname{End}(F_X)$, then for any $w(x_1,\dots,x_n)\in\mathcal{L}(\mathcal{K})$,

$$\sigma w(x_1,\dots,x_n)=w(\sigma x_1,\dots,\sigma x_n)\in\mathcal{L}(\mathcal{K})$$

and so $\mathcal{L}(\mathcal{K})$ is fully invariant in F_X.□

Theorem 12.8 *Let X be a nonempty set and let $w(x_1,\dots,x_n)$ be a word over X'. If (K_X,κ) is $\mathcal{K}$-universal on X, then the following are equivalent:*
1) $w(\kappa x_1,\dots,\kappa x_n)=1$ in K_X
2) $w(x_1,\dots,x_n)$ is a law of $\mathcal{K}$-groups.□

We have said that there are classes of groups for which relatively free groups do not exist. For example, the class of all finite groups is such a class, as we will see later. There is one very important type of class for which relatively free groups do exist, however.

Definition *Let X be a nonempty set and let $\mathcal{L}=\{w_i\mid i\in I\}$ be a subset of the concrete free group F_X. The* **equational class** *(or* **variety***) with* **laws** *$\mathcal{L}$ is the class $\mathcal{E}(\mathcal{L})$ of all groups G for which each $w_i\in\mathcal{L}$ is identically 1 on G.*□

Note that an equational class is closed under subgroup, quotient and direct product. For equational classes, we can construct relatively free groups using our construction of the concrete free group.

Theorem 12.9 *Let $\mathcal{K}=\mathcal{E}(\mathcal{L})$ be an equational class of groups with laws $\mathcal{L}$.*
1) For any group G, the **verbal subgroup**

$$\mathcal{L}(G)=\langle w(a_1,\dots,a_n)\mid w\in\mathcal{L},a_i\in G\rangle$$

is fully invariant in G.
2) If F_X is the concrete free group on X, then the pair

$$(K_X=F_X/\mathcal{L}(F_X),\kappa=\pi\circ j)$$

where π is projection modulo $\mathcal{L}(F_X)$ and $j\colon X \to F_X$ is inclusion, is $\mathcal{K}$-universal for X. We refer to $\mathcal{K}_X$ as the **concrete $\mathcal{K}$-free group**.

Proof. Write $F = F_X$. For part 1), if $\sigma \in \operatorname{End}(G)$, then for any $a_i \in G$,

$$\sigma w(a_1, \ldots, a_n) = w(\sigma a_1, \ldots, \sigma a_n) \in \mathcal{L}(G)$$

and so $\mathcal{L}(G)$ is fully invariant in G. For part 2), to see that $K = K_X$ is a $\mathcal{K}$-group, if $w(x_1, \ldots, x_n) \in \mathcal{L}$, then for any $a_i \in F$,

$$w(a_1\mathcal{L}(F), \ldots, a_n\mathcal{L}(F)) = w(a_1, \ldots, a_n)\mathcal{L}(F) = \mathcal{L}(F)$$

and so K_X satisfies the laws in $\mathcal{L}$.

To see that $(\mathcal{K}_X, \pi \circ j)$ is $\mathcal{K}$-universal, referring to Figure 12.2, let $f\colon X \to H$, where H is a $\mathcal{K}$-group.

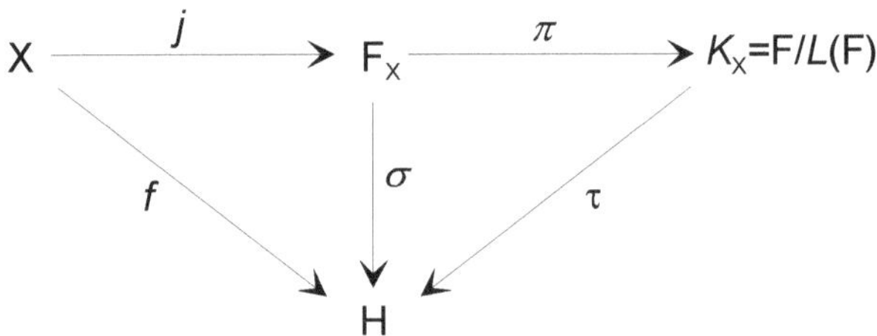

Figure 12.2

Then f can be lifted uniquely to a homomorphism $\sigma\colon F \to H$ satisfying $\sigma \circ j = f$. But if $w(x_1, \ldots, x_n) \in \mathcal{L}$ and $u_i \in F_X$, then

$$\sigma w(u_1, \ldots, u_n) = w(\sigma u_1, \ldots, \sigma u_n) = 1$$

Hence, $w(u_1, \ldots, u_n) \in \ker(\sigma)$ and so $\mathcal{L}(F) \leq \ker(\sigma)$. Thus, the universality of quotients implies that σ can be lifted uniquely to a homomorphism $\tau\colon F/\mathcal{L}(F) \to H$ for which $\tau \circ \pi = \sigma$ and so

$$\tau \circ \pi \circ j = \sigma \circ j = f$$

As to uniqueness, if

$$\tau' \circ \pi \circ j = \tau'' \circ \pi \circ j = f$$

then the uniqueness of σ implies that $\tau' \circ \pi = \tau'' \circ \pi$ and the uniqueness of τ implies that $\tau' = \tau''$.□

More on Equational Classes

If $\mathcal{K}$ is a class of groups, then the equational class $\mathcal{E}(\mathcal{L}(\mathcal{K}))$ is the class of all groups that satisfy the laws of $\mathcal{K}$-groups. Of course, a $\mathcal{K}$-group satisfies the laws of $\mathcal{K}$-groups and so $\mathcal{K} \subseteq \mathcal{E}(\mathcal{L}(\mathcal{K}))$. The interesting issue is that of equality, that is, for which classes $\mathcal{K}$ is it true that the laws of $\mathcal{K}$-groups hold *only* for $\mathcal{K}$-groups? The answer is that this happens if and only if $\mathcal{K}$ is an equational class.

For if $\mathcal{K} = \mathcal{E}(\mathcal{M})$ is the equational class for a set $\mathcal{M}$ of laws over X', then $\mathcal{M} \subseteq \mathcal{L}(\mathcal{K})$ and so

$$\mathcal{E}(\mathcal{L}(\mathcal{K})) \subseteq \mathcal{E}(\mathcal{M}) = \mathcal{K} \subseteq \mathcal{E}(\mathcal{L}(\mathcal{K}))$$

whence $\mathcal{E}(\mathcal{L}(\mathcal{K})) = \mathcal{K}$.

Theorem 12.10 *A class $\mathcal{K}$ of groups satisfies*

$$\mathcal{K} = \mathcal{E}(\mathcal{L}(\mathcal{K}))$$

that is, the laws of $\mathcal{K}$-groups hold only for $\mathcal{K}$-groups, if and only if $\mathcal{K}$ is an equational class.□

Example 12.11 Let $\mathcal{K}$ be the class of all finitely-generated groups. If $w(x_1, \ldots, x_n)$ is a law of $\mathcal{K}$, then $w(x_1, \ldots, x_n) = 1$ in all finitely-generated groups and therefore in all groups. Hence, $\mathcal{E}(\mathcal{L}(\mathcal{K}))$ is the class of all groups and so $\mathcal{K}$ is not an equational class.□

The following theorem makes it relatively easy to tell when a class $\mathcal{K}$ is an equational class.

Theorem 12.12 (Birkhoff) *The following are equivalent for a class $\mathcal{K}$ of groups:*
1) $\mathcal{K}$ is an equational class
2) $\mathcal{K}$ is closed under subgroup, quotient and direct product
3) $\mathcal{K}$ is closed under quotient and subdirect product.
Proof. It is clear that 1) implies 2), which implies 3). To show that 3) implies 1), we show that $\mathcal{E}(\mathcal{L}(\mathcal{K})) \subseteq \mathcal{K}$. Let $G \in \mathcal{E}(\mathcal{L}(\mathcal{K}))$, that is, G satisfies the laws of $\mathcal{K}$-groups. For the proof, we work with quotients of free groups.

Now, if $\text{card}(Y) \geq \text{card}(G)$, there is an $N \trianglelefteq G$ for which $G \approx F_Y/N$ and so F_Y/N satisfies the laws of $\mathcal{K}$-groups. Hence, if $w(y_1, \ldots, y_n) \in \mathcal{L}_Y(\mathcal{K})$, then $w(y_1, \ldots, y_n) \in N$ and so $\mathcal{L}_Y(\mathcal{K}) \leq N$. Hence,

$$G \approx \frac{F_Y}{N} \approx \frac{F_Y}{\mathcal{L}_Y(\mathcal{K})} \bigg/ \frac{N}{\mathcal{L}_Y(\mathcal{K})}$$

and the proof would be complete if we knew that $F_Y/\mathcal{L}_Y(\mathcal{K})$ was a $\mathcal{K}$-group. In fact, the same argument works for any $I \leq \mathcal{L}_Y(\mathcal{K})$, since then $I \leq N$ and so

$$G \approx \frac{F_Y}{N} \approx \frac{F_Y}{I} \bigg/ \frac{N}{I}$$

Thus, we only need to find an $I \leq \mathcal{L}_Y(\mathcal{K})$ for which F_Y/I is a $\mathcal{K}$-group.

Now, if $w \in F_Y$ is *not* a law of $\mathcal{K}$-groups, then there is a $\mathcal{K}$-group K_w that violates w, that is, for which

$$w(k_{w,1},\ldots,k_{w,n_w})\neq 1$$

for some $k_{w,i}\in K_w$. Moreover, if $\mathrm{card}(Y)\geq\mathrm{card}(K_w)$, then there is an isomorphism $\tau_w\colon F_Y/N_w\approx K_w$ for which $\tau_w(y_iN_w)=k_{w,i}$ for all i. Hence, F_Y/N_w also violates w in the sense that

$$w(y_1,\ldots,y_{n_w})\notin N_w$$

and so

$$I:=\bigcap_{w\notin\mathcal{L}(\mathcal{K})}N_w\leq\mathcal{L}_Y(\mathcal{K})$$

But Theorem 12.3 implies that the quotient F_Y/I is isomorphic to a subdirect product of $\mathcal{F}$ and is therefore a $\mathcal{K}$-group.□

We can now relate equational classes and existence of relatively free groups.

Theorem 12.13 (Birkhoff) *The following are equivalent for a nontrivial class $\mathcal{K}$ of groups:*
1) $\mathcal{K}$ is an equational class, that is, $\mathcal{E}(\mathcal{L}(\mathcal{K}))=\mathcal{K}$
2) $\mathcal{K}$ is closed under quotient and for every nonempty set X, there is a $\mathcal{K}$-universal pair (K_X,κ).

Proof. We have seen that 1) implies 2). For the converse, we show that 2) implies $\mathcal{E}(\mathcal{L}(\mathcal{K}))\subseteq\mathcal{K}$. Let $G\in\mathcal{E}(\mathcal{L}(\mathcal{K}))$.

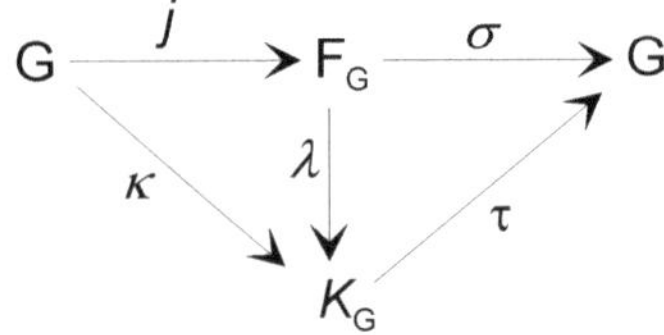

Figure 12.3

Referring to Figure 12.3, let $j\colon G\to F_G$ be the inclusion map into the concrete free group F_G. We lift two maps. The $\mathcal{K}$-universal map κ can be lifted to a unique homomorphism $\lambda\colon F_G\to K_G$ for which $\lambda j=\kappa$. Moreover, λ is surjective since κG generates K_G. Also, the identity map $\iota\colon G\to G$ can be lifted to a unique epimorphism $\sigma\colon F_G\twoheadrightarrow G$ for which $\sigma j=\iota$.

To see that $\ker(\lambda)\leq\ker(\sigma)$, if $w(x_1,\ldots,x_n)$ is a word over X' and if $w(ja_1,\ldots,ja_n)\in\ker(\lambda)$ for $a_i\in G$, then

$$w(\kappa a_1,\ldots,\kappa a_n)=\lambda w(ja_1,\ldots,ja_n)=1$$

in K_G and so Theorem 12.8 implies that $w(x_1,\ldots,x_n)\in\mathcal{L}(\mathcal{K})$. Hence in G,

$$\sigma w(ja_1, \ldots, ja_n) = w(\sigma ja_1, \ldots, \sigma ja_n) = w(a_1, \ldots, a_n) = 1$$

and so $w(ja_1, \ldots, ja_n) \in \ker(\sigma)$, whence $\ker(\lambda) \leq \ker(\sigma)$.

Hence, σ induces an epimorphism $\tau: K_G \twoheadrightarrow G$ defined for each $k \in K_G$ by $\tau k = \sigma\lambda^{-1}(k)$ and so $G \approx K_G/\ker(\tau)$ and since the latter is a $\mathcal{K}$-group, so is G.□

Free Abelian Groups

The class of abelian groups is an equational class, with law $[x, y] = 1$. Thus, free abelian groups exist. In fact, if G is a group, then the verbal subgroup is

$$\mathcal{L}(G) = \langle [a, b] \mid a, b \in G \rangle$$

which is just the commutator subgroup G' of G. Hence, Theorem 12.9 implies that if F_X is the concrete free group on X, then the group

$$A_X = F_X / F'_X$$

is free abelian.

Definition *Let X be a nonempty set. If F_X is the concrete free abelian group on X, then $A_X = F_X/F'_X$ is the* **concrete free abelian group** *on X. It is customary to think of A_X as the group F_X with the additonal condition that the elements of X commute.*□

Theorem 12.14 *If X is a nonempty set, then the free abelian group A_X satisfies*

$$A_X \approx \boxplus_{x \in X} \langle x \rangle \approx \boxplus_{x \in X} \mathbb{Z}$$

Proof. The function $f: X \to \boxplus \langle x \rangle$ defined by

$$f(x)(y) = \begin{cases} x & \text{if } y = x \\ 1 & \text{if } y \neq x \end{cases}$$

for all $y \in X$ can be lifted uniquely to a homomorphism $\tau: A_X \to \boxplus \langle x \rangle$ for which

$$\tau(x_1^{e_1} \cdots x_n^{e_n}) = f(x_1)^{e_1} \cdots f(x_n)^{e_n}$$

and so

$$\tau(x_1^{e_1} \cdots x_n^{e_n})(y) = \begin{cases} x_k^{e_k} & \text{if } y = x_k \\ 1 & \text{if } y \notin \{x_1, \ldots, x_n\} \end{cases}$$

Now, τ is surjective, since if $\alpha \in \boxplus \langle x \rangle$ has support $\{x_1, \ldots, x_n\}$ and $\alpha(x_k) = x_k^{e_k}$, then $\alpha = \tau(x_1^{e_1} \cdots x_n^{e_n})$. Also, τ is injective, since if $\tau(x_1^{e_1} \cdots x_n^{e_n}) = 0$, then $x_k^{e_k} = 1$ for all k and so $e_k = 0$ for all k. Thus, τ is an isomorphism.□

The following theorem explains the term *basis* used for the set X in the free abelian group A_X.

Definition *A subset S of the free abelian group A_X is* **independent** *in A_X if*

$$s_i \in S, \quad s_1^{e_1} \cdots s_n^{e_n} = \epsilon, \quad s_i \neq s_j \quad \Rightarrow \quad e_i = 0 \text{ for all } i \qquad \square$$

Theorem 12.15 *Let X be a nonempty subset of an abelian group A. The following are equivalent:*

1) *A is a free abelian group with basis X*
2) *X is independent in A and generates A*
3) *Except for the order of the factors, every nonidentity element $a \in A$ has a unique expression of the form*

$$a = x_1^{e_1} \cdots x_n^{e_n} \text{ for } x_i \neq x_j, e_k \neq 0, n \geq 1$$

Proof. If 1) holds, we have seen that X generates A and if

$$w = x_1^{e_1} \cdots x_n^{e_n} = \epsilon$$

for $x_i \neq x_j$, then $e_i = 0$ for all i, since otherwise w would be a reduced word equivalent to the shorter word ϵ. Hence, 2) holds. If 2) holds, then every element of A has at least one such expression. But if $a \in A$ has two distinct expressions:

$$a = x_1^{e_1} \cdots x_n^{e_n} = y_1^{f_1} \cdots y_m^{f_m}$$

where we may assume that $x_n \neq y_m$ (or we can cancel), then

$$x_1^{e_1} \cdots x_n^{e_n} y_m^{-f_m} \cdots y_1^{-f_1} = \epsilon$$

violates independence. Hence, 3) holds.

Finally, to see that 3) implies 1), suppose that $f: X \to G$, where G is an abelian group. If a mediating morphism τ does exist, then $\tau(x) = f(x)$ for all $x \in X$ and so τ is unique, since X generates A. Define a map $\tau: A \to G$ by

$$\tau(x_1^{e_1} \cdots x_n^{e_n}) = f(x_1)^{e_1} \cdots f(x_n)^{e_n}$$

which is well defined since the expressions $x_1^{e_1} \cdots x_n^{e_n}$ are unique up to order but G is abelian. The map τ is easily seen to be a homomorphism.$\square$

The next theorem says that, up to isomorphism, there is only one free (or free abelian) group of each cardinal rank.

Theorem 12.16 *Let X and Y be nonempty sets.*

1) a) $A_X \approx A_Y \Leftrightarrow |X| = |Y|$
   b) *If S generates A_X, then* $|S| \geq |X|$
2) a) $F_X \approx F_Y \Leftrightarrow |X| = |Y|$

b) If S generates F_X, then $|S| \geq |X|$

Proof. Suppose first that $|X| = |Y|$ and let $f: X \to Y$ be a bijection. Extend the range of f so that $f: X \to F_Y$ (or $f: X \to A_Y$). Then there is a unique mediating morphism $\tau: F_X \to F_Y$ (or $\tau: A_X \to A_Y$) for which

$$\tau(x) = f(x)$$

for all $x \in X$. To see that τ is injective, if $w = x_1^{e_1} \cdots x_n^{e_n}$ and $f(x_i) = y_i$, then

$$\tau(w) = \tau(x_1^{e_1} \cdots x_n^{e_n}) = f(x_1)^{e_1} \cdots f(x_n)^{e_n} = y_1^{e_1} \cdots y_n^{e_n}$$

Hence, $\tau(w) = \epsilon$ implies $e_i = 0$ for all i and so $w = \epsilon$. Also, τ is surjective since $f: X \to Y$ is surjective. Hence, $F_X \approx F_Y$ (and $A_X \approx A_Y$).

For the converse of part 1), let A represent either A_X or A_Y. We use additive notation. The quotient $A/2A$ is elementary abelian of exponent 2 and is therefore a vector space over $\mathbb{Z}_2$. Moreover, if S generates the group A, then

$$S/2A = \{s + 2A \mid s \in S\}$$

generates the vector space $A/2A$ over $\mathbb{Z}_2$ and if S is independent in A, then $S/2A$ is linearly independent in $A/2A$, since an equation of the form

$$(s_1 + 2A) + \cdots + (s_n + 2A) = 2A$$

for $n > 0$ and $s_i \neq s_j$ in S implies that

$$s_1 + \cdots + s_n = 2(e_1 t_1 + \cdots + e_m t_m)$$

for $t_i \in S$ and $e_i \in \mathbb{Z}$, which is not possible.

It follows that $X/2A_X$ is a basis for $A_X/2A_X$ and so

$$\dim(A_X/2A_X) = |X|$$

and similarly for Y. Thus, since $A_X \approx A_Y$ implies $A_X/2A_X \approx A_Y/2A_Y$, we have

$$|X| = \dim(A_X/2A_X) = \dim(A_Y/2A_Y) = |Y|$$

and if S generates A_X, then

$$|S| \geq |S/2A_X| \geq |X|$$

This completes the proof of part 1).

For part 2), if $F_X \approx F_Y$, then

$$A_X \approx F_X/F_X' \approx F_Y/F_Y' \approx A_Y$$

and so part 1) implies that $|X| = |Y|$. Finally, if S generates F_X, then

$S/F'_X = \{sF'_X \mid s \in S\}$ generates $A_X = F_X/F'_X$ and so

$$|S| \geq |S/F'_X| \geq |X/F'_X| = |X| \qquad \square$$

Theorem 12.17 *Let A_X be free abelian on X. Then all independent sets have cardinality at most $|X|$.*

Proof. It is sufficient to prove the result for $A = \boxplus_{x \in X} \mathbb{Z}_x$ where $\mathbb{Z}_x = \mathbb{Z}$ for all x. The set $V = \boxplus_{x \in X} \mathbb{Q}_x$ is a vector space over the rational field $\mathbb{Q}$ and it is easy to see that a subset

$$\mathcal{B} = \{v_i \mid i \in I\} \subseteq A$$

is dependent in A if and only if $\mathcal{B}$ is linearly dependent over $\mathbb{Q}$. But in the vector space V, all sets of cardinality greater than $|X|$ are linearly dependent over $\mathbb{Q}$ and therefore also over $\mathbb{Z}$.$\square$

The Nielsen–Schreier Theorem says that every subgroup of a free group is free and so the subgroups of free groups are very restricted. (Nielsen proved this result for finitely-generated groups in 1921 and Schreier generalized it to all groups in 1927.)

Theorem 12.18

1) *Any subgroup of a free group is free.*
2) *Any subgroup S of a free abelian group A_X is free abelian and* $\text{rk}(S) \leq \text{rk}(A)$.

Proof. We omit the difficult proof of part 1) and refer the interested reader to Robinson [26]. For part 2), we may assume that $A_X = \boxplus_{x \in X} \langle x \rangle$ and that X is well ordered. Since the elements of S have finite support, for $f \in S$, we can let $i(f)$ be the largest index x for which $f(x) \neq 1$.

For each $x \in X$, consider the set

$$I_x = \{f(x) \mid f \in S, \, i(f) \leq x\}$$

Then $I_x \leq \langle x \rangle$ and so $I_x = \langle f_x(x) \rangle$ for some $f_x \in S_x$. We show that S is free on the set

$$\mathcal{B} = \{f_x \mid x \in X, f_x(x) \neq 1\}$$

If $\mathcal{B}$ does not span S, among those elements of S not in the span of $\mathcal{B}$, choose an element g for which $y = i(g)$ is the smallest possible. Since $g(y) \in I_y = \langle f_y(y) \rangle$, it follows that $g(y) = f_y^k(y)$ for some nonzero $k \in \mathbb{Z}$. Then

$$(gf_y^{-k})(x) = g(x)f_y^{-k}(x) = 1 \text{ for all } x \geq y$$

and so $i(gf_y^{-k}) < y$, which implies that gf_y^{-k} is in the span of $\mathcal{B}$. But then

$$g = (gf_y^{-k})f_y^k$$

is also in the span of $\mathcal{B}$, a contradiction. Thus, $\mathcal{B}$ spans S.

Also, $\mathcal{B}$ is independent, since if

$$f_{x_1}^{e_1} \cdots f_{x_n}^{e_n} = 1$$

where $x_i < x_j$ for $i < j$, then applying this to x_n gives

$$f_{x_n}^{e_n}(x_n) = 1$$

and so $e_n = 0$. Similarly, $e_i = 0$ for all i and so $\mathcal{B}$ is independent. Hence, Theorem 12.15 implies that S is free over $\mathcal{B}$. Also, it is clear that $|\mathcal{B}| \leq |X|$ and so $\text{rk}(S) \leq \text{rk}(A)$.□

Applications of Free Groups

Sometimes free groups can help produce complements.

Theorem 12.19 *If $\sigma: G \twoheadrightarrow F_X$ is an epimorphism, where F_X is free on X, then $K = \ker(\sigma)$ is complemented in G.*
Proof. Define a function $\tau: X \to G$ by letting τx be a fixed member of $\sigma^{-1}(x)$. Since F_X is free on X, there is a unique homomorphism $\tau: F_X \to G$ that extends τ on X. Since for any $x \in X$,

$$\sigma \circ \tau(x) = x$$

it follows that $\sigma \circ \tau = \iota$. In other words, τ is a right inverse of σ and so Theorem 5.23 implies that $\ker(\sigma)$ is complemented in G.□

With the help of free groups, we can provide an example of a finitely-generated *nonabelian* group with a subgroup that is not finitely generated. We have already proved (Theorem 2.21) that if G is an n-generated *abelian* group, then every subgroup of G can be generated by n or fewer elements.

Theorem 12.20 *Let $X = \{x, y\}$ and let F_X be the 2-generated free group on X. Let $G = \langle S \rangle$, where*

$$S = \{y^k x y^{-k} \mid k > 0\}$$

Then G is isomorphic to the free group F_Z on a countably infinite set $Z = \{z_1, z_2, \ldots\}$ and so is not finitely generated.
Proof. Consider the function $f: Z \to G$ defined by $f(z_k) = y^k x y^{-k}$. Then there is a unique mediating morphism $\tau: F_Z \to G$ for which $\tau(z_k) = y^k x y^{-k}$. It is clear that τ is surjective.

In addition, if

$$\tau(z_{i_1}^{e_{i_1}} \cdots z_{i_m}^{e_{im}}) = \epsilon$$

where $i_k \neq i_{k+1}$, $e_{i_k} \neq 0$ and $m \geq 1$, then

$$y^{i_1} x^{e_{i_1}} y^{-i_1+i_2} x^{e_{i_2}} y^{-i_2+i_3} x^{e_{i_3}} \cdots x^{e_{i_{m-1}}} y^{-i_{m-1}+i_m} x^{e_{im}} y^{-i_m} = \epsilon$$

in $G \leq F_X$. Since the left-hand side can be reduced to ϵ using only removal steps, it follows that $i_k = i_{k+1}$ for all k and so $m = 1$ and

$$y^{i_1} x^{e_{i_1}} y^{-i_1} = \epsilon$$

But the left-hand side of this equation reduces to ϵ by removal steps if and only if $e_1 = 0$, which is false. Hence, τ is injective and therefore an isomorphism.□

We can also provide an example of a group G with a subgroup $H \leq G$ for which $aHa^{-1} < H$.

Theorem 12.21 *Let F_X be the free group on $X = \{x, y\}$. Then F_X has a subgroup H for which $xHx^{-1} < H$.*
Proof. Let H consist of the empty word ϵ and the set of all words of the form

$$w = x^{n_1} y^{k_1} x^{n_2} y^{k_2} \cdots x^{n_r} y^{k_r} x^{n_{r+1}}$$

with $r \geq 1$, $k_i \neq 0, n_1 \geq 0, n_{r+1} \leq 0$, $n_i \neq 0$ for $2 \leq i \leq r$ and $\sum n_i = 0$. Note that $y \in H$. We leave it to the reader to show that H is a subgroup of F_X. It is clear that

$$xwx^{-1} = x^{n_1+1} y^{k_1} x^{n_2} y^{k_2} \cdots x^{n_r} y^{k_r} x^{n_{r+1}-1} \in H$$

and so $xHx^{-1} \leq H$. However, $xwx^{-1} \neq y$ for all $w \in H$ since if $w \neq \epsilon$, then xwx^{-1} has length at least 3. Hence, $xHx^{-1} < H$.□

Presentations of a Group

One way to define a group is to list all of the elements of the group and then give the group's *multiplication table*, which shows explicitly how to multiply all pairs of elements of the group. Then it is necessary to verify the defining axioms of a group: associativity, identity and inverses. For example, the cyclic group $C_3(a)$ is

$$C_3 = \{1, a, a^2\}$$

with multiplication table

	1	a	a^2
1	1	a	a^2
a	a	a^2	1
a^2	a^2	1	a

Of course, we generally abbreviate this description by writing

$$C_3 = \{a^i \mid 0 \leq i \leq 2\}, \quad a^i a^j = a^{(i+j) \bmod 3}$$

It is routine in this case to check the group axioms.

On the other hand, it is tempting to define a group by giving a set of *generators* for the group along with some properties satisfied by these generators. The issue then becomes one of deciding whether there is a group that has these generators with these properties.

To illustrate, consider the following description of a group G:

$$G = \langle a, b \rangle, a^2 = 1, b^4 = 1, abab = 1$$

This description gives a nonempty set $X = \{a, b\}$ of generators for G and certain *relations* on G, that is, equations of the form $w = 1$ where w is a word over the generators and their inverses. Note that there is nothing in the description above that precludes the possibility that $a = b$.

In order to guarantee that such a group exists, we require that all relations must have the form $w = 1$. The left-hand side w is referred to as a *relator*. Expressions such as $w \neq 1$ are not permitted. However, it is customary to take the liberty of writing a relation in the form $w = v$ as a more intuitive version of $wv^{-1} = 1$. For example, the last relation above can be written

$$ba = a^{-1}b^{-1} \quad \text{or} \quad ba = ab^3$$

In view of the precise nature of relations, given any nonempty generating set X and any set $\mathcal{R}$ of relations, there is always one group that is generated by X and satisfies the relations $\mathcal{R}$: It is the trivial group, where each generator is taken to be the identity.

On the other hand, the best hope for getting a *nontrivial* group generated by X and satisfying the relations $\mathcal{R}$ is to start with the concrete free group F_X and factor out by the *smallest* normal subgroup N required in order to satisfy the given relations, that is, the normal closure of the relators of $\mathcal{R}$.

With respect to the example above, the quotient group

$$G_1 = \frac{F_{\{x,y\}}}{\langle x^2, y^4, xyxy\rangle_{\text{nor}}}$$

has generators $a = xN$ and $b = yN$, where $N = \langle x^2, y^4, xyxy\rangle_{\text{nor}}$ that satisfy the relations given for a group G.

Thus, we are lead to the concept of a free presentation of a group, given in the next definition.

Definition *Let F_X be free on X. An epimorphism $\sigma\colon F_X \twoheadrightarrow G$ is called a* **free presentation** *of G. The set X is called a set of* **generators** *for the presentation and if* $\ker(\sigma) = \langle \mathcal{R}\rangle_{\text{nor}}$ *so that*

$$G \approx \frac{F_X}{\langle \mathcal{R}\rangle_{\text{nor}}}$$

the set $\mathcal{R}$ is called a set of **relators** *of the presentation. In this case, we write*

$$G \approx \langle X \mid \mathcal{R}\rangle$$

If $r \in \mathcal{R}$ is a relator, then the equation $r = 1$ is called a **relation**.□

We will often refer to a free presentation of G simply as a presentation of G. It is common to say that the group G itself has presentation $\langle X \mid \mathcal{R}\rangle$ when

$$G \approx \frac{F_X}{\langle \mathcal{R}\rangle_{\text{nor}}}$$

and that G is **defined by generators and relations**. Note, however, that it is the set σX that actually generates G. A presentation $\langle X \mid \mathcal{R}\rangle$ is **finite** if $X \cup \mathcal{R}$ is a finite set. Finally, we will often blur the distinction between a relator and the corresponding relation, using whichever is more convenient at the time.

We can form a more concrete version of the group defined by generators X and relations $\mathcal{R}$ as follows. If $G = \langle X\rangle$ and if F_X is the concrete free group on X, then a word over X' has two contexts: as an element of F_X and as an element of G. Moreover, there is a unique epimorphism $\sigma\colon F_X \twoheadrightarrow G$ defined by specifying that $\sigma x = x$ for all $x \in X$. This map can be thought of simply as a *change of context* and it is convenient to give it this name officially.

Definition *Let $G = \langle X\rangle$ and let F_X be the concrete free group on X.*

1) *We call the unique epimorphism $\sigma\colon F_X \twoheadrightarrow G$ defined by $\sigma x = x$ for all $x \in X$ the* **change of context map** *associated to G and X.*
2) *If the change of context map $\sigma\colon F_X \twoheadrightarrow G$ has kernel $N = \langle \mathcal{R}\rangle_{\text{nor}}$, so that the induced map $\tau\colon F_X/N \to G$ defined by*

$$\tau(xN) = x$$

is an isomorphism, we say that G has **concrete presentation** *$\langle X \mid \mathcal{R}\rangle$ and write $G = \langle X \mid \mathcal{R}\rangle$.*□

It is clear that if $G = \langle X \mid \mathcal{R}\rangle$ with $N = \langle \mathcal{R}\rangle_{\text{nor}}$ and if $w(x_1, \ldots, x_n)$ is a word over X', then

$$\begin{aligned} w(x_1, \ldots, x_n) = 1 \text{ in } G \quad &\Leftrightarrow \quad w(x_1N, \ldots, x_nN) = N \text{ in } F_X/N \\ &\Leftrightarrow \quad w(x_1, \ldots, x_n) \in N \text{ in } F_X \end{aligned}$$

Every Group Has a Presentation

The change of context map $\sigma\colon F_G \twoheadrightarrow G$ shows that every group has a concrete presentation. Moreover, the kernel of this presentation is essentially the multiplication table for G. To be more specific, if $a, b, c \in G$ and $c = ab$, then $abc^{-1} = 1$ in G and so abc^{-1} must be factored out of F_G. So let

$$N = \langle abc^{-1} \mid a, b, c \in G, c = ab \text{ in } G\rangle \leq F_G$$

It is easy to see that N is a fully invariant subgroup of F_G and that $N \leq \ker(\sigma)$. For the reverse inclusion, if $w(a_1, \ldots, a_n) \in \ker(\sigma)$, then $w(a_1, \ldots, a_n) = 1$ in G and so $w(a_1, \ldots, a_n) \in N$. Thus, $N = \ker(\sigma)$ and so $G = \langle G \mid \mathcal{R}\rangle$, where

$$\mathcal{R} = \{abc^{-1} \mid a, b, c \in G, c = ab \text{ in } G\}$$

Note also that if G is finite, then so is $\mathcal{R}$.

Theorem 12.22 *Every group G has concrete presentation $\langle G \mid \mathcal{R}\rangle$, where*

$$\mathcal{R} = \{abc^{-1} \mid a, b, c \in G, c = ab \textit{ in } G\}$$

Moreover, if G is finite, then $\langle G \mid \mathcal{R}\rangle$ is a finite presentation of G.□

The concrete presentation $\langle G \mid \mathcal{R}\rangle$ is rather large and we can improve upon it in general.

Theorem 12.23 *If $G \approx \langle X \mid \mathcal{R}\rangle$, then $G = \langle Y \mid \mathcal{S}\rangle$, where $|Y| \leq |X|$ and $|\mathcal{R}| \leq |\mathcal{S}|$. In particular, G has a finite presentation if and only if it has a finite concrete presentation.*

Proof. Let $\mu\colon F_X \twoheadrightarrow G$ be a free presentation of G with kernel $N = \langle \mathcal{R}\rangle_{\text{nor}}$. Then the set $Y = \mu X$ generates G. Let $\sigma\colon F_Y \twoheadrightarrow G$ be the change of context map. If $\mathcal{S}$ is the set of relators in Y obtained from $\mathcal{R}$ by replacing each occurrence of $x \in X$ by μx, then the following are equivalent:

$$\begin{aligned} w(\mu x_1, \ldots, \mu x_n) &\in \ker(\sigma) \\ w(\mu x_1, \ldots, \mu x_n) &= 1 \text{ in } G \\ w(x_1, \ldots, x_n) &\in \langle \mathcal{R} \rangle_{\text{nor}} \\ w(\mu x_1, \ldots, \mu x_n) &\in \langle \mathcal{S} \rangle_{\text{nor}} \end{aligned}$$

and so $\ker(\sigma) = \langle \mathcal{S} \rangle_{\text{nor}}$ and $G = \langle Y \mid \mathcal{S} \rangle$.□

Finitely Presented Groups

A group is **finitely presented** if it has a finite presentation.

Theorem 12.24 *If G has a finite presentation and if X is a generating set for G, then G has a finite presentation of the form*

$$\langle x_1, \ldots, x_n \mid r_1, \ldots, r_m \rangle$$

where $x_i \in X$.
Proof. Let $\langle Y \mid \mathcal{S} \rangle$ be a finite concrete presentation of G, with

$$Y = \{y_1, \ldots, y_u\} \quad \text{and} \quad \mathcal{S} = \{s_i(y_1, \ldots, y_u) \mid i = 1, \ldots, v\}$$

Then there is a finite subset $X_0 = \{x_1, \ldots, x_n\} \subseteq X$ for which $Y \subseteq \langle X_0 \rangle$ and so X_0 generates G and we can write

$$x_i = \xi_i(y_1, \ldots, y_u); \quad i = 1, \ldots, n$$

and

$$y_j = \lambda_j(x_1, \ldots, x_n); \quad j = 1, \ldots, u$$

Let

$$H = \langle x_1, \ldots, x_n \mid \mathcal{R} \rangle$$

where $\mathcal{R}$ is the set of relators formed from the relations

$$s_i(\lambda_1(x_1, \ldots, x_n), \ldots, \lambda_u(x_1, \ldots, x_n)) = 1; \quad i = 1, \ldots, v$$

and

$$x_j = \xi_j(\lambda_1(x_1, \ldots, x_n), \ldots, \lambda_u(x_1, \ldots, x_n)); \quad j = 1, \ldots, n$$

Since each of these relations holds in G, the subgroup $\langle \mathcal{R} \rangle_{\text{nor}}$ is contained in the kernel of the change of context epimorphism $\sigma: F_{X_0} \twoheadrightarrow G$ and so σ induces an epimorphism $\sigma': H \twoheadrightarrow G$ defined by $\sigma'(x_i) = x_i$.

To see that σ' is injective, if

$$\sigma'(w(x_1, \ldots, x_n)) = 1$$

then

$$w(x_1, \ldots, x_n) = 1$$

in G, whence

$$w(\xi_1(y_1, \ldots, y_u), \ldots, \xi_n(y_1, \ldots, y_u)) = 1$$

in G and so

$$w(\xi_1(y_1, \ldots, y_u), \ldots, \xi_n(y_1, \ldots, y_u)) \in \langle \mathcal{S} \rangle_{\text{nor}}$$

in F_Y. Replacing each y_j by $\lambda_j(x_1, \ldots, x_n)$ implies that

$$w(x_1, \ldots, x_n) \in \langle \mathcal{R} \rangle_{\text{nor}}$$

in H, that is, $w(x_1, \ldots, x_n) = 1$ in H. Hence, σ' is an isomorphism and so G has presentation $\langle x_1, \ldots, x_n \mid \mathcal{R} \rangle$.□

Theorem 12.25 *Let G be a group and let $N \trianglelefteq G$. If N and G/N are finitely presented, then so is G.*
Proof. Let

$$N = \langle x_1, \ldots, x_n \mid r_1, \ldots, r_u \rangle$$

and let

$$G/N = \langle y_1 N, \ldots, y_m N \mid s_1, \ldots, s_v \rangle$$

If $X = \{x_1, \ldots, x_n\}$ and $Y = \{y_1, \ldots, y_m\}$ and $A = X \cup Y$, then G is generated by A. As to relators, we have

$$s_i(y_1 N, \ldots, y_n N) = 1 \quad \Leftrightarrow \quad s_i(y_1, \ldots, y_n) = w_i(x_1, \ldots, x_n)$$

where $w_i(x_1, \ldots, x_n)$ is a word in the x_i's and x_i^{-1}'s. Also, N is normal if and only if

$$y_j x_i y_j^{-1} = v_{i,j}(x_1, \ldots, x_n) \quad \text{and} \quad y_j^{-1} x_i y_j = z_{i,j}(x_1, \ldots, x_n)$$

for words $v_{i,j}$ and $z_{i,j}$. The following set $\mathcal{R}$ of relations captures the relations r_i and s_j as well as the fact that $N \trianglelefteq G$:

$$\begin{aligned} r_i(x_1, \ldots, x_n) &= 1 \text{ for } i = 1, \ldots, u \\ s_i(y_1, \ldots, y_n) &= w_i(x_1, \ldots, x_n) \text{ for } i = 1, \ldots, v \\ y_j x_i y_j^{-1} &= v_{i,j}(x_1, \ldots, x_n) \text{ for } i = 1, \ldots, n; j = 1, \ldots, m \\ y_j^{-1} x_i y_j &= z_{i,j}(x_1, \ldots, x_n) \text{ for } i = 1, \ldots, n; j = 1, \ldots, m \end{aligned}$$

Let $X' = \{x'_1, \ldots, x'_n\}$, $Y' = \{y'_1, \ldots, y'_m\}$ and $A' = X' \cup Y'$ and let

$$H = \langle A' \mid \mathcal{R}'_{A \to A'} \rangle$$

If $N' = \langle x'_1, \ldots, x'_n \rangle \leq H$, then the relators in $\mathcal{R}'$ show that $N' \trianglelefteq H$.

Let $f: F_{A'} \to G$ be the unique epimorphism for which $fx'_i = x_i$ and $fy'_j = y_j$. Since $\langle \mathcal{R}' \rangle_{\text{nor}} \le \ker(f)$, there is a unique epimorphism $g: H \twoheadrightarrow G$ for which

$$gx'_i = fx'_i = x_i \quad \text{and} \quad gy'_i = fy'_i = y_i$$

Moreover, the restriction $g: N' \twoheadrightarrow N$ is an isomorphism, since if

$$g(w(x'_1, \ldots, x'_n)) = 1$$

then

$$w(x_1, \ldots, x_n) = 1$$

in N and so $w(x_1, \ldots, x_n) \in \langle \mathcal{R}' \rangle_{\text{nor}}$ in F_X. It follows that

$$w(x'_1, \ldots, x'_n) \in \langle \mathcal{R}'_{A \to A'} \rangle_{\text{nor}}$$

in $F_{A'}$ and so $w(x'_1, \ldots, x'_n) = 1$ in N'. Hence, if $K = \ker(g)$, then

$$K \cap N' = \{1\}$$

Our goal is to show that $K = \{1\}$.

Since $g(N') = N$, the epimorphism $g: H \twoheadrightarrow G$ induces an epimorphism $g': H/N' \twoheadrightarrow G/N$ for which

$$g'(y'_j N') = g(y'_j)N = y_j N$$

for all j. Moreover, g' is an isomorphism, since if

$$g'(w(y'_1 N', \ldots, y'_m N'))) = w(y_1 N, \ldots, y_m N)) = N$$

in G/N, then

$$w(y_1 N, \ldots, y_m N)) \in \langle s_1(y_1 N, \ldots, y_m N), \ldots, s_u(y_1 N, \ldots, y_m N) \rangle_{\text{nor}}$$

in the free group on $\{y_1 N, \ldots, y_m N\}$. Hence,

$$w(y'_1 N', \ldots, y'_m N')) \in \langle s_1(y'_1 N', \ldots, y'_m N'), \ldots, s_u(y'_1 N', \ldots, y'_m N') \rangle_{\text{nor}}$$

in the free group on $\{y'_1 N', \ldots, y'_m N'\}$, which implies that

$$w(y'_1, \ldots, y'_m) \in N'$$

that is,

$$w(y'_1 N', \ldots, y'_m N')) = N'$$

in H/N'. Thus, g' is an isomorphism. Finally, if $1 \ne k \in K$, then $k \notin N'$ and so $kN' \ne N'$, whence

$$N = g(k)N = g'(kN') \ne N$$

It follows that $K = \{1\}$, that is, $g\colon H \approx G$ is an isomorphism and so G is finitely presented.□

Combinatorial Group Theory

Before looking at other examples, let us return briefly to the question of whether a given presentation $\langle X \mid \mathcal{R}\rangle$ defines a nontrivial group. From one point of view, this question has a rather surprising answer. It can be shown that no *algorithm* can ever exist that determines whether or not an arbitrary set of generators and relations defines a nontrivial group! Nor is there any algorithm that determines whether the group defined by an arbitrary finite presentation is finite or infinite.

Definition *A* **decision problem** *is a problem that has a yes or no answer, such as whether or not a given word in F_X is the identity in a group G.*

1) *A decision problem is* **decidable** *or* **solvable** *if there is an algorithm, called a* **decision procedure**, *that stops after a finite number of steps and returns "yes" when the answer is yes and "no" when the answer is no. A decision problem is* **undecidable** *or* **unsolvable** *if it is not decidable.*
2) *A decision problem is* **semidecidable** *or* **semisolvable** *if there is an algorithm that stops after a finite number of steps and returns "yes" when the answer is yes. However, the algorithm need not stop if the answer is no.*□

The **word problem** for a group G with presentation $\langle X \mid \mathcal{R}\rangle$, first formulated in 1911 by Max Dehn, is the problem of deciding whether or not an arbitrary word over X' is the identity element of G (or, equivalently, whether or not two arbitrary words over X' are the same element of G). It has been shown that there exist individual groups G with finite presentations for which the word problem is unsolvable.

On the other hand, there are large classes of groups for which the word problem is solvable. For example, the word problem is solvable for all free groups (in view of Theorem 12.4), for all finite groups and for all finitely-generated abelian groups. In fact, it is an active area of current research in group theory to study classes of groups for which the word problem can be solved.

On the other hand, the word problem for finitely-presented groups is semidecidable. For if $\langle X \mid \mathcal{R}\rangle$ is a finite presentation of a group G and if $w \in G$, then since $\mathcal{R}$ is a finite set, there is an algorithm that checks all of the elements of $\langle \mathcal{R}\rangle_{\text{nor}}$ one-by-one looking for w. If $w = 1$ in G, then this algorithm will eventually encounter w. The problem is that the algorithm will not terminate if $w \neq 1$ and so this is not a decision procedure. One way to mitigate this problem is to intermix the steps of this algorithm with the steps of another algorithm that stops if $w \neq 1$, assuming that such an algorithm exists.

For example, the following defines a class of group for which such an algorithm does exist.

Definition *A group G is* **residually finite** *if for any $1 \neq a \in G$, there is a normal subgroup $N \trianglelefteq G$ for which $a \notin N$ and G/N is finite.* $\square$

Theorem 12.26 *The word problem is solvable for the class of all finitely-presented residually finite groups.*

Proof. Let $G = \langle X \mid \mathcal{R} \rangle$ be a finite presentation of a residually finite group G and let $w \in G$. It is possible to enumerate all finite groups by constructing multiplication tables. Also, for a given finite group F, there are only a finite number of group homomorphisms from G to F, since all such homomorphisms are uniquely determined by the functions $f: X \to F$. Consider the following algorithm:

1) Compute the next finite group F.
2) For each group homomorphism $\sigma: G \to F$, stop the algorithm if $\sigma w \neq 1$.

Now, if $w \neq 1$ in G, then since G is residually finite, there is an $N \trianglelefteq G$ for which $w \notin N$ and G/N is finite. Hence, the canonical projection $\pi: G \to G/N$, which is encountered in step 2) above, satisfies $\pi w \neq 1$ and so the algorithm will stop if $w \neq 1$. Thus, we can intermix this algorithm with the aforementioned algorithm to get a decision procedure for the word problem for G.$\square$

The issues discussed above fall under the auspicies of an area of algebra known as **combinatorial group theory**.

On the Order of a Presented Group

Since a relation cannot be a *nonequality*, there is no way to specify the *order* of an element or subgroup of a group by relations. Thus, for example, among the following descriptions of a group, only the first description is a presentation:

1) $G_1 = \langle a, b \mid a^2 = 1, b^4 = 1, abab = 1 \rangle$

2) $G_2 = \langle a, b \rangle, o(G_2) = 8, a^2 = 1, b^4 = 1, abab = 1$

3) $G_3 = \langle a, b \rangle, o(a) = 2, o(b) = 4, abab = 1$

Since the relation $abab = 1$ is equivalent to the commutativity relation

$$ba = ab^3$$

each of the groups above has underlying set

$$S = \{a^i b^j \mid 0 \leq i \leq 1, 0 \leq j \leq 3\}$$

and so $o(G_k) \leq 8$ for $k = 1, 2, 3$. Moreover, since $o(G_2) = 8$ implies $o(a) = 2$ and $o(b) = 4$, any group satisfying 2) also satisfies 3). Conversely, $o(G_3) = 4$ or 8, but if $o(G_3) = 4$, then $G_3 = \langle b \rangle$ and $a = b^2$, which does not satisfy $abab = 1$. Hence, $o(G_3) = 8$ and so 2) and 3) describe the same group.

Thus, if we show that a group G fitting description 2) or 3) exists, then G has order 8 and *satisfies* the relations $\mathcal{R}$ given by 1). Hence, $\mathcal{R}$ is contained in the kernel of the change of context map $\sigma\colon F_{\{a,b\}} \to G$ and so

$$8 \geq (F_{\{a,b\}} : \langle \mathcal{R} \rangle_{\text{nor}}) \geq (F_{\{a,b\}} : \ker(\sigma)) = |G| = 8$$

It follows that $\ker(\sigma) = \langle \mathcal{R} \rangle_{\text{nor}}$ and so G has presentation $\langle X \mid \mathcal{R} \rangle$.

Theorem 12.27 *Suppose that a group with presentation $\langle X \mid \mathcal{R} \rangle$ has order at most $n < \infty$. Then any group G of order n generated by X and satisfying the relations in $\mathcal{R}$ has presentation $\langle X \mid \mathcal{R} \rangle$.*

Proof. Let $\sigma\colon F_X \twoheadrightarrow G$ be the change of context epimorphism. Since $\langle \mathcal{R} \rangle_{\text{nor}} \leq \ker(\sigma)$, we have

$$n \geq (F_X : \langle \mathcal{R} \rangle_{\text{nor}}) \geq (F_X : \ker(\sigma)) = |G| = n$$

from which it follows that $\ker(\sigma) = \langle \mathcal{R} \rangle_{\text{nor}}$ and so

$$G \approx \frac{F_X}{\langle \mathcal{R} \rangle_{\text{nor}}}$$

□

Referring to our previous example, since the dihedral group D_8 fits description 2) and has order 8, we have

$$D_8 \approx \langle a, b \mid a^2 = 1, b^4 = 1, abab = 1 \rangle$$

More generally, one of the simplest presentations with two generators is

$$H = \langle x, y \mid x^n = 1, y^m = 1, yx = xy^t \rangle$$

for some $0 < t < m$. The commutativity relation shows that

$$H = \{x^i y^j \mid 0 \leq i < n \text{ and } 0 \leq j < m\}$$

and so $o(H) \leq mn$. Moreover, one can prove by induction that for $0 \leq k < m$ and $0 \leq j < n$,

$$y^k x^j = x^j y^{kt^j}$$

and so

$$(x^i y^k)(x^j y^\ell) = x^{i+j} y^{kt^j + \ell} \tag{12.28}$$

However, we can define a group G whose underlying set consists of the mn distinct formal symbols

$$G = \{x^i y^j \mid 0 \le i < n \text{ and } 0 \le j < m\}$$

with product defined by (12.28). This product is associative, since

$$[(x^i y^k)(x^j y^\ell)](x^u y^v) = (x^{i+j} y^{kt^j+\ell})(x^u y^v) = x^{i+j+u} y^{(kt^j+\ell)t^u+v}$$

and

$$(x^i y^k)[(x^j y^\ell)(x^u y^v)] = x^i y^k (x^{j+u} y^{\ell t^u+v}) = x^{i+j+u} y^{kt^{j+u}+\ell t^u+v}$$

and inverses exist, since

$$(x^i y^k)^{-1} = x^{n-i} y^{m-kt^{n-i}}$$

Hence, G is a group of size nm that satisfies $\mathcal{R}$ and so G has presentation $\langle X \mid \mathcal{R}\rangle$ and $o(H) = mn$.

Theorem 12.29

1) The presentation

$$\langle X \mid \mathcal{R}\rangle = \langle a, b \mid a^n = 1, b^m = 1, ba = ab^t\rangle$$

where $0 < t < m$ defines the group

$$G = \{a^i b^j \mid 0 \le i < n \text{ and } 0 \le j < m\}$$

where $o(G) = mn$, $o(a) = n$, $o(b) = m$ and

$$(a^i b^k)(a^j b^\ell) = a^{i+j} b^{kt^j+\ell}$$

Moreover, any group of order mn that is generated by X and satisfies the relations $\mathcal{R}$ has presentation $\langle X \mid \mathcal{R}\rangle$.

2) The presentation

$$\langle Y \mid \mathcal{S}\rangle = \langle c, d \mid c^n = 1, d^m = 1, dc = c^s d\rangle$$

defines the group

$$H = \{c^i d^j \mid 0 \le i < n \text{ and } 0 \le j < m\}$$

where $o(H) = mn$, $o(c) = n$, $o(d) = m$ and

$$(c^i d^k)(c^j d^\ell) = c^{i+js^k} d^{k+\ell}$$

Moreover, any group of order mn that is generated by Y and satisfies the relations $\mathcal{S}$ is defined by $\langle Y \mid \mathcal{S}\rangle$.□

Let us now consider some examples of presentations.

Dihedral Groups

Since the dihedral group $D_{2n} = \langle \sigma, \rho \rangle$ of order $2n$ satisfies the relations

$$\mathcal{R} = \{\sigma^2 = 1, \rho^n = 1, \rho\sigma = \sigma\rho^{n-1}\}$$

Theorem 12.29 implies that D_{2n} has presentation $\langle \{\sigma, \rho\} \mid \mathcal{R} \rangle$ and multiplication table

$$(\sigma^i \rho^k)(\sigma^j \rho^\ell) = \sigma^{i+j} \rho^{k(n-1)^j + \ell}$$

We leave it as an exercise to show that D_{2n} is also presented by

$$\langle Y \mid \mathcal{S} \rangle = \langle x, y \mid x^2 = 1, y^2 = 1, (xy)^n = 1 \rangle$$

Thus, two rather different looking presentations can be **equivalent**, that is, can present the same group.

Quaternion Group

To find a presentation for the quaternion group, note that $Q = \langle i, j \rangle$ satisfies the relations

$$\mathcal{R} = \{i^4 = 1, i^2 = j^2, ji = i^3 j\}$$

and so we need only show that any group presented by $\langle X \mid \mathcal{R} \rangle$ has order at most 8. If G has presentation

$$\langle X \mid \mathcal{R} \rangle = \langle x, y \mid x^4 = 1, y^2 = x^2, yx = x^3 y \rangle$$

then

$$G = \{x^s y^t \mid 0 \leq s, t \leq 3\}$$

However, since $y^2 = x^2$, we see that

$$G = \{x^s y^t \mid 0 \leq s \leq 3, 0 \leq t \leq 1\}$$

and so $o(G) \leq 8$. Thus, $Q \approx \langle X \mid \mathcal{R} \rangle$.

Dicyclic Groups

Consider the presentation

$$\langle X \mid \mathcal{R} \rangle = \langle x, y \mid x^{2n} = 1, y^2 = x^n, yx = x^{-1} y \rangle$$

If

$$H = \langle X \mid \mathcal{R} \rangle$$

then using the fact that $y^2 = x^n$, we have

$$H = \{x^i y^j \mid 0 \le i \le 2n-1, 0 \le j \le 1\}$$

and so $o(H) \le 4n$.

A double induction shows that for $0 \le k < 4$ and $0 \le j < 2n$,

$$y^k x^j = x^{(-1)^k j} y^k = \begin{cases} x^j y^k & k \text{ even} \\ x^{2n-j} y^k & k \text{ odd} \end{cases}$$

and so

$$(x^i y^k)(x^j y^\ell) = x^{i+(-1)^k j} y^{k+\ell} \tag{12.30}$$

However, we can define a group G by choosing two distinct symbols x and y and setting

$$G = \{x^i y^j \mid 0 \le i < 2n, j = 0, 1\}$$

with product defined by (12.30). This product is associative:

$$[(x^i y^k)(x^j y^\ell)](x^u y^v) = (x^{i+(-1)^k j} y^{k+\ell})(x^u y^v) = x^{i+(-1)^k j+(-1)^{k+\ell} u} y^{k+\ell+v}$$

and

$$(x^i y^k)[(x^j y^\ell)(x^u y^v)] = x^i y^k (x^{j+(-1)^\ell u} y^{\ell+v}) = x^{i+(-1)^k (j+(-1)^\ell u)} y^{k+\ell+v}$$

and inverses exist:

$$(x^i y^k)^{-1} = x^{(-1)^k (2n-i)} y^{4-k}$$

Hence, G is a group of size $4n$ that satisfies the relations $\mathcal{R}$ and so $G \approx \langle X \mid \mathcal{R} \rangle$ and $o(H) = 4n$. Any group with presentation $\langle X \mid \mathcal{R} \rangle$ is called a **dicyclic group** of order $4n$.

A special case of the dicyclic group is when $2n$ has the form 2^{n-1}, in which case the presentation is

$$\langle X \mid \mathcal{R} \rangle = \langle x, y \mid x^{2^{n-1}} = 1, y^2 = x^{2^{n-2}}, yx = x^{-1} y \rangle$$

A group with this presentation is called a **generalized quaternion group**. When $n = 3$, this is

$$\langle X \mid \mathcal{R} \rangle = \langle x, y \mid x^4 = 1, y^2 = x^2, yx = x^{-1} y \rangle$$

which is the presentation for the quaternion group Q.

The Symmetric Group

Recall from Theorem 6.5 that the symmetric group S_n is generated by the $n-1$ adjacent transpositions $t_k = (k\, k+1)$ for $k = 1, \ldots, n-1$. Note also that the t_k's satisfy the relations

$$t_k^2 = 1, \quad t_k^{t_{k-1}} = t_{k-1}^{t_k}, \quad t_k t_j = t_j t_k \text{ for } j - k \neq \pm 1$$

Now let

$$X = \{x_1, \ldots, x_{n-1}\}$$

and let $\mathcal{R}$ be the relations

$$x_k^2 = 1, \quad x_{k-1}x_k x_{k-1} = x_k x_{k-1} x_k, \quad x_k x_j = x_j x_k \text{ for } j - k \neq \pm 1$$

Let $G = \langle X \mid \mathcal{R} \rangle$. Since S_n is generated by the elements $t_k = (k\, k+1)$ and satisfies the relations $\mathcal{R}$ with x_k replaced by t_k, Theorem 12.27 implies that if $o(G) \leq n!$, then $S_n \approx \langle X \mid \mathcal{R} \rangle$. We prove the former by induction on n.

If $n = 2$, then $G = \langle x_1 \rangle$ where $x_1^2 = 1$ and so $G = \{1, x_1\}$ has order 2. Assume that the result holds for the subgroup $H = \langle x_1, \ldots, x_{n-2} \rangle$ with relations $\mathcal{R}$. We show that every $a \notin H$ can be written in the form

$$a = x_k x_{k+1} \cdots x_{n-1} w$$

for $1 \leq k \leq n-1$, where $w \in H$. Let us refer to a substring $x_i x_{i+1} \cdots x_{n-1}$ as being in *proper order*. Then

$$a = u(x_i x_{i+1} \cdots x_{n-1}) v$$

for some $i \leq n-1$, where $v \in H$. If u is the empty string, then we are done. Otherwise, let $u = u' x_j$. Here are the possibilities:

1) If $j \leq i-2$, then x_j commutes with all factors to its right and so

$$a = u'(x_i x_{i+1} \cdots x_{n-1}) x_j v$$

where $x_j v \in H$.

2) If $j = i-1$, then

$$a = u'(x_{i-1} x_i x_{i+1} \cdots x_{n-1}) v$$

3) If $j = i$, then

$$a = u' x_i (x_i x_{i+1} \cdots x_{n-1}) v = u'(x_{i+1} \cdots x_{n-1}) v$$

4) If $j \geq i+1$, then

$$\begin{aligned} a &= u' x_j (x_i x_{i+1} \cdots x_{n-1}) v \\ &= u'(x_i \cdots x_j x_{j-1} x_j \cdots x_{n-1}) v \\ &= u'(x_i \cdots x_{j-1} x_j x_{j-1} \cdots x_{n-1}) v \\ &= u'(x_i \cdots x_{j-1} x_j \cdots x_{n-1}) x_{j-1} v \end{aligned}$$

where $x_{j-1} v \in H$.

Thus, in all cases, we can reduce the length of the substring appearing to the left of the substring in proper order by one symbol. Repeated application brings a to the desired form

$$a = x_k x_{k+1} \cdots x_{n-1} w$$

It follows that $o(G) = n \cdot o(H) \le n!$.

Theorem 12.31 *The symmetric group S_n has presentation $\langle X \mid \mathcal{R} \rangle$, where*

$$X = \{x_1, \ldots, x_{n-1}\}$$

and let $\mathcal{R}$ consist of the relations

$$x_k^2 = 1, \quad x_{k-1} x_k x_{k-1} = x_k x_{k-1} x_k, \quad x_k x_j = x_j x_k \textit{ for } j - k \ne \pm 1 \qquad \square$$

We close by noting that the relations above are equivalent to

$$x_k^2 = 1, \quad (x_{k-1} x_k)^3 = 1, \quad (x_k x_j)^2 = 1 \textit{ for } j \le k - 2$$

Exercises

1. An equational class $\mathcal{K}$ is *abelian* if all members of the class are abelian groups. Characterize abelian equational classes.
2. Let G be the dicyclic group of order $4n$, $n > 1$, with presentation

$$\langle X \mid \mathcal{R} \rangle = \langle x, y \mid x^{2n} = 1, y^2 = x^n, yx = x^{2n-1} y = x^{-1} y \rangle$$

   a) G has exactly one involution z.
   b) $Z(G) = \langle z \rangle$
   c) $G/Z(G) \approx D_{2n}$
3. Let X be a nonempty set and let (F_X, κ) be universal for X. Let $w(x_1, \ldots, x_n)$ be a word over X' and let $f: \{x_1, \ldots, x_n\} \to \{y_1, \ldots, y_n\}$ be an injection, where $y_i \in X$. Prove that

$$w(\kappa x_1, \ldots, \kappa x_n) = 1 \quad \Leftrightarrow \quad w(\kappa y_1, \ldots, \kappa y_n) = 1$$

4. Let F be the free group on the set $X = \{x_1, \ldots, x_n\}$. Show that F has a subgroup of index m for all $1 \le m \le n$.
5. Let F be the free group on $X = \{x, y\}$ and let $G = \langle a \rangle \boxtimes \langle b \rangle$ be the direct product of two infinite cyclic groups. The function $f: X \to G$ defined by $fx = (a, 1)$, $fy = (1, b)$ induces a unique mediating morphism $\tau: F \to G$. What is the kernel of τ?
6. Let F_X be the free group on X. Prove the following:
   a) If $Y \subseteq X$ is nonempty, then $F_Y \le F_X$.
   b) If $Y \subseteq X$ is nonempty, then $F_Y \cap F_{X \setminus Y} = \{1\}$. In particular, if $x \in X \setminus Y$, then $x \notin F_Y$.
7. Characterize the abelian groups that are free groups.
8. Let F be a free group and let $H \le F$ have finite index. Show that H intersects every nontrivial subgroup of F nontrivially.

9. Let F be free on the disjoint union $X \cup Y$ of nonempty sets. Prove that $F/\langle Y \rangle_{\text{nor}}$ is free on X.
10. Show that if $N \trianglelefteq G$ and G/N is free, then $G = N \rtimes H$ for some $H \le G$.
11. Prove that if $|X| > 1$, then the free group F_X is centerless.
12. Let F be a free group. Suppose that $\sigma: A \twoheadrightarrow B$ is an epimorphism and that $\tau: F \to B$ is a homomorphism. Prove that there is a $\lambda: F \to A$ for which $\sigma \circ \lambda = \tau$. This is called the **projective property** of free groups.
13. If $\mathcal{K}$ is a class of groups, then a group G is **residually $\mathcal{K}$** or a **residually $\mathcal{K}$-group** if for any $1 \ne a \in G$, there is a normal subgroup $N_a \trianglelefteq G$ for which $a \notin N_a$ and G/N_a is a $\mathcal{K}$-group.
    a) Prove that a group G is residually $\mathcal{K}$ if and only if it is isomorphic to a subdirect product of $\mathcal{K}$-groups.
    b) Prove that if $\mathcal{N} = \{N_i \mid i \in I\}$ is a family of normal subgroups of a group G and if G/N_i is a $\mathcal{K}$-group for all $i \in I$, then $G/\bigcap N_i$ is residually $\mathcal{K}$.
14. Prove that the presentation

$$\langle Y \mid S \rangle = \langle c, d \mid c^n = 1, d^m = 1, dc = c^s d \rangle$$

defines the group

$$H = \{c^i d^j \mid 0 \le i < n \text{ and } 0 \le j < m\}$$

where $o(H) = mn$, $o(c) = n$, $o(d) = m$ and

$$(c^i d^k)(c^j d^\ell) = c^{i+js^k} d^{k+\ell}$$

15. Show that D_{2n} is presented by

$$\langle Y \mid S \rangle = \langle x, y \mid x^2 = 1, y^2 = 1, (xy)^n = 1 \rangle$$

16. Let σ and ρ be distinct symbols. Let

$$D = \{\rho^i, \sigma\rho^i \mid i \in \mathbb{Z}\}$$

with product defined by the properties $\sigma^2 = 1, \rho\sigma = \sigma\rho^{-1}$. Thus,

$$\rho^i \sigma = \sigma \rho^{-i}$$

a) Show that D is a group and that D is presented by

$$P_1 = \langle x, y \mid x^2 = 1, yx = xy^{-1} \rangle$$

b) Show that D is also presented by

$$P_2 = \langle x, y \mid x^2 = 1, y^2 = 1 \rangle$$

Any group presented by P_1 or P_2 is called an **infinite dihedral group**.

17. Let $G = \langle X \mid \mathcal{R} \rangle$ be a finite presentation of G, where $X = \{x_1, \ldots, x_n\}$ and $\mathcal{R}' = \{r_1, \ldots, r_m\}$ and $m < n$. Show that G is an infinite group as follows.

a) Reduce the problem to the abelian case as follows. Let A_X be the free abelian group on X. Show that there is an epimorphism from $F_X/\langle \mathcal{R}' \rangle_{\text{nor}}$ to $A_X/\langle \mathcal{R}' \rangle$.
b) Show that $A_X/\langle \mathcal{R}' \rangle$ is infinite.

Chapter 13
Abelian Groups

In this chapter, we study abelian groups. We will write abelian groups using additive notation. One of our main goals is to provide a complete solution to the classification problem for finitely-generated abelian groups. That is, we will describe all finitely-generated abelian groups up to isomorphism.

Perhaps the most natural place to begin is to observe that the elements of finite order in an abelian group A form a subgroup of A, since

$$o(ab) = \mathrm{lcm}(o(a), o(b))$$

Let us remind the reader that this is not the case in a general group. For example, in $GL(2, \mathbb{C})$, let

$$A = \begin{pmatrix} 0 & -1 \\ 1 & 0 \end{pmatrix} \quad \text{and} \quad B = \begin{pmatrix} 0 & 1 \\ -1 & -1 \end{pmatrix}$$

Then A and B have finite order but AB has infinite order.

Definition *Let A be an abelian group. An element $a \in A$ that has finite order is called a* **torsion element**. *The subgroup A_{tor} of all torsion elements in A is called the* **torsion subgroup** *of A. A group that has no nonzero torsion elements is said to be* **torsion free** *and a group all of whose elements are torsion elements is said to be* **torsion**.□

Of course, a finite group is a torsion group, but the converse is not true: Consider the direct product $\mathbb{Z}_2^{\aleph_0}$ of an infinite number of copies of $\mathbb{Z}_2$.

The quotient A/A_{tor} is easily seen to be torsion free and it would be nice if A_{tor} was always a direct summand of A, that is, if

$$A = A_{\mathrm{tor}} \bowtie B$$

for some B, A, since then $B \approx A/A_{\mathrm{tor}}$ would be torsion free and we would have a nice decomposition of any abelian group. Unfortunately, this is not the case. To show this, we require a definition.

Definition *Let A be an abelian group. An element $a \in A$ is* **divisible** *by an integer n if there is an element $b \in A$ for which $a = nb$. A group A is* **divisible** *if every element is divisible by every nonzero integer.*$\square$

Theorem 13.1 *The torsion subgroup of an abelian group need not be complemented.*

Proof. Let $A = \boxtimes \mathbb{Z}_p$ be the external direct product of the abelian groups $\mathbb{Z}_p$, taken over all primes p. For $a \in A$, we use the notation a_p in place of $a(p)$. The torsion subgroup A_{tor} is the subgroup of all elements with finite support. If $A = A_{\text{tor}} \bowtie B$, then $B \approx A/A_{\text{tor}}$, so this prompts us to look for an isomorphism-invariant property that holds in A/A_{tor} but not in B. This property is divisibility.

Specifically, we will show the following:

1) No nonzero element of A (and hence B) is divisible by all primes p.
2) There are nonzero elements of A/A_{tor} that are divisible by all primes p.

Since the only element of $\mathbb{Z}_p$ that is divisible by p is 0, if $a \in A$ is divisible by p, then $a_p = 0$. Hence, if a is divisible by all primes, it follows that $a = 0$.

Now let $a \in A$ be the element for which $a_p = 1$ for all p. For a given prime p, to say that $a = pb$ for some $b \in A$ is to say that $pb_q = 1$ for all primes q. But if $q \neq p$, then p has an inverse r_p in the field $\mathbb{Z}_q$. Hence, if b is defined by $b_p = r_p$ and $b_p = 0$, then

$$(a - pb)_p = a_p - pb_p = 1$$

and for all $q \neq p$,

$$(a - pb)_q = a_q - pb_q = 0$$

and so $a - pb \in A_{\text{tor}}$.$\square$

Despite the negative nature of the previous result, we will show that if A is a *finitely-generated* abelian group, then A_{tor} is complemented. This is a key to the structure theorem for finitely-generated abelian groups.

An Abelian Group as a $\mathbb{Z}$-Module

An abelian group A has a natural scalar multiplication defined upon it, namely, multiplication by the integers: If $\alpha \in \mathbb{Z}$ and $a \in A$, we set

$$\alpha a = \begin{cases} 0 & \text{if } \alpha = 0 \\ \underbrace{a + \cdots + a}_{\alpha \text{ terms}} & \text{if } \alpha > 0 \\ -(-\alpha)a & \text{if } \alpha < 0 \end{cases} \tag{13.2}$$

Under this operation, an abelian group A is a $\mathbb{Z}$-module, as defined below.

Definition *Let R be a commutative ring with identity. An* R-**module** *(or a* **module over** R*) is an abelian group M, together with a scalar multiplication, denoted by juxtaposition, that assigns to each pair $(r, u) \in R \times M$, an element $rv \in M$. Furthermore, the following properties must hold for all $r, s \in R$ and $u, v \in M$:*

$$\begin{aligned} r(u + v) &= ru + rv \\ (r + s)u &= ru + su \\ (rs)u &= r(su) \\ 1u &= u \end{aligned}$$

The ring R is called the **base ring** *of M and the elements of R are called* **scalars**. □

Note that an abelian group A is a $\mathbb{Z}$-module and, conversely, a $\mathbb{Z}$-module is nothing more than an abelian group, since the scalar multiplication of a $\mathbb{Z}$-module M must be the operation defined in (13.2). Moreover, the subgroups of the abelian group M are the submodules of the module M and the group homomorphisms between the abelian groups M to N are the linear (module) maps between the $\mathbb{Z}$-modules M and N.

The Classification of Finitely-Generated Abelian Groups

We solved the classification problem for finite abelian groups in Theorem 5.7. Also, Theorem 12.14 solves the classification problem for free abelian groups. Using these theorems, we can now solve the classification problem for finitely-generated abelian groups. The first step is to note the following.

Theorem 13.3 *A finitely-generated abelian group A is torsion free if and only if it is free.*

Proof. We leave proof that if A is free, then it is torsion free as an exercise. For the converse, let $S = \{v_1, \ldots, v_n\}$ be a generating set for the torsion-free abelian group A. The proof is based on the fact that since A is torsion free, it is a torsion-free $\mathbb{Z}$-module. Moreover, for any $a \in A$, the multiplication map $\mu_a \colon A \to A$ defined by $\mu_a x = ax$ is a $\mathbb{Z}$-module automorphism of A.

Let $\{u_1,\ldots,u_k\}$ be a maximal linearly independent subset of S. Of course, if $k=n$, then S is a basis for A and so A is free. Assume otherwise and let

$$S=\{u_1,\ldots,u_k,v_1,\ldots,v_{n-k}\}$$

For each v_i, the set $\{u_1,\ldots,u_k,v_i\}$ is linearly dependent and so there exist $a_i\in\mathbb{Z}$ for which

$$a_iv_i\in\langle u_1,\ldots,u_k\rangle$$

If $a=\alpha_1\cdots\alpha_{n-k}$, then

$$\mu_aA=a\langle u_1,\ldots,u_k,v_1,\ldots,v_{n-k}\rangle\subseteq\langle u_1,\ldots,u_k\rangle$$

and since the latter is a free abelian group, Theorem 12.18 implies that μ_aA is also free and therefore so is A.□

If A is a finitely-generated abelian group A, then A_{tor} is a subgroup of A and the quotient A/A_{tor} is torsion-free and finitely generated and so is free. Since the canonical projection map $\pi\colon A\to A/A_{\text{tor}}$ is an epimorphism, Theorem 12.19 implies that A_{tor} has a complement:

$$A=F\bowtie A_{\text{tor}}$$

where $F\approx A/A_{\text{tor}}$ is free and finitely-generated. Moreover, Theorem 12.16 implies that F has finite rank and since A_{tor} is finitely generated (Theorem 2.21), torsion and abelian, it is finite.

As to uniqueness of this decomposition, if

$$A=F'\bowtie T$$

where F' is free and T is torsion, then clearly, $T\leq A_{\text{tor}}$. But if $a\in A_{\text{tor}}$ and $a=t+f$ where $t\in T$ and $f\in F'$ and so $f=a-t$ is torsion, whence $f=0$ and $a\in T$. Thus, $T=A_{\text{tor}}$. It follows that F and F' are both complements of A_{tor} and hence are isomorphic.

We can now state the fundamental theorem of finitely-generated abelian groups.

Theorem 13.4 (The fundamental theorem of finitely-generated abelian groups) *Let A be a finitely-generated abelian group, with torsion subgroup A_{tor}.*

1) Then

$$A=F\bowtie A_{\text{tor}}$$

where F is free of finite rank r and A_{tor} is finite. As to uniqueness, if

$$A=F'\bowtie T$$

where F' is free and T is torsion, then $T = A_{\text{tor}}$ and $\text{rk}(F) = \text{rk}(F')$. The number $r = \text{rk}(F)$ is called the **free rank** *of A.*

2) **(Invariant factor decomposition)** *A is the direct sum of a finite number of cyclic subgroups*

$$A = \langle x_1 \rangle \bowtie \cdots \bowtie \langle x_r \rangle \bowtie \langle u_1 \rangle \bowtie \cdots \bowtie \langle u_n \rangle$$

where $o(x_i) = \infty$ and $o(u_i) = \alpha_i \geq 2$ and

$$\alpha_n \mid \alpha_{n-1} \mid \cdots \mid \alpha_1$$

The orders α_i are called the **invariant factors** *of A.*

3) **(Primary cyclic decomposition)** *If*

$$\alpha_k = p_1^{e_{k,1}} \cdots p_m^{e_{k,m}}$$

then

$$\begin{aligned} A = {} & \langle x_1 \rangle \bowtie \cdots \bowtie \langle x_r \rangle \\ & \bowtie [\langle u_{1,1} \rangle \bowtie \cdots \bowtie \langle u_{1,k_1} \rangle] \bowtie \cdots \bowtie [\langle u_{m,1} \rangle \bowtie \cdots \bowtie \langle u_{m,k_m} \rangle] \end{aligned}$$

where $o(x_i) = \infty$ and $o(u_{i,j}) = p_i^{e_{i,j}}$ and

$$e_{i,1} \geq e_{i,2} \geq \cdots \geq e_{i,k_i} \geq 1$$

The numbers $p_i^{e_{i,j}}$ are called the **elementary divisors** *of A.*

3) *The multiset $\{\alpha_i\}$ of invariant factors and the multiset $\{p_i^{e_{i,j}}\}$ of elementary divisors are uniquely determined by the group A.*□

Projectivity and the Right-Inverse Property

A diagram of the form

$$A \xrightarrow{\sigma} B \xrightarrow{\tau} C$$

where A, B and C are groups and σ and τ are group homomorphisms is **exact** if

$$\text{im}(\sigma) = \ker(\tau)$$

It is customary to regard the figure

$$A \xrightarrow{\sigma} B \longrightarrow 0$$

as exact and to omit the second homomorphism, since it must be the zero map. Thus, this figure says precisely that σ is surjective. Similarly, the figure

$$0 \longrightarrow A \xrightarrow{\sigma} B$$

says precisely that σ is injective.

According to Theorem 5.23, a group homomorphism $\sigma: G \to G'$ has a right inverse σ_R if and only if it is surjective and $\ker(\sigma)$ is complemented:

$$G = \ker(\sigma) \rtimes K$$

for some $K \le G$ and in this case,

$$G = \ker(\sigma) \rtimes \operatorname{im}(\sigma_R)$$

Let us say that an abelian group P has the **right-inverse property** if every epimorphism $\sigma: A \twoheadrightarrow P$, where A is abelian, has a right inverse. This is illustrated in Figure 13.1.

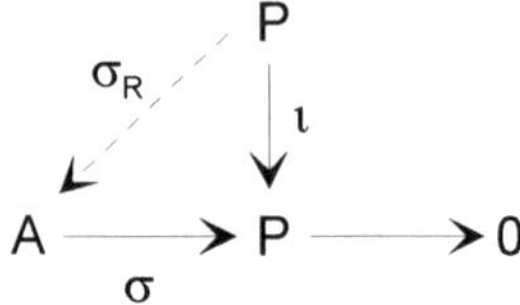

Figure 13.1

An apparently stronger property is given in the following definition.

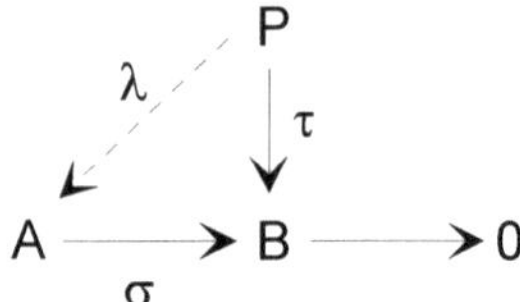

Figure 13.2

Definition *An abelian group P is* **projective** *if, referring to Figure 13.2, for any epimorphism $\sigma: A \twoheadrightarrow B$ of abelian groups and any homomorphism $\tau: P \to B$, there is a homomorphism $\lambda: P \to A$ for which*

$$\sigma \circ \lambda = \tau$$

In this case, we say that τ can be **range-lifted** *to λ.*□

While the projective property appears to be stronger than the right-inverse property, the two properties are actually equivalent. The following theorem is the main result on projective abelian groups.

Theorem 13.5 *Let G be an abelian group. The following are equivalent:*
1) *G is a free abelian group.*
2) *G is projective.*
3) *G has the right-inverse property, that is, any surjection $\tau: A \to G$, where A is abelian, has a right inverse.*
4) *If $\sigma: A \twoheadrightarrow G$, where A is abelian, then $\ker(\sigma)$ is a direct summand of A.*

Proof. We have seen that 3) and 4) are equivalent. Assume that 1) holds and let $G = F_X$ be free on X. Let $\sigma: A \to B$ be surjective and let $\tau: F_X \to B$. Then for each $x \in X$, there is an $a_x \in A$ for which $\sigma a_x = \tau x$. Define a function $f: X \to A$ by $f(x) = a_x$. Since F_X is free, there is a unique homomorphism $\lambda: F_X \to A$ for which $\lambda x = a_x$. Then

$$\sigma \circ \lambda(x) = \sigma a_x = \tau x$$

and so $\sigma \circ \lambda = \tau$ on X and therefore on F_X. Hence, $G = F_X$ is projective and 2) holds. It is clear that 2) implies 3).

Finally, suppose that 3) holds. The identity map $\iota: G \to G$ can be lifted to a homomorphism $\sigma: F_G \to G$ where F_G is the free abelian group with basis G. Of course, σ is surjective and so 3) implies that σ has a right inverse $\sigma_R: G \to F_G$. Hence, $\ker(\sigma)$ is complemented, that is,

$$F_G = \ker(\sigma) \bowtie S$$

But $\sigma: S \to G$ is an isomorphism and so G is isomorphic to a direct summand of a free abelian group and is therefore free abelian by Theorem 12.18. Hence, 1) holds.□

Injectivity and the Left-Inverse Property

We have seen that a monomorphism $\sigma: A \hookrightarrow B$ has a left-inverse σ_L if and only if $\operatorname{im}(\sigma)$ has a complement K in B, in which case $K \approx \ker(\sigma_L)$. Dual to the right-inverse property is the **left-inverse property**: An abelian group E has the left-inverse property if every monomorphism $\sigma: E \hookrightarrow B$ to an abelian group B has a left inverse.

Dual to the projective property is the injective property.

Definition *An abelian group E is* **injective** *if, referring to Figure 13.3, for any embedding $\sigma: A \hookrightarrow B$ and any homomorphism $\tau: A \to E$ there is a homomorphism $\lambda: A \to E$ for which*

$$\lambda \circ \sigma = \tau$$

In this case, we say that $\tau: A \to E$ can be **domain-lifted** *to $\lambda: B \to E$ by $\sigma: A \to B$.*□

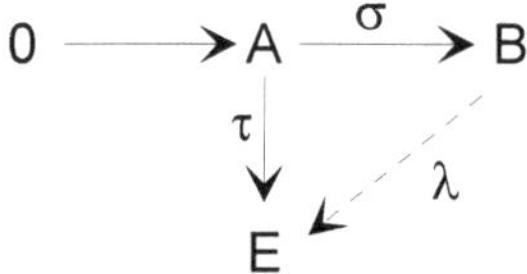

Figure 13.3

Baer [3] proved that if this condition holds in the special case where $B = \mathbb{Z}$ is the group of integers and $A \leq \mathbb{Z}$ and $\sigma: A \to \mathbb{Z}$ is the inclusion map, then the condition holds in general and E is injective.

Theorem 13.6 *An abelian group E is injective if and only if it satisfies* **Baer's criterion**: *Any homomorphism $\tau: I \to E$, where I is a subgroup of the integers $\mathbb{Z}$ can be extended to a homomorphism $\lambda: \mathbb{Z} \to E$ on $\mathbb{Z}$.*

Proof. We wish to show that any homomorphism $\tau: A \to E$ can be domain-lifted to $\lambda: B \to E$ by any monomorphism $\sigma: A \hookrightarrow B$. Note that τ can be domain-lifted by σ to $\tau \circ \sigma^{-1}: \sigma A \to E$, since $\sigma: A \to \sigma A$ is an isomorphism. This is shown in Figure 13.4.

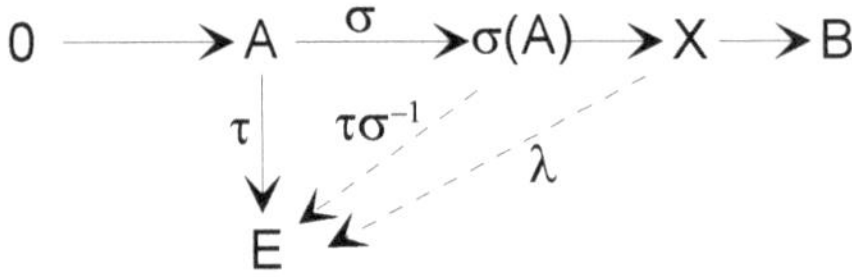

Figure 13.4

Suppose we have domain-lifted τ by σ to $\lambda: X \to B$, where $\sigma A \leq X \leq B$, that is,

$$\tau = \lambda \circ \sigma$$

If $X < B$, then for any $a \in X \setminus B$, we can lift λ by σ to a map on $X + \langle a \rangle$ as follows. We need only define λ on $\langle a \rangle = \mathbb{Z}a$. But λ is already defined on $\mathbb{Z}a \cap X = Ia$ for some $I \leq Z$. So if $\lambda_1: I \to E$ is defined by

$$\lambda_1(\alpha) = \lambda(\alpha a)$$

then Baer's criterion implies that λ_1 can be extended to $\lambda_2: \mathbb{Z} \to E$. Then the map $\overline{\lambda}: X + \langle a \rangle \to E$ defined by

$$\overline{\lambda}(x + ra) = \lambda(x) + \lambda_2(r)a$$

for any $r \in \mathbb{Z}$ is well defined since if $x + ra = y + sa$, then $x - y = (s - r)a$ and so $s - r \in I$, which implies that

$$\lambda_1(s - r) = \lambda((s - r)a) = \lambda(x - y)$$

and so

$$\overline{\lambda}(x + ra) = \lambda(x) + \lambda_2(r)a = \lambda(y) + \lambda_2(s)a = \overline{\lambda}(y + sa)$$

Moreover, since $\lambda \circ \sigma = \tau$ and $\text{im}(\sigma) \leq \text{dom}(\lambda)$, it follows that $\overline{\lambda} \circ \sigma = \tau$.

This discussion prompts us to apply Zorn's lemma. Let $\mathcal{S}$ be the collection of all pairs (X, λ), where $\sigma(A) \subseteq X \subseteq B$ and λ is a lifting of τ by σ to a map on X. Then $\mathcal{S}$ is nonempty since we may take $X = \sigma A$. Order $\mathcal{S}$ by setting $(X, \mu) \leq (Y, \lambda)$ if $X \subseteq Y$ and $\lambda|_Y = \mu$.

If $\mathcal{C} = \{(X_i, \mu_i)\}$ is a chain in $\mathcal{S}$, let $U = \bigcup X_i$. If $x \in X_i \cap X_j$, then one of μ_i and μ_j is an extension of the other and so $\mu_i x = \mu_j x$. Hence, we may define μ by $\mu x = \mu_i x$ for any i satisfying $x \in X_i$. Then (U, μ) is an upper bound for $\mathcal{C}$ and so Zorn's lemma implies that $\mathcal{S}$ has a maximal element (M, λ). But if $M < B$, then there is a further lifting of λ, which contradicts the maximality of (M, λ) and so $M = B$.$\square$

We can now present our main theorem on injective groups.

Theorem 13.7 *Let G be an abelian group. The following are equivalent:*
1) *G is injective.*
2) *G is divisible.*
3) *G has the left-inverse property, that is, every monomorphism $\sigma: G \hookrightarrow B$ to an abelian group B has a left inverse.*
4) *If $\sigma: G \hookrightarrow A$, where A is abelian, then* $\operatorname{im}(\sigma)$ *is a direct summand of A.*

Proof. We know that 3) and 4) are equivalent. Assume first that G is injective. For $g \in G$ and $n \in \mathbb{Z}$ we seek $h \in G$ for which $g = nh$. The map $\tau: \langle n \rangle \to E$ defined by $\tau(rn) = rg$ for all $r \in \mathbb{Z}$ can be domain-lifted by the inclusion map $j: \langle n \rangle \to \mathbb{Z}$, that is,

$$\lambda \circ j = \tau$$

Hence,

$$g = \tau(n) = \lambda \circ j(n) = \lambda(n) = n\lambda(1)$$

and so $h = \lambda(1)$. Thus, G is divisible and 1) implies 2).

To see that 2) implies 1), we show that 2) implies the Baer criterion. Let $\tau: \langle k \rangle \to G$ where $\langle k \rangle \leq \mathbb{Z}$. To show that τ can be extended to $\mathbb{Z}$, define $\lambda_a: \mathbb{Z} \to G$ by $\lambda_a(n) = na$, for $a \in G$. Then λ_a extends τ if $\lambda_a(k) = \tau(k)$, that is, if $\tau(k) = ka$. But since G is divisible, there is an $a \in G$ for which this holds. Hence, 2) implies 1).

It is clear that 1) implies 3). Finally, suppose that E has the left-inverse property. We wish to show that E is injective. The story of the proof is shown in Figure 13.5.

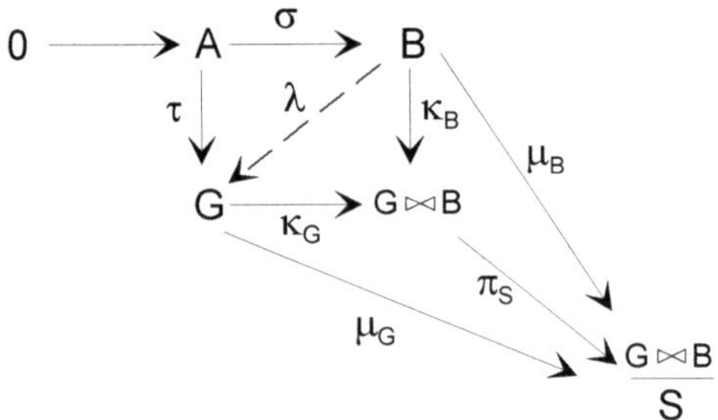

Figure 13.5

We seek the map $\lambda: B \to G$ for which $\lambda \circ \sigma = \tau$. A first attempt might be to consider the direct sum $G \bowtie B$, with canonical injections κ_G and κ_B. Since $\kappa_G: G \to G \bowtie B$ is an injection, it has a left inverse $(\kappa_G)_L$ and we can take $\lambda = (\kappa_G)_L \circ \kappa_B$. However, it may not be the case that $\lambda \circ \sigma = \tau$.

On the other hand, perhaps we can factor the direct sum $G \bowtie B$ by a subgroup S, with projection map π_S in such a way that the compositions

$$\mu_B = \pi_S \circ \kappa_B: B \to (G \bowtie B)/S$$

and

$$\mu_G = \pi_S \circ \kappa_G: G \to (G \bowtie B)/S$$

satisfy

$$\mu_B \circ \sigma = \mu_G \circ \tau$$

where μ_G is left-invertible. In this case, if $\lambda = (\mu_G)_L \circ \mu_B: B \to G$, then

$$\lambda \circ \kappa = (\mu_G)_L \circ \mu_B \circ \sigma = (\mu_G)_L \circ \mu_G \circ \tau = \tau$$

as desired. But the condition $\mu_B \circ \sigma = \mu_G \circ \tau$ is

$$\pi_S \circ \kappa_B \circ \sigma = \pi_S \circ \kappa_G \circ \tau$$

that is,

$$(0, \sigma a) + S = (\tau a, 0) + S$$

for all $a \in A$ and so if

$$S = \{(\tau a, -\sigma a) \mid a \in A\}$$

then this equation holds. Also, μ_G is injective since $g \in \ker(\mu_G)$ implies that $(g, 0) \in S$ and so $(g, 0) = (\tau a, -\sigma a)$. Hence, $\sigma a = 0$ and so $a = 0$, whence $g = 0$.□

Exercises

1. Let A be a torsion-free abelian group. Prove that if $a \in A$ is divisible by $n \in \mathbb{Z}$, then the "quotient" $b \in A$ is unique.
2. Prove that a subset $\mathcal{B}$ of an abelian group A is a basis if and only if for every $a \in A$, there are unique elements $b_1, \ldots, b_n \in \mathcal{B}$ and unique scalars $\alpha_1, \ldots, \alpha_n \in \mathbb{Z}$ for which
$$a = \alpha_1 b_1 + \cdots + \alpha_n b_n$$
3. Let A be a free abelian group. Show that it is not necessarily true that any linearly independent set of size $\mathrm{rk}(A)$ is a basis for A.
4. Let A be a free abelian group of finite rank and let $A = B \bowtie C$. Prove that $\mathrm{rk}(A) = \mathrm{rk}(B) + \mathrm{rk}(C)$.
5. Prove that any abelian group A is isomorphic to a quotient of a free abelian group.
6. A subgroup $H \leq G$ of an additive abelian group G is **pure** if for any $h \in H$ and $m \in \mathbb{Z}$,
$$h = mg \text{ for } g \in G \quad \Rightarrow \quad h = mh' \text{ for some } h' \in H$$
   a) Prove that if G is an abelian group, then the set of periodic elements of G is pure.
   b) Prove that any direct summand of an abelian group is pure.
   c) Find a nonpure subgroup of the cyclic group $\mathbb{Z}_{n^2}$.
7. Prove that an abelian group A is finitely generated if and only if it is isomorphic to a quotient of a free abelian group of finite rank.
8. Let A be a free abelian group of finite rank n. Let $S = \{s_1, \ldots, s_n\}$ be a generating set for A. Prove that S is a basis for A. *Hint*: Let $X = \{x_1, \ldots, x_n\}$ be a basis for A and define the map $\tau: A \to A$ by $\tau(x_i) = s_i$ and extending to a surjective homomorphism. What about $\ker(\tau)$?
9. Let F be a free abelian group of rank n. Let H be a subgroup of F of rank $k < n$. Prove that G/H contains an element of infinite order.
10. Show that, in general, a basis for a subgroup of a free abelian group cannot be extended to a basis for the entire group.
11. Prove that any free abelian group is torsion free.
12. Let G be a torsion-free abelian group. Suppose that G has a subgroup F that is free and has finite index. Prove that G is free abelian.
13. Let A be a finite abelian group.
   a) Prove that if $pA = \{0\}$, then A is a vector space over the field $\mathbb{Z}_p = \mathbb{Z}/p\mathbb{Z}$.
   b) Prove that for any subgroup S of A the set
$$S^{(p)} = \{v \in S \mid pv = 0\}$$

is also a subgroup of A and if $A = S \bowtie T$, then

$$A^{(p)} = S^{(p)} \bowtie T^{(p)}$$

14. Let F be a free abelian group of rank n. Show that $\operatorname{Aut}(F)$ is isomorphic to the group of all $n \times n$ matrices with determinant equal to ± 1.
15. Let $\mathbb{Q}^+$ be the multiplicative group of positive rational numbers.
    a) Show that $\mathbb{Q}^+$ is isomorphic to the additive group $\mathbb{Z}[x]$ of polynomials over the integers. *Hint*: Use the fundamental theorem of arithmetic.
    b) Show that the multiplicative group $\mathbb{Q}^+$ of positive rationals is a free abelian group of countably infinite rank.
16. Prove that the quasicyclic group $\mathbb{Z}(p^\infty)$ is divisible.
17. Prove that a free abelian group is not divisible.
18. a) Show that the rational numbers $\mathbb{Q}$ are not finitely generated as an abelian group under addition.
    b) Show that $\mathbb{Q}$ is divisible and therefore not free by an earlier exercise.
    c) Show that $\mathbb{Q}$ is torsion free.

    Thus, a torsion-free abelian group need not be free.
19. Prove that if $A = \boxtimes A_i$, where A and A_i are abelian groups, then A is injective if and only if A_i is injective for all i.
20. Let A be an abelian group. Show that the set A_{div} of all divisible elements is a subgroup of A. Show that A_{div} is a direct summand of A.
21. Let D be a divisible group. Prove that D_{tor} is divisible.
22. Let A be a finitely generated abelian group. For any subgroup S of A let

    $$A_{(p)} = \{a \in A \mid a \text{ is a } p\text{-element}\}$$

    Show that $A_{(p)} \sqsubseteq A$. Describe the order of $A_{(p)}$ in terms of the elementary divisors of A.
23. How can one tell from the elementary divisors of a finite abelian group when that group is cyclic?
24. Use one of the decomposition theorems to prove that a finite abelian group A has a subgroup of order k for every $k \mid o(A)$.
25. Find, up to isomorphism, all finite abelian groups of order 1000. Which are cyclic?
26. Prove that every abelian group of order 426 is cyclic.
27. Let p be a prime. Let

    $$A = \langle u_1 \rangle \bowtie \cdots \bowtie \langle u_m \rangle$$

    where $o(u_i) = p^{f_i}$ and let

    $$B = \langle v_1 \rangle \bowtie \cdots \bowtie \langle v_n \rangle$$

    where $o(v_i) = p^{e_i}$. Assume that $A \leq B$.
    a) Prove that $m \leq n$.

b) Prove that

$$f_m \leq e_n, f_{m-1} \leq e_{n-1}, \ldots, f_1 \leq e_{n-m+1}$$

References

[1] Allen, W., Complex Groups, *The American Mathematical Monthly*, Vol. 67, No. 7 (1960) 637–641.

[2] Aschbacher, M., The Status of the Classification of Finite Simple Groups, *Notices of the AMS*, Volume 51, Number 7, 2004.

[3] Baer, R., Situation der Untergruppen und Struktur der Gruppe, *S.-B. Heidelberg. Akad. Math.-Nat. Klasse* 2 (1933) 12–17.

[4] Brauer, R. and Fowler, K. A., On Groups of Even Order, *Annals of Mathematics*, 2nd Ser., Vol. 62, No. 3 (1955) 565–583.

[5] Cassidy, P. J., Products of commutators are not always commutators: An example, *The American Mathematical Monthly*, Vol. 86, No. 9 (1979) 772.

[6] Cauchy, A.-L., Mémoire sur les arrangements que l'on peut former avec des lettres données et sur les permutations ou substitutions à l'aide desquelles on passe d'un arrangement à un autre. Exercises d'analyse et de physique mathématique 3 (1845) 151–252. (Reprinted in: Oeuvres complètes, vol. 13, second series. Gauthier-Villars, Paris (1932) pp. 171–282.)

[7] Cayley, A., On the theory of groups as depending on the symbolic equation $\theta^n = 1$, Phil. Mag., 7 (4) (1854), 40–47.

[8] Christensen, C., Complementation in groups, Math. Zeitschr. 84 (1964) 52–69.

[9] Christensen, C., Groups with complemented normal subgroups, J. London Math. Soc., 42 (1967) 208–216.

[10] Feit, W. and Thompson, J. G., A Solvability Criterion for Finite Groups and Some Consequences, *Proc. Nat. Acad. Sci. USA* 48 (1962) 968–970.

[11] Feit, W. and Thompson, J. G., Solvability of Groups of Odd Order, *Pacific J. Math* 13 (1963) 775–1029.

[12] Frattini, G., Intorno alla generazione dei gruppi di operazioni, *Rend. Atti Accad. Lincei*, (4) 1 (1885) 281-285, 455–457.

[13] Frobenius, G., Berliner Sitzungsberichte, (1895) 995.

[14] Fuchs, L., *Abelian Groups*, Pergammon Press, 1960.

[15] Guralnick, R., Expressing group elements as commutators, *Rocky Mountain J. Math.*, Vol. 10, No. 3 (1980), 651–654.

[16] Hall, M., *The Theory of Groups*, AMS Chelsea, 1976.

[17] Hall, P., Complemented groups, *J. London Math. Soc.*, 12, (1937), 201–204.

[18] Hirshon, R., On Cancellation in Groups, *American Mathematical Monthly*, Vol. 76, No. 9 (1969) 1037–1039.

[19] Kappe, L.-C. and Morse, R., On commutators in groups, *Proceedings of Groups St. Andrews 2005 LMS Lecture Series*, also available online at http://faculty.evansville.edu/rm43/publications/commutatorsurvey.pdf.

[20] Kappe, L.-C. and Morse, R., On commutators in p-groups, *J. Group Theory*, Vol. 8, No. 4 (2005), 415–429.

[21] Kertész, A., On groups every subgroup of which is a direct summand, *Publ. Math. Debrecen*, 2 (1952), 74–75.

[22] MacDonald, I. D., Commutators and their products, *Amer. Math. Monthly*, Vol. 93, No. 6 (1986), 440–444.

[23] McKay, J. H., Another proof of Cauchy's group theorem, *American Mathematical Monthly* 66, 1959, 119.

[24] Meo, M., The mathematical life of Cauchy's group theorem, *Historia Mathematica* 31 (2004) 196–221.

[25] Miller, G. A., The regular substitution groups whose orders is less than 48, *Quarterly Journal of Mathematics*, 28, (1896), 232–284.

[26] Robinson, D., *A Course in the Theory of Groups*, Springer, 1996.

[27] Roman, S., *Field Theory*, Second Edition, Springer, 2006.

[28] Roman, S., *Advanced Linear Algebra*, Third Edition, Springer, 2007.

[29] Schreier, O., Über den Jordan-Hölderschen Satz, Abh. math. Sem. Univ. Hamburg 6 (1928), 300-302.

[30] Schmidt, R., *Subgroup Lattices of Groups*, De Gruyter Expositions in Mathematics 14, 1994.

[31] Spiegel, E., Calculating commutators in groups. *Math. Mag.*, Vol. 49, No. 4 (1976), 192–194.

[32] Sylow, L., Théorèmes sur les groupes de substitutions, *Math. Ann.* 5 (1872), 584–594.

[33] Weber, H., *Lehrbuch der Algebra*, Vol 2. Braunschweig, second edition, 1899.

[34] Weigold, J., On Direct Factors in Groups, *J. London Math Soc.*, 35 (1960), 310–320.

[35] Wielandt, H., Eine Verallgemeinerung der invarianten Untergruppen, *Math. Z.*, 45 (1939), 209–244.

[36] Weinstein, M., Examples of Groups, Second Edition, Polygonal Publishing, 2000.

[37] Zassenhaus, H. J., Qum Satz von Jordan-Hölder-Schreier, *Abh. Math. Semin. Hamb. Univ.* 10 (1934) 106–108.

References on the Burnside Problem

[38] Adjan, S.I., *The Burnside problem and identities in groups*, Translated from the Russian by John Lennox and James Wiegold, Ergebnisse der

Mathematik und ihrer Grenzgebiete [Results in Mathematics and Related Areas], 95 (1979), Springer-Verlag.

[39] Burnside, W., On an Unsettled Question in the Theory of Discontinuous Groups, *Quart. J. Pure Appl. Math.* 33 (1902) 230–238.

[40] Golod, E. S., On Nil-Algebras and Residually Finite -Groups, *Isv. Akad. Nauk SSSR Ser. Mat.* 28 (1964) 273–276.

[41] Hall, M., Solution of the Burnside Problem for Exponent Six, *Ill. J. Math.* 2 (1958) 764–786.

[42] Ivanov, S., On the Burnside Problem on Periodic Groups, *Bulletin of the American Mathematical Society*, Volume 27, Number 2 (1992) 257–260.

[43] Levi, F. and van der Waerden, B. L., Über eine besondere Klasse von Gruppen, *Abh. Math. Sem. Univ. Hamburg* 9 (1933) 154–158.

[44] Novikov, P. S. and Adjan, S. I., Infinite Periodic Groups I, II, III., *Izv. Akad. Nauk SSSR Ser. Mat.* 32 (1968) 212–244, 251–524, and 709–731.

[45] Sanov, I. N., Solution of Burnside's problem for exponent four, *Leningrad State Univ. Ann. Math.* Ser. 10 (1940) 166–170.

List of Symbols

$\leq$: subgroup
$\prec$: $a \prec b$ means that b covers a
$:=$ the item on the left is defined by the item on the right
$=:$ the item on the right is defined by the item on the left
$\sqcup$: disjoint union of sets
$\times$: cartesian product
$\bowtie$: internal direct sum; $\oplus$ in the abelian case
$\boxtimes$: external direct product
$\boxplus$: external direct sum
$\rtimes$: semidirect product
$K \rtimes H \bmod J$: $K \trianglelefteq G, G = HK, H \cap K = J$
$\sqsubseteq$: $H \sqsubseteq G$ means that H is characteristic in G
$\sqsubset$: $H \sqsubset G$ means that H is characteristic and proper in G
$\hookrightarrow$: denotes an embedding (injective map)
$[u]$: the equivalence class containing u
(k, n): If k and n are integers, this is $\gcd(k, n)$
ACC: ascending chain condition
BCC: both chain conditions
DCC: descending chain condition
$\mathcal{DS}(G)$: the set of direct summands in G
$C_n(a)$: the cyclic group $\langle a \rangle$ of order n
$\mathrm{nc}(X, G)$: the normal closure of X in G
ι: identity map
$I_n = \{1, \ldots, n\}$
Y_p for Sylows so we do not conflict with symmetric group.
$\Phi(G)$: the Frattini subgroup of G
$(G : H)$: index of H in G
$\gcd(a, b)$: greatest common divisor of a and b
$\mathrm{lcm}(a, b)$: least common multiple of a and b
$[H, K]$: the commutator subgroup of H and K
H°: the normal interior of H
$H \bullet K$: The set product HK where H and K are essentially disjoint.
$H_{(i)}$: $\bigvee\{H_j \mid j \neq i\}$

$H_{(k)}$: a term in the sequence of normal closures of $H \leq G$
$a^M = \{a^m \mid m \in M\}$
$\mathrm{conj}_G(H)$: the set of conjugates of H by G
$\mathrm{Fix}_X(G)$: the set of elements of X fixed by the action of G
H^G: the normal closure of H in G
$\Gamma_G(H) = [H, G]$
$\zeta_G(H) = X$ where $X/H = Z(G/H)$
$\mathcal{C}[G]$: the set of commutators of G
$\mathcal{K}(G)$: the set of all normal subgroups of G that belong to class $\mathcal{K}$
$G/\mathcal{K}$: the set of all normal subgroups N of G for which G/N belongs to class $\mathcal{K}$
$\mathcal{L}(\mathcal{K})$: The set of laws of $\mathcal{K}$ groups
$\mathrm{sub}(G)$: the lattice of subgroups of a group G
$\mathrm{sub}_d(G)$: the subgroups of a finite p-group G of order p^d
Ω-$\mathrm{sub}(G)$: the lattice of Ω-subgroups of a group G
$\mathrm{nor}(G)$: the lattice of normal subgroups of a group G
$\mathrm{nor}_d(G)$: the normal subgroups of a finite p-group G of order p^d
Ω-$\mathrm{nor}(G)$: the lattice of normal Ω-subgroups of a group G
$\mathrm{sub}(N; G)$: the family of all subgroup of G containing N
$\mathrm{subn}_\Omega(H; G)$: the family of all Ω-subnormal subgroups of an Ω-group G that contain H
$\mathrm{subn}_\Omega(G)$: the family of all Ω-subnormal subgroups of an Ω-group G
$\mathrm{Syl}_p(G)$: the Sylow p-subgroups of G
$\mathrm{Syl}_p(S; G)$: the Sylow p-subgroups of G that contain S
$\mathrm{supp}(f)$: the support of f
$s(H, G)$: the length of the sequence of normal closures of H in G
$\mathcal{S}_d(G)$: for a p-group G, the set of subgroups of order p^d
$\mathcal{N}_d(G)$: for a p-group G, the set of normal subgroups of order p^d
$\wp(S)$: the power set of a set S
$\exists\mathrm{CompSer}_\Omega(G)$: G has an Ω-composition series
$\exists\mathrm{CompSer}_\Omega(H; K)$: G has a composition series from H to K
$\mathbb{Z}_n^*$: $\{a \in \mathbb{Z}_n \mid (a, n) = 1\}$
SDR: system of distinct representatives
$\mathrm{hom}(G, H)$: The set of all homomorphisms from G to H
$\wr$: the restricted wreath product
$\wr\wr$: the complete wreath product
$\wr\wr_r$: the regular wreath product
$\trianglelefteq$: normal
$\triangleleft$: normal and proper
$\trianglelefteq\trianglelefteq$: subnormal
$\triangleleft\triangleleft$: subnormal and proper
$X' = X \sqcup X^{-1}$, where X is a nonempty set

Index

MIX
Papier aus verantwortungsvollen Quellen
Paper from responsible sources
FSC® C105338

If you have any concerns about our products,
you can contact us on
ProductSafety@springernature.com

In case Publisher is established outside the EU,
the EU authorized representative is:
Springer Nature Customer Service Center GmbH
Europaplatz 3, 69115 Heidelberg, Germany

Printed by Libri Plureos GmbH
in Hamburg, Germany